de Gruyter Expositions in Mathematics 24

de Gruyter Expositions in Mathematics

1 The Analytical and Topological Theory of Semigroups, *K. H. Hofmann, J. D. Lawson, J. S. Pym* (Eds.)

2 Combinatorial Homotopy and 4-Dimensional Complexes, *H. J. Baues*

3 The Stefan Problem, *A. M. Meirmanov*

4 Finite Soluble Groups, *K. Doerk, T. O. Hawkes*

5 The Riemann Zeta-Function, *A. A. Karatsuba, S. M. Voronin*

6 Contact Geometry and Linear Differential Equations, *V. E. Nazaikinskii, V. E. Shatalov, B. Yu. Sternin*

7 Infinite Dimensional Lie Superalgebras, *Yu. A. Bahturin, A. A. Mikhalev, V. M. Petrogradsky, M. V. Zaicev*

8 Nilpotent Groups and their Automorphisms, *E. I. Khukhro*

9 Invariant Distances and Metrics in Complex Analysis, *M. Jarnicki, P. Pflug*

10 The Link Invariants of the Chern-Simons Field Theory, *E. Guadagnini*

11 Global Affine Differential Geometry of Hypersurfaces, *A.-M. Li, U. Simon, G. Zhao*

12 Moduli Spaces of Abelian Surfaces: Compactification, Degenerations, and Theta Functions, *K. Hulek, C. Kahn, S. H. Weintraub*

13 Elliptic Problems in Domains with Piecewise Smooth Boundaries, *S. A. Nazarov, B. A. Plamenevsky*

14 Subgroup Lattices of Groups, *R. Schmidt*

15 Orthogonal Decompositions and Integral Lattices, *A. I. Kostrikin, P. H. Tiep*

16 The Adjunction Theory of Complex Projective Varieties, *M. C. Beltrametti, A. J. Sommese*

17 The Restricted 3-Body Problem: Plane Periodic Orbits, *A. D. Bruno*

18 Unitary Representation Theory of Exponential Lie Groups, *H. Leptin, J. Ludwig*

19 Blow-up in Quasilinear Parabolic Equations, *A. A. Samarskii, V. A. Galaktionov, S. P. Kurdyumov, A. P. Mikhailov*

20 Semigroups in Algebra, Geometry and Analysis, *K. H. Hofmann, J. D. Lawson, E. B. Vinberg* (Eds.)

21 Compact Projective Planes, *H. Salzmann, D. Betten, T. Grundhöfer, H. Hähl, R. Löwen, M. Stroppel*

22 An Introduction to Lorentz Surfaces, *T. Weinstein*

23 Lectures in Real Geometry. *F. Broglia* (Ed.)

Evolution Equations and Lagrangian Coordinates

by

A. M. Meirmanov
V. V. Pukhnachov
S. I. Shmarev

Walter de Gruyter · Berlin · New York 1997

Authors

A. M. Meirmanov
Departamento de Matemática
Universidade da Beira Interior
R. Marquês d'Ávila e Bolama
6200 Covilhã
Portugal

V. V. Pukhnachov
Lavrentyev Institute
of Hydrodynamics
The Russian Academy
of Sciences
Siberian Division
630090 Novosibirsk,
Russia

S. I. Shmarev
Lavrentyev Institute
of Hydrodynamics
The Russian Academy of Sciences
Siberian Division
630090 Novosibirsk, Russia
and
Departamento de Matématicas
Faculdad de Ciencias
Universidad de Oviedo
33007 Oviedo, Spain

1991 Mathematics Subject Classification: 35-02, 35K55, 80A22, 76S05, 58G35
Keywords: nonlinear parabolic equations, Stefan-type problems, nonlocal equivalence transformation, asymptotic behaviour of solutions

♾ Printed on acid-free paper which falls within the guidelines of the ANSI to ensure permanence and durability.

Library of Congress Cataloging-in-Publication Data

Evolution equations and Lagrangian coordinates / by A. M. Meirmanov, V. V. Pukhnachev, S. I. Shmarev.
p. cm. – (De Gruyter expositions in mathematics ; 24)
Includes bibliographical references and index.
ISBN 3-11-014875-7 (alk. paper)
1. Evolution equations – Numerical solutions. 2. Lagrangian equations – Numerical solutions. 3. Mathematical physics. I. Meĭrmanov, A. M. (Anvarbek Mukatovich) II. Pukhnachev, V. V. III. Shmarev, S. I. (Sergei I.), 1958– . IV. Series.
QC20.7.E88E96 1997
515'.353–dc21 96-54 303
CIP

Die Deutsche Bibliothek – Cataloging-in-Publication Data

Meirmanov, Anvarbek M.:
Evolution equations and Lagrangian coordinates / by A. M. Meirmanov ; V. V. Pukhnachov ; S. I. Shmarev. – Berlin ; New York : de Gruyter, 1997
(De Gruyter expositions in mathematics ; 24)
ISBN 3-11-014875-7
NE: Puchnačev, Vladislav, V.:; Šmarev, Sergej I.:; GT

Typeset using the authors' TEX files: I. Zimmermann, Freiburg
Printing: Arthur Collignon GmbH, Berlin. Binding: Lüderitz & Bauer GmbH, Berlin.
Cover design: Thomas Bonnie, Hamburg.

Preface

This book presents recent results of the investigation of evolution equations by means of the method of Lagrangian coordinates.

There are two basic methods for describing the motions of continua, those of Euler and Lagrange. In the first method all characteristics of the motion are considered as functions of time and independent space variables in a coordinate system not connected with the medium itself. In the approach of Lagrange, the observer follows the particles' paths, e.g. all parameters of the motion are assumed to be functions of time and the initial positions of the particles. Each of the two approaches has its own merits. In particular, the method of Lagrange is sometimes preferable to deal with the motions of continua which involve free boundaries, a priori unknown.

As was noticed some ten years ago, one may profit from this mechanical ideology to study abstract evolution equations. Given an evolution equation of divergent form, whatever its physical applications may be, it may always be viewed as the mass balance law of some motion of a continuous medium written in Euler variables. This fictitious motion may then be given the Lagrangian description which is of different mathematical nature than the original equation. The study of the Lagrangian counterpart proves very informative and if the formal correspondence between the solutions of the starting equation and its "twin" has been somehow established, we advance in the study of both.

These ideas are developed in several directions: we outline their algebraic and numerical aspects, and use them to study free boundary problems. The contents are structured into four chapters.

Chapter 1 is devoted to the study of Verigin's problem. This problem occurs in the mathematical modeling of joint motions of two immiscible liquids through a porous medium. The process gives rise to the free boundary that separates the regions occupied by the liquids. The complete investigation of this problem in the one-dimensional formulation is presented, the description of the physical process and the governing mathematical models included. For the deduced boundary value problems the existence and uniqueness theorems are proven. The self-similar solutions are constructed and it is shown that the self-similar solutions of Stefan's problem yield the similarity solutions of the limiting Verigin problem.

In Chapter 2 we discuss algebraic properties of the coordinate transformation employed. Special symmetry properties of second and higher-order parabolic equations are derived. In particular, we describe a class of equations possessing the nonlocal symmetries which turn out to be the classical Lie symmetries after the passage to Lagrangian coordinates. The classes of equations are stated 1) which are invariant with respect to

Lagrangian coordinate transformation; 2) that can be linearized or simplified via this change of variables. In the latter case we construct special exact solutions generating new explicit solutions of the original equations.

Lagrangian coordinates prove to be a favorable framework to obtain new classes of exact solutions for certain boundary value problems. We present exact solutions to the one-phase Stefan problem and to the Cauchy–Verigin problem. Of special interest are those exact solutions of Stefan's problem which describe the process of phase disappearance. The method is also extended to some systems of equations and to equations with complex variables.

Chapter 3 contains a systematic study of the Cauchy problem for a number of equations contained in the class of equations usually referred to as the generalized porous medium equation or the advection-diffusion equation. Here we are interested principally in the regularity and propagation properties of the interfaces separating the regions where solutions are strictly positive from those where they are identically zero. Although an appropriate mathematical theory of such problems already exists, we employ the Lagrangian coordinate transformation to get detailed information on the interface properties. We establish conditions on data providing $C^{1,\alpha}$ and C^{∞}-smoothness of the occurring interfaces, the asymptotic representations for the interfaces are derived, and it is shown that, under certain restrictions, these asymptotics are differentiable. For a subclass of the generalized porous medium equation the effect of instantaneous loss of initial smoothness is discovered: whatever the input data, the space derivative of the solution instantly becomes discontinuous and, moreover, the parameters of this "jump of discontinuity" do not depend on the nature of the data.

In the concluding sections of this chapter we demonstrate an application of this approach to the study of unbounded solutions for parabolic equations. A special non-classical variant of Stefan's problem with degeneracy at the free boundary is discussed in the final section.

Chapter 4 collects results of the investigation of multi-dimensional free boundary problems for parabolic equations. The first of the problems under consideration is the classical one-phase Stefan problem. The Lagrangian formulation of this problem is presented. After the passage to Lagrangian coordinates we obtain, instead of a single parabolic equation, a system of evolution equations (which need not be parabolic). The dimension of this system equals the number of independent spatial variables in the original equation. The well-posedness of the linear model of this problem is studied. For the complete problem, the existence of self-similar solutions is proven. These self-similar solutions can be interpreted as describing the process of ice melting. The initial temperature is assumed to be constant along each ray exiting from the top of some cone, the temperature is positive inside the cone and zero otherwise.

By their very nature, Lagrangian coordinates can always be introduced in a number of ways and we benefit from this arbitrariness to treat the equation of polytropic gas filtration (the porous medium equation). For this equation an analog of the system of mass Lagrangian coordinates is proposed in which the equivalent system of equations adopts a canonical form. Next, we study Boussinesq's equation of filtration theory. The group analysis of Boussinesq's equation is given and two types of self-similar solutions are constructed and investigated. This study serves the groundwork for the construction of new nonlocal invariant solutions to Boussinesq's equation. We present an example of the numerical treatment of these self-similar solutions and propose a method to linearize Boussinesq's equation at its solution.

The final section of the chapter discusses local smoothness properties of the free boundaries in solutions of the N-dimensional porous medium equation. We prove that the interfaces preserve the initial regularity in the space variables and become infinitely differentiable in time.

The results and presentation of Chapter 1 are due to A. M. Meirmanov. The material of Chapter 2 was prepared by V. V. Pukhnachov. Chapter 3 was written by S. I. Shmarev, except for Sections 26–27 presented by V. V. Pukhnachov. Sections 28, 33–34 of Chapter 4 were written by V. V. Pukhnachov, Sections 31–32 and 35 by S. I. Shmarev, Sections 29–30 are due to A. M. Meirmanov and V. V. Pukhnachov.

The idea of writing a book was born when all three authors were collaborators of the Lavrentyev Institute of Hydrodynamics, Siberian Division of Russian Academy of Sciences, Novosibirsk. The original motivation was to survey the results which had already been obtained up to that time by the Lagrangian coordinates method. However, the selection of material and preparation of the text have taken much more time than was expected and the final version of the book contains a number of new results obtained while this monograph was in progress.

At different stages of work on this book the authors were supported by different institutions. Primarily, in 1991 S. I. Shmarev was supported with a Research Grant by Lavrentyev Institute of Hydrodynamics, during 1993–1994 he was given a scholarship of FICYT (Principado de Asturias, Spain) at the Mathematical Department of the University of Oviedo. The work of V. V. Pukhnachov was partially supported with the grant EP1/931S by the Japan Society for the Promotion of Science, and with the grant NR 6000 by the International Science Foundation. The authors would like to express here their deepest gratitude to these institutions and their members for encourage and support.

Our pleasant obligation is to thank here the colleages whose aid, discussions, and criticism were invaluable in making this book a reality. Especially, we would like to express our sincere gratitude to V. S. Belonosov, A. P. Chupakhin, J. I. Diaz, T. Nagai, L. V. Ovsyannikov, C. Rogers, J. L. Vazquez.

Contents

Chapter I

The Verigin problem

1 Review of results 1

2 Filtration in a porous soil 2

3 Formulation of the problem 4

4 Self-similar solutions; Stefan's problem as a limit case of Verigin's problem . . 6

5 One-dimensional problem: main statements and formulation of results 9

6 Proofs of Theorems 5.1 and 5.2: Verigin's problem with given mass flux on the known boundaries and the Cauchy–Verigin problem 12

6.1 Equivalent problem in a fixed domain 12

6.2 Solutions to approximate problems 17

6.3 Uniform estimates for solutions of approximation problems 21

6.4 Passage to the limit 26

6.5 Existence of classical solutions to original problems 27

6.6 Uniqueness of solution to problem (E_1) 28

6.7 Global in time solvability of problem (E_1) 30

7 Proof of Theorem 5.3: Verigin's problem with the given pressure on the known boundaries 33

7.1 Equivalent problem (L_3). Existence of solution 33

7.2 Uniqueness of solution to problem (L_3) 37

Chapter II

Equivalence transformations of evolution equations

8 Main ideas. A historical survey 42

9 Reciprocity transformations of second-order equations 47

10 Hidden symmetry of evolution equations 51

11 Linearization by means of Lagrangian coordinates 62

12 Lagrange-invariant equations 65

13 Equations with spherical and cylindrical symmetries 69

13.1 Gradient-nonlinear equations 69

13.2 Equation of heat conduction in solid hydrogen 73

13.3 The nonlinear Schrödinger equation 75

14 Equivalence transformations for higher-order equations and systems of equations 79

14.1 The Kingston–Rogers theorem 79

14.2 Third-order L-invariant equations 82

14.3 Reciprocity transformations for systems of one-dimensional evolution equations 84

14.4 Evolution equations in complex domains 87

15 A remarkable equation of nonlinear heat conduction 89

16 The one-phase Stefan problem: explicit solutions with functional arbitrariness 94

16.1 Statement of the problem 94

16.2 Solvability conditions 96

16.3 Global in time explicit solutions 98

16.4 Explicit solutions describing the process of phase disappearance . . . 100

16.5 Explicit solutions of the Cauchy–Verigin problem 101

Chapter III

One-dimensional parabolic equations

17 Introduction 104

17.1 An example 104

17.2 Finite speed of propagation. Interfaces 106

17.3 Preliminaries and results 108

18 Lagrangian coordinates in one-dimensional evolution equations 121

18.1 Equations in divergence form 122

18.2 Some equations not in divergence form 126

18.3 A commentary 129

19 Analysis of the problem in Lagrangian terminology 130

19.1 Regularization 130
19.2 Level curves. Positivity properties of solutions 132
19.3 Solvability of the problem in Lagrangian formulation 136
20 Uniform estimates. The inverse transformation 137
21 Some starting properties of the interface 143
22 Estimates for the time derivative and the higher-order derivatives 146
22.1 Local variables 146
22.2 Preliminary estimation of the time derivative 146
22.3 An improved estimate 151
22.4 Estimating the higher-order derivatives 154
23 The interface regularity 158
23.1 Global $C^{1+\alpha}$-smoothness 158
23.2 Infinite differentiability of the moving interface 160
24 Regularity of interfaces for a generalized porous medium equation 161
24.1 Solvability of Lagrangian statement 161
24.2 Further estimates 166
24.3 The inverse change of variables 169
24.4 Estimating the time derivative 170
24.5 Continuous differentiability of the right-side interface 175
24.6 Left-side interface 177
24.7 Infinite differentiability of the interfaces 177
24.8 A nonlocal equation in population dynamics 179
25 Axially symmetrical solutions of the porous medium equation:
long-time asymptotic behavior 185
25.1 Lagrangian statement 185
25.2 Solvability of problem (L). Regularization. Invertibility of the
coordinate transformation 187
25.3 Similarity solutions 191
25.4 Power nonlinearity. Differentiable asymptotics 194
25.5 Equations with arbitrary nonlinearity 200
26 A non-classical problem for a degenerate parabolic equation:
uniqueness of unbounded solutions 204

27 The Stefan problem with degeneracy at the free boundary: example of exact solution . . . 207

Chapter IV

Parabolic equations in several space dimensions

28 Review of results . . . 210
29 Lagrangian coordinates in the one-phase Stefan problem . . . 212
29.1 Formulation of the problem . . . 212
29.2 Linearization . . . 214
30 Correctness of the linear model . . . 217
31 Similarity solutions of the Stefan problem . . . 222
31.1 Self-similar statement . . . 222
31.2 Definitions. The main result . . . 224
31.3 Auxiliary propositions . . . 226
31.4 Poisson's equation in weighted classes . . . 231
32 Solvability of the nonlinear problem . . . 237
32.1 Newton's method . . . 237
32.2 The linear model: potential and solenoidal parts . . . 238
32.3 Solenoidal component . . . 240
32.4 Potential component . . . 241
32.5 The nonlinear problem . . . 243
33 Canonical Lagrangian coordinates . . . 244
34 Boussinesq's equation in filtration theory . . . 250
34.1 Group analysis of Boussinesq's equation in Lagrangian coordinates . 250
34.2 Similarity solutions of first and second type . . . 253
34.3 Existence of self-similar solutions . . . 255
34.4 Computation of self-similar solutions. An example . . . 257
34.5 How could one linearize the Boussinesq equation? . . . 266
35 Local regularity of interfaces . . . 270
35.1 Lagrangian coordinates . . . 270
35.2 Newton's method . . . 275

35.3 The linear model 279
35.4 The potential part 283
35.5 Solution to the nonlinear problem 286
35.6 Regularity of the solution 288

Bibliography 295

Notation 309

Index 310

Chapter I

The Verigin problem

1 Review of results

Verigin's problem is a mathematical problem describing the advancement of the surface of contact between two immiscible compressible liquids in the process of joint filtration through a porous soil. This problem was originally considered in the self-similar statement by N. N. Verigin in 1954, [172]. There the sought functions were the common boundary $\Pi(t)$ of two domains $\Omega_1(t)$ and $\Omega_2(t)$ occupied by liquids "1" and "2", respectively, and the pressure in each of the liquids. The pressure satisfied the heat equation in the domains $\Omega_1(t)$ and $\Omega_2(t)$ and was assumed continuous along with its conormal derivative across the contact surface $\Pi(t)$. The boundary $\Pi(t)$ itself was found from the additional condition of coincidence of the velocity of the liquid particle at the contact surface with the velocity of this surface.

Most likely, this problem was given the name "the Verigin problem" in works by L. I. Kamynin [81, 80]. As opposed to [172], in these works the pressure satisfied a general linear strictly parabolic equation in each domain $\Omega_1(t)$ and $\Omega_2(t)$. In papers [81, 80] the existence of a classical solution for the one-dimensional problem was proven in the class of input data (the initial and boundary conditions) that provided monotonicity of the sought (free) boundary.

In the literature, this problem is often referred to as Muskat's model, [54, 45, 46]. It should be mentioned, however, that Muskat himself considered the problem of motion of incompressible fluids separated by an interface. In 1969, W. Fulks and R. Guenter [54] stated the existence and uniqueness of a classical solution locally in time, and L. Evans [45, 46] in 1977/1978 proved the existence of a solution for the one-dimensional problem globally in time. In the above-mentioned papers the pressure satisfied the linear heat equation in each domain $\Omega_1(t)$ and $\Omega_2(t)$.

In its complete statement, when the mass balance equation and Darcy's law were considered simultaneously, the one-dimensional Verigin problem was studied by A. M. Meirmanov [105]. There were proven the uniqueness of classical solutions corresponding to different types of boundary conditions at the known boundaries. In addition, as opposed to the other above-mentioned works, in [105] some physically reasonable restrictions were found on the life span of a classical solution, for example, the time of displacement of one of the liquids by another.

For the multi-dimensional Verigin problem, the existence of a classical solution for small times was proven by E. V. Radkevich in 1985 [140]. The derived sufficient conditions for the existence of a solution are close to the necessary conditions in the sense that in the spherically symmetric case they coincide with the conditions obtained by A. M. Meirmanov, which define the restrictions on the existence time of a classical solution. Paper [140] also gives conditions for the existence of a classical solution of the dynamical angle problem for the Verigin–Muskat problem, i. e. in the case when the free boundary $\Pi(t)$ and the fixed boundary intersect.

2 Filtration in a porous soil

Let us consider the process of filtration of a compressible liquid in a porous soil occupying the domain $\Omega \subset \mathbb{R}^3$. The mathematical model describing this motion is based on the mass balance law in the differential form (the continuity equation)

$$m\frac{\partial \rho}{\partial t} + \operatorname{div}(\rho\vec{v}) = 0, \tag{2.1}$$

and Darcy's law

$$\vec{v} = -\frac{k}{\mu}\nabla p + \vec{f}. \tag{2.2}$$

Here m is the porosity of the soil, ρ is the density of the filtrating liquid and $\vec{v}$ is its velocity, p is the pressure in the liquid, k is the filtration coefficient, μ is the viscosity of the filtrating liquid, and $\vec{f}$ is the vector of the mass force density.

In the system of equations (2.1)–(2.2) the unknowns are the density ρ, the pressure p, and the velocity $\vec{v}$ – five functions in total. All other parameters of the model are supposed given. As a rule, m, k, μ are positive constants, and the vector $\vec{v}$ is a known function of the space variables $\vec{x}$, time t, and, perhaps, the sought functions p and ρ.

System (2.1)–(2.2) is formally incomplete — to define five parameters one has only four scalar equations. To complete the model, one has to have recourse to the hydrodynamic axioms. In what follows we restrict ourselves to the simplest case of isentropic motion where the density is a known function of the pressure:

$$\rho = \varphi(p). \tag{2.3}$$

To yield a unique solution, equations (2.1)–(2.3) are completed by the boundary condition on $\Gamma = \partial\Omega$ for $t > 0$ and the initial condition

$$\rho(x, 0) = \rho_0(x), \quad x \in \Omega. \tag{2.4}$$

As a rule, on Γ either the pressure

$$p = p^0(x, t), \quad x \in \Gamma,\ t > 0, \tag{2.5}$$

or the mass consumption are given:

$$\rho \cdot (\vec{v}, \vec{n}) = \rho v_n = g(x, t), \quad x \in \Gamma, \ t > 0. \tag{2.6}$$

In (2.6) $\vec{n}$ stands for the outer normal vector to the hypersurface Γ.

The system of equations (2.1)–(2.3) with the initial condition (2.4) and one of the boundary conditions (2.5) or (2.6) is a complete mathematical model describing the process of filtration of the compressible liquid in the porous soil. It is easy to see that this system is equivalent to the single scalar equation

$$m \frac{\partial}{\partial t} (\varphi(p)) = \operatorname{div} \left(\frac{k}{\mu} \varphi(p) \nabla p - \varphi(p) \vec{f} \right) \tag{2.7}$$

for the pressure p.

If

$$\varphi'(p) > 0, \tag{2.8}$$

then equation (2.7) is parabolic. In particular, for

$$\varphi(p) = p^{1/\gamma}, \quad \gamma = \text{const.} > 0, \tag{2.9}$$

equation (2.7) is equivalent to the well-known nonlinear filtration equation

$$\frac{\partial \rho}{\partial t} = \operatorname{div} \left(\frac{k\gamma}{m\mu} \rho^{\gamma} \nabla \rho - \rho \vec{f} \right). \tag{2.10}$$

The motions of continua where the sought characteristics satisfy the differential equations of mechanics in the ordinary sense (e.g. possess the required set of continuous derivatives) are usually called the *continuous motions*. Following this, we shall call a *continuous filtration* a filtration process for which the functions ρ, $\vec{v}$, p are continuously differentiable.

It is well known that the continuous motions do not exhaust the class of motions of continua. That is why, besides the continuous motions, the *generalized motions* are studied where the sought characteristics need not be differentiable and satisfy the conservation laws only in the integral form.

Also, the continuous filtration does not contain the entire class of admissible solutions of the problem describing the process of filtration of a fluid in a porous soil. Proceeding by analogy, to deal with the filtration problem one can define correctly the class of liquid motions where the sought functions ρ and $\vec{v}$ are only integrable and satisfy the mass conservation law in the integral form

$$\int_Q (m\rho \cos(\vec{\nu}, t) + \rho \cdot (\vec{v}, \vec{n}) \sin(\vec{\nu}, t))\, d\sigma = 0. \tag{2.11}$$

Here Q is an arbitrary volume with piecewise-smooth boundary $\partial\Omega$, lying in the four-dimensional domain of points $(\vec{x}, t)$, $x \in \Omega$ and $t > 0$; $\vec{\nu}$ is the unit normal vector to $\partial\Omega$ and $\vec{n}$ is the unit outer normal vector to the truncation of the boundary ∂Q by the hyperplane $\{t = \text{const.}\}$. The detailed proof of relation (2.11) can be found in [122].

We shall term a *generalized filtration* a liquid filtration for which the sought characteristics satisfy the mass balance law in the form (2.11) and Darcy's law (2.2).

It is easy to see that such a definition of a generalized filtration is equivalent to the introduction of the concept of weak solution for the parabolic equation (2.7). Thus, if η is an arbitrary function, sufficiently smooth, compactly supported in Q, and vanishing as $t = T$, then the weak solution of equation (2.7) in the domain $\Omega_T = \Omega \times (0, T)$ is a function $p(\vec{x}, t) \in W_q^{1,0}(\Omega_T)$, $q \geq 1$, satisfying the integral identity

$$\int_Q \left\{ -m\frac{\partial \eta}{\partial t} + \nabla\eta \left(\frac{k}{\mu}\nabla p - \vec{f} \right) \right\} \varphi(p)\, d\Omega\, dt = \int_\Omega [m\varphi(p)\eta]|_{t=0}\, d\Omega. \tag{2.12}$$

This definition of the generalized solution is only one of many possible. In all definitions of this sort the key point is the choice of the function space which the solution $p(\vec{x}, t)$ must belong to. This may depend on many factors and, generally speaking, each concrete problem corresponds to its own definition of the generalized solution.

3 Formulation of the problem

Verigin's problem refers to the process of filtration of two different immiscible liquids having a surface of contact. To simplify the presentation, from now on we assume that the vector of the mass forces is equal to zero. Let the liquids occupy the domains $\Omega_1(t)$ and $\Omega_2(t)$ such that $\overline{\Omega_1(t)} \cup \overline{\Omega_2(t)} = \Omega$, and in each domain $\Omega_1(t)$ and $\Omega_2(t)$ the pressure p, the density ρ, and the velocity vector $\vec{v}$ satisfy the continuity equation (2.1) and Darcy's law (2.2) in the ordinary sense but, generally speaking, with different relations (2.3) and different viscosities

$$\mu = \mu_i, \quad \varphi = \varphi_i(p) \quad \text{in } \Omega_i(t),\ i = 1, 2.$$

The pressure and the conormal component of the velocity vector are continuous on the contact surface $\Pi(t)$ (the free boundary)

$$[p] = 0, \tag{3.1}$$

$$\left[\vec{v} \cdot \vec{n}\right] = 0. \tag{3.2}$$

Here $\vec{n}$ is the unit normal vector to the hypersurface $\Pi(t)$, and the symbol $[u]$ here and elsewhere denotes the jump of the function u across $\Pi(t)$:

$$[u] = \lim_{\substack{x \to x_0 \\ x \in \Omega_2(t)}} u(x,t) - \lim_{\substack{x \to x_0 \\ x \in \Omega_1(t)}} u(x,t), \quad x_0 \in \Pi(t).$$

Furthermore, on $\Pi(t)$ the conormal component of the velocity is proportional to the normal velocity of this boundary in the direction of the normal $\vec{n}$:

$$v_n \equiv \vec{v} \cdot \vec{n} = mV_n. \tag{3.3}$$

If the hypersurface is given by the equation $H(x,t)=0$, then

$$\vec{n} = \nabla H / |\nabla H|, \qquad V_n = -\frac{\partial H}{\partial t}\frac{1}{|\nabla H|}. \tag{3.4}$$

The mathematical model is completed by posing the boundary condition (2.5) or (2.6) on the boundary $\Gamma = \partial\Omega$, the initial condition (2.4), and by fixing the initial position of the free boundary

$$\Pi(0) = \Pi_0. \tag{3.5}$$

There are different methods to derive the boundary conditions (3.1)–(3.3) on the sought boundary $\Pi(t)$. All of them are based, to various extents, on conservation laws in the integral form. In our opinion, the most natural method is to derive the model viewing the filtration of two different liquids as the *generalized filtration* of a single liquid which possesses different viscosities and dependences (2.3) between the density and the pressure in the domains $\Omega_1(t)$ and $\Omega_2(t)$. The filtration is continuous everywhere outside the hypersurface

$$\Pi_T = \bigcup_{t=0}^{T} \Pi(t)$$

of the four-dimensional domain $\Omega_T = \Omega \times (0, T)$, i.e., outside Π_T the functions ρ, p, $\vec{v}$ are continuously differentiable and satisfy equations (2.1), (2.2) in the ordinary sense. In continuum mechanics such a motion is usually called *motion with strong discontinuity* [122]. Likewise, Verigin's problem can be termed *generalized filtration with strong discontinuity*.

A byproduct of this definition is given by the conditions on the surface of contact discontinuity

$$[\rho(v_n - mV_n)] = 0, \tag{3.6}$$

$$[p] = 0. \tag{3.7}$$

Condition (3.6) follows from the mass conservation law (2.11), and (3.7) is implied by Darcy's law. The derivation of these conditions is much the same as in continuum mechanics [122].

Now, if we assume additionally that $\Pi(t)$ is a contact surface, e.g., there is no exchange of particles across this surface, then condition (3.6) implies conditions (3.2) and (3.3). In this case the limiting values of the conormal components of the velocity at $\Pi(t)$ inside the domains $\Omega_i(t)$ coincide and are proportional to the velocity of the boundary in the direction of the chosen normal:

$$v_n = mV_n.$$

Note that in filtration theory the latter condition means that the hypersurface Π_T is always constituted by the same particles and is a surface of contact discontinuity. The difference between this condition and the well-known counterpart condition in continuum mechanics is explained by the fact that in the filtration theory the equation of the particles' trajectories has the form

$$m\frac{d\vec{x}}{dt} = \vec{v}(\vec{x}, t). \tag{3.8}$$

4 Self-similar solutions; Stefan's problem as a limit case of Verigin's problem

The purpose of this section is to derive the one-phase Stefan problem from Verigin's problem. Before starting the formal derivation we consider self-similar solutions of Verigin's problem and show that these solutions converge to self-similar solutions of Stefan's problem when the viscosity of one of the liquids tends to zero.

Let us consider the one-dimensional Verigin problem when the domains $\Omega_1(0)$, $\Omega_2(0)$ are the half-spaces $\{x > 0\}$, $\{x < 0\}$, respectively, and the initial pressures are constant therein (and different).

In this case Verigin's problem possesses a self-similar solution where the sought boundary $\Pi(t)$ is given by the equation $x = h_* \sqrt{t}$, $h_* =$ const., and in each of the domains $\Omega_i(t)$ the pressure depends only on the variable $\xi = x/\sqrt{t}$.

For the sake of simplicity, let us suppose that in each liquid the dependence (2.3) has the form

$$\rho = \varphi(p) \equiv e^p.$$

Then the function

$$u(\xi) = \exp(p(x, t))$$

satisfies in each of the domains

$$Q_1^* = \{\xi : -\infty < \xi < h_*\}, \quad Q_2^* = \{\xi : h_* < \xi < \infty\}$$

the equation

$$\frac{d^2u}{d\xi^2} + 2a\xi\frac{du}{d\xi} = 0, \tag{4.1}$$

where

$$a = \frac{m}{4k}\mu_i \quad \text{as } \xi \in Q_i^*, \ i = 1, 2.$$

In the self-similar statement of Verigin's problem, the sought functions are the boundary $\xi = h_*$ between the domains Q_1^*, Q_2^* and the function $u(\xi)$.

The boundary conditions for the solutions of equation (4.1) in each of the domains Q_i^* and the equation for defining the unknown point h_* are formed by the initial conditions for the density (or the pressure) and conditions (3.1)–(3.3) on the free boundary. Thus, the initial conditions for the pressure transform into the conditions at infinity

$$\lim_{\xi \to -\infty} u(\xi) = u_1, \quad \lim_{\xi \to \infty} u(\xi) = u_2. \tag{4.2}$$

Condition (3.1) corresponds to the condition

$$[u] = 0 \quad \text{as } \xi = h_* \tag{4.3}$$

and conditions (3.2)–(3.3) have to be replaced by

$$\frac{1}{\mu_1}\frac{du}{d\xi}(h_* - 0) = \frac{1}{\mu_2}\frac{du}{d\xi}(h_* + 0), \tag{4.4}$$

$$\frac{1}{\mu_1}\frac{du}{d\xi}(h_* - 0) = -\frac{m}{2}h_*. \tag{4.5}$$

The proof of the solvability of problem (4.1)–(4.5) is fairly easy. It relies on the simplest variant of the fixed point theorem. Let us fix an arbitrary number h and solve equation (4.1) in the domains $Q_1 = \{\xi : -\infty < \xi < h\}$, $Q_2 = \{\xi : h < \xi < \infty\}$ with conditions (4.2)–(4.4) at infinity and as $\xi = h$. We have:

$$u = u_1 + \frac{u_2 - u_1}{\Lambda}\mu_1 \exp\left(a_1 h^2\right)\int_{-\infty}^{\xi} \exp\left(-a_1 s^2\right) ds, \quad \xi < h, \tag{4.6}$$

$$u = u_2 + \frac{u_1 - u_2}{\Lambda}\mu_2 \exp\left(a_2 h^2\right)\int_{\xi}^{\infty} \exp\left(-a_2 s^2\right) ds, \quad \xi > h, \tag{4.7}$$

$$\Lambda = \mu_1 \exp\left(a_1 h^2\right)\int_{-\infty}^{h} \exp\left(-a_1 s^2\right) ds + \mu_2 \exp\left(a_2 h^2\right)\int_{h}^{\infty} \exp\left(-a_2 s^2\right) ds. \tag{4.8}$$

This solution depends continuously on the parameter h. The equation for defining h_* is given by condition (4.5). The left-hand side of condition (4.5) taken at $\xi = h - 0$ equals

$$I(h) \equiv \frac{u_2 - u_1}{\Lambda}.$$

This is a continuous function of the variable h. Moreover, $I(h)$ doesn't change sign and $|I(h)| \to \infty$ when $h \to \pm\infty$. The right-hand side of (4.5) is a linear function of h. Evidently, the graph of this function has to intersect the graph of $I(h)$ at least at one point $h = h_*$. Such a point h_* defines the domains Q_1^*, Q_2^*, and the solution $u(\xi)$ to problem (4.1)–(4.5) is then given by formulas (4.6)–(4.8) with $h = h_*$.

Let us observe how this solution behaves in dependence on $\mu_2 > 0$. The function $I(h)$ is uniformly bounded with respect to $\mu_2 > 0$. Hence, for all $\mu_2 > 0$ the function $h_*(t)$ is uniformly bounded, whence the free boundary $\xi = h_*(\mu_2)$ can advance only a finite distance. Letting $\mu_2 \to 0$ in formulas (4.6)–(4.7), we get that $h_*(\mu_2)$ converges to h, while $u(\xi, \mu_2)$ tends to the function $\tilde{u}(\xi)$, where

$$\tilde{u}(\xi) \equiv u_2 \quad \text{for } \xi > \tilde{h},$$

$$\tilde{u}(\xi) \equiv u_1 + (u_2 - u_1)\frac{\int_{-\infty}^{\xi} \exp\left(-a_1 s^2\right) ds}{\int_{-\infty}^{h} \exp\left(-a_1 s^2\right) ds} \quad \text{as } \xi < \tilde{h}. \tag{4.9}$$

In turn, $\tilde{h}$ is a solution of the equation

$$\frac{1}{\mu_1}\frac{d\tilde{u}}{d\xi}(\tilde{h}-0) = -\frac{m}{2}\tilde{h} \tag{4.10}$$

coinciding with (4.5).

Thus, the limit function $\tilde{u}(\xi)$ solves the following problem: to find a function $\tilde{u}(\xi)$ and a parameter $\tilde{h}$ such that $\tilde{u}(\xi)$ satisfies in the sought domain $\{-\infty < \xi < \tilde{h}\}$ the equation

$$\frac{d^2\tilde{u}}{d\xi^2} + 2a_1\xi\frac{d\tilde{u}}{d\xi} = 0, \tag{4.11}$$

the conditions

$$\lim_{\xi\to-\infty}\tilde{u}(\xi) = u_1, \quad \tilde{u}(h-0) = u_2, \tag{4.12}$$

and the additional condition (4.10) on the unknown (sought) boundary.

The limiting problem (4.10)–(4.12) is exactly the Stefan problem in the self-similar statement. Hence, if the viscosity in one of the liquids tends to zero, the self-similar solution of Verigin's problem tends to the self-similar solution of Stefan's problem.

It is not difficult to present the formal deduction of Stefan's problem from Verigin's problem. It amounts to setting $\mu_2 = 0$ in the formulation of Darcy's law in the domain $\Omega_2(t)$ written in the form

$$\nabla p = -\frac{\mu_2}{k}\vec{v},$$

(assuming, of course, that the velocity $\vec{v}$ is bounded). This gives

$$p(x,t) \equiv p^0(t), \quad x \in \Omega_2(t), \quad t > 0.$$

Given a constant pressure p_* at the exterior boundary of $\Omega_2(t)$, everywhere in $\Omega_2(t)$

$$p(x,t) \equiv p_* \quad \text{as } t > 0.$$

For the pressure $p(x,t)$, in the domain $\Omega_1(t)$ we have the following problem:

$$m\frac{\partial\varphi(p)}{\partial t} = \operatorname{div}\left(\frac{k}{\mu}\varphi(p)\nabla p\right), \quad x \in \Omega_1(t), \tag{4.13}$$

$$p = p_*, \quad \frac{k}{\mu}\frac{\partial p}{\partial n} = -mV_n, \quad x \in \Pi(t),\ t > 0, \tag{4.14}$$

$$p = p^0(x,t), \quad x \in \Gamma = \partial\Omega,\ t > 0, \tag{4.15}$$

$$\Pi(0) = \Pi_0, \quad p(x,0) = p_0(x), \quad x \in \Omega_1(0). \tag{4.16}$$

Problem (4.13)–(4.16) is the one-phase Stefan problem. For instance, if $\varphi(p) = \exp(p)$, then the "temperature" $u = \exp(p)$ in the domain $\Omega_1(t)$ obeys the heat equation where the coefficient of thermal capacity is equal to k/μ, and the latent specific heat of melting in Stefan's condition (4.14) is equal to $m\exp(p)$.

5 One-dimensional problem: main statements and formulation of results

Let $\Omega = (x_1, x_2) \subset \mathbb{R}$ be a finite or infinite interval occupied by the filtrating liquid. If $x = h(t)$ is the equation of the sought contact boundary of two viscous immiscible liquids "1" and "2", then the domain occupied by liquid "1" at the instant $t = 0$ corresponds to the interval $\Omega_1(t) = (x_1, h(t))$, while the domain occupied by liquid "2" corresponds to the interval $\Omega_2(t) = (h(t), x_2)$.

Let us set

$$\Omega_{1,T} = \bigcup_{t=0}^{T} \Omega_1(t), \quad \Omega_{2,T} = \bigcup_{t=0}^{T} \Omega_2(t), \quad T > 0.$$

The pressure p and the density ρ satisfy in the domains $\Omega_{1,T}$, $\Omega_{2,T}$ the equation

$$m\frac{\partial \rho}{\partial t} = \frac{\partial}{\partial x}\left(\frac{k}{\mu}\rho\frac{\partial p}{\partial x}\right) \tag{5.1}$$

and are connected by the relation

$$\rho = \varphi_i(p), \quad (x, t) \in \Omega_{i,T}, \quad i = 1, 2. \tag{5.2}$$

If $x_1 > -\infty$, then either

$$p = p_1(t) \quad \text{as } x = x_1, t > 0, \tag{5.3}$$

or

$$\varphi_1(p)\frac{\partial p}{\partial x} = -\frac{\mu_1}{k}g_1(t) \quad \text{as } x = x_1, t > 0. \tag{5.4}$$

Analogously, if $x_2 < \infty$, then either

$$p = p_2(t) \quad \text{as } x = x_2, t > 0, \tag{5.5}$$

or

$$\varphi_2(p)\frac{\partial p}{\partial x} = -\frac{\mu_2}{k}g_2(t) \quad \text{as } x = x_2, t > 0. \tag{5.6}$$

On the sought contact boundary $x = h(t)$

$$p(h(t) - 0, t) = p(h(t) + 0, t), \quad t > 0, \tag{5.7}$$

$$\frac{1}{\mu_1}\frac{\partial p}{\partial x}(h(t) - 0, t) = \frac{1}{\mu_2}\frac{\partial p}{\partial x}(h(t) + 0, t), \quad t > 0, \tag{5.8}$$

$$\frac{1}{\mu_1}\frac{\partial p}{\partial x}(h(t) - 0, t) = -\frac{m}{k}\frac{dh(t)}{dt}, \quad t > 0. \tag{5.9}$$

Finally, at the initial instant

$$h(0) = h_0, \quad p = p_0(x), \quad x \in \Omega. \tag{5.10}$$

Without any loss of generality, the constants m and k can be taken to be unity. If needed, one may always use for this purpose the rescaling transformation for the pressure p and time t:

$$t \to \frac{t}{m}, \qquad p \to kp.$$

With respect to the functions $\varphi_i(t)$, $g_i(t)$, $i = 1, 2$, $p_j(t)$, $j = 0, 1, 2$, we assume the following:

$$\text{(U.1)} \quad \begin{cases} \varphi_1, \varphi_2 \in C^3(-\infty, \infty), \quad \{\varphi_1', \varphi_2'\} \geq \nu > 0; \\ \text{there exists a constant } p_\infty \text{ such that} \\ \{\varphi_1, \varphi_2\} > 0 \quad \text{as } s > p_\infty; \end{cases}$$

$$\text{(U.2)} \quad \begin{cases} g_{1,2} \in C^{(2+\alpha)/2}[0, \infty), \quad \alpha \in (0, 1); \\ p_1, p_2 \in C^{(3+\alpha)/2}[0, \infty), \quad h_0 \in (x_1, x_2); \\ p_0 \in C^{3+\alpha}(\Omega_1(0)) \cap C^{3+\alpha}(\Omega_2(0)); \quad \{p_1, p_2, p_0\} > p_\infty; \\ \text{If } |x_i| < \infty, \text{ then } p_0 \in C^{3+\alpha}\left(\overline{Q}\right), \text{ where } Q \subset \Omega_i(0) \text{ is} \\ \text{an arbitrary domain bounded away from} \\ \text{the point } h_0 \text{ by a positive distance.} \end{cases}$$

(U.3) If $|x_1| < \infty$, then at the points $x = x_i$, $i = 1, 2$, for the boundary conditions (5.4) or (5.6) and the initial condition (5.10) the first-order compatibility conditions are fulfilled [97].

(U.4) If $|x_1| < \infty$, then at the points $x = x_i$, $i = 1, 2$, for the boundary conditions (5.3) or (5.5) and the initial condition (5.10) the first-order compatibility conditions are fulfilled.

(U.5) Let $\mu_0(x) = \mu_i$ as $x \in \Omega_i(0)$, $i = 1, 2$. Then the functions $p_0(x)$ and

$$v_0(x) = -\frac{1}{\mu_0(x)} \frac{dp_0}{dx}$$

belong to the space $W_2^1(\Omega)$.

We will study the following initial-boundary value problems:

(E_1) to find a solution of the problem (5.1), (5.2), (5.4), (5.6), (5.7)–(5.10);

(E_2) if $x_1 = -\infty$, $x_2 = \infty$, to find a solution of the problem (5.1), (5.2), (5.7)–(5.10); this problem we also term the *Cauchy–Verigin problem*;

(E_3) to find a solution of the problem (5.1), (5.2), (5.3), (5.5), (5.7)–(5.10).

Denote by $\mu(x, t)$ the function equal to μ_i in $\Omega_{i,T}$ and set

$$v(x, t) = -\frac{1}{\mu(x, t)} \frac{\partial p(x, t)}{\partial x}.$$

Let us recall that the notation $L_p(0, T; B)$, $p \geq 1$, denotes the space of functions $u(x, t)$ belonging to the Banach space B at the truncations $\{t = \text{const.}\}$ and such that their norms in the space B, as functions of time t, belong to $L_p(0, T)$. Throughout the chapter we adhere to the traditional notation for the function spaces and their norms introduced, for example, in [97, Ch. 2].

Let us formulate the main results.

Theorem 5.1. *Let* $|x_i| < \infty$, $i = 1, 2$, *and conditions* (U.1)–(U.5) *be fulfilled. Then on a small enough time interval* $(0, T_*)$ *problem* (E_1) *has a unique classical solution such that*

$$h \in C^{(2+\alpha)/2}[0, T], \quad p \in C^{\beta,\beta/2}\left(\overline{\Omega_T}\right) \cap C^{3+\alpha,(3+\alpha)/2}\left(\overline{Q}\right),$$

$$v \in C^{\beta,\beta/2}\left(\overline{\Omega_T}\right) \cap L_\infty\left(0, T; W_2^1(Q)\right),$$

where $\beta \in (0, 1]$, $0 < T < T_*$, *and* $Q \subset \Omega_T$ *is an arbitrary domain such that the distance between the contact boundary* $\{x = h(t)\}$ *and* Q *remains positive throughout.*

The instant T_* is defined by two factors: either one of the domains $\Omega_1(T_*)$ or $\Omega_2(T_*)$ degenerates into a point, i.e., the free boundary $\{x = h(t)\}$ arrives at one of the boundaries $\{x = x_i\}$, $i = 1, 2$, meaning the total displacement of one of the liquids by another, or the density of one of the liquids becomes zero or infinity.

If $g_1(t) \geq 0$, $g_2(t) \leq 0$, then $T_* = \infty$. In any case, $T_* \leq T_0$ where T_0 is the minimal time of the degeneracy of one of the domains $\Omega_1(t)$ and $\Omega_2(t)$:

$$\int_{x_1}^{h_0} \varphi(p_0(x))dx + \int_0^{T_0} g_1(t)dt = 0,$$

or

$$\int_{h_0}^{x_2} \varphi_2(p_0(x))dx - \int_0^{T} g_2(t)dt = 0.$$

Theorem 5.2. *Let* $x_1 = -\infty$, $x_2 = \infty$ *and conditions* (U.1), (U.2), (U.5) *hold. Then problem* (E_2) *has a unique classical solution*

$$h \in C^{(\beta+2)/2}[0, T], \quad p \in C^{\beta,\beta/2}\left(\overline{\Omega_T}\right) \cap C^{3+\alpha,(3+\alpha)/2}(\overline{Q}),$$

$$v \in C^{\beta,\beta/2}\left(\overline{\Omega_T}\right) \cap L_\infty\left(0, T; W_2^1(Q)\right), \quad 0 < T < \infty,$$

where the domain Q *is described in the conditions of Theorem* 5.1.

Theorem 5.3. *Let* $|x_i| < \infty$, $i = 1, 2$, *and conditions* (U.1)–(U.5) *hold. Then on a small enough time interval* $(0, T_*)$ *problem* (E_3) *has a unique classical solution*

$$h \in C^{(2+\beta)/2}[0, T], \quad p \in C^{\beta,\beta/2}\left(\overline{\Omega_T}\right) \cap C^{3+\alpha,(3+\alpha)/2}\left(\overline{Q}\right),$$

$$v \in C^{\beta,\beta/2}\left(\overline{\Omega_T}\right) \cap L_\infty\left(0, T; W_2^1(Q)\right), \quad 0 < T < T_*,$$

where the domain Q *is described in the conditions of Theorem* 5.1.

The instant T_* is the moment of the total displacement of one of the liquids by the other

$$(x_2 - h(T_*))\,(p(T_*) - x_1) = 0.$$

Remark 5.1. The condition $g_1 \geq 0$, $g_2 \leq 0$ in Theorem 5.1 has a transparent physical meaning: the liquid can be only "pumped in" through both lateral boundaries, therefore the density must increase, and the free boundary $\{x = h(t)\}$ will always be bounded away from the lateral boundaries.

Remark 5.2. It is not clear what may happen after the total displacement of one liquid by another. It was implicitly assumed that if there was an influx of a liquid through the boundary $x = x_i$, then it was the "i"-th liquid that flowed in. If at some time instant the domain Ω is filled only by one liquid, then the next moment when another liquid may appear should be specially indicated. The study of the problem arising is beyond the scope of this book.

6 Proofs of Theorems 5.1 and 5.2: Verigin's problem with given mass flux on the known boundaries and the Cauchy–Verigin problem

6.1 Equivalent problem in a fixed domain. The assumption that the sought boundary is the contact surface and, besides, at each time instant is constituted by the same particles allows one to introduce new variables rendering this boundary independent of time.

There are two principal methods of describing the motions of continua, those of Euler and Lagrange. The first method is mainly used in gas and fluid mechanics. There the independent variables are time t and the points of the physical space x. The second method is often used in the theory of rigid deformable bodies. Here the independent variables are time t and the initial position of the particle ξ. Thus, the Euler description distinguishes the points of physical space, while the approach of Lagrange differentiates between the points of the medium.

The connection between the Euler and Lagrange variables is performed by means of the phase flow of the system of equations

$$\frac{dx}{dt} = v(x,t), \qquad x|_{t=0} = \xi, \tag{6.1}$$

defining the trajectories of the particles of the medium. Here $v(x,t)$ is the velocity field given in the Euler variables. Let us recall that if

$$x = X(\xi, t)$$

is a solution of system (6.1) depending on $\xi \in \Omega_0$ as a parameter, then the phase flow of the system of equations (6.1) is the mapping $\xi \mapsto X_t(\xi) = x$ of the domain Ω_0 onto the domain Ω_t, where $X_t(\xi) \equiv X(\xi, t)$.

Once Verigin's problem is posed in the Lagrange variables, the free boundary becomes time-independent. Thus, in the Lagrange variables the boundary of contact between two liquids is known beforehand. Let us benefit from this fact to reformulate the problem in the new variables. It will be convenient to introduce the *mass Lagrangian coordinate* by the formula

$$y = \int_{h(t)}^{x} \rho(s, t)\, ds. \tag{6.2}$$

Since the density ρ is assumed strictly positive, the change of variables defined by (6.2) is one-to-one.

Let

$$x = X(\xi, t) \tag{6.3}$$

be the transformation inverse to (6.2). Differentiating the identity

$$y \equiv \int_{h(t)}^{X(y,t)} \rho(s, t)\, ds$$

in y and t, we get

$$1 = \rho(x, t)\frac{\partial X}{\partial y}(y, t), \tag{6.4}$$

$$0 = \rho(x, t)\frac{\partial X}{\partial t} - \rho(h, t)\frac{dh}{dt} + \int_{h(t)}^{x} \frac{\partial \rho(s, t)}{\partial t}\, ds.$$

Let us make use of the continuity equation (2.1) for the density ρ to replace the derivative $\partial\rho/\partial t$ by $\operatorname{div}(\rho v)$

$$\int_{h(t)}^{x} \frac{\partial}{\partial s}[\rho v]\, ds = \rho(x, t)v(x, t) - \rho(h, t)v(h, t),$$

whence

$$0 = \rho(x, t)\left(\frac{\partial X}{\partial t} - v(x, t) - \rho(h, t)\left(\frac{dh}{dt} - v(h, t)\right)\right).$$

Applying now condition (3.3), we get an equivalent form of equation (6.1)

$$\frac{\partial X}{\partial t} = v(X, t). \tag{6.5}$$

Now, let us make use of Darcy's law

$$v = \frac{1}{\mu}\frac{\partial p}{\partial x}(x, t)$$

with the derivative in x recalculated in the new variables. If

$$P(y, t) = p(x, t), \tag{6.6}$$

then, with the use of (6.4),

$$\frac{\partial p}{\partial x} = \frac{\partial P}{\partial y} : \frac{\partial X}{\partial y} = \rho \frac{\partial P}{\partial y}.$$

Thus,

$$\frac{\partial X}{\partial t} = -\frac{\rho}{\mu}\frac{\partial P}{\partial y}.$$

Let us differentiate in y the expression obtained and then make use once again of equality (6.4). We have

$$\frac{\partial}{\partial t}\left(-\frac{1}{\rho}\right) = \frac{\partial}{\partial y}\left(-\frac{\rho}{\mu}\frac{\partial P}{\partial y}\right)$$

or, since $\rho = \varphi(P)$,

$$\frac{\partial}{\partial t}\left(-\frac{1}{\varphi(P)}\right) = \frac{\partial}{\partial y}\left(\frac{\varphi(P)}{\mu}\frac{\partial P}{\partial y}\right). \tag{6.7}$$

Thus, the pressure $P(y,t)$ satisfies the quasilinear parabolic equation (6.7) in each of the domains $G_{1,T}$, $G_{2,T}$ (which are the images of $\Omega_{1,T}$, $\Omega_{2,T}$ under the mapping (6.2), $0 < t < T$).

Also, one has to bear in mind that $\varphi = \varphi_i(P)$, $\mu = \mu_i$ as $(y,t) \in G_{i,T}$, $i = 1, 2$: in liquids "1" and "2" the viscosities and the dependences of the densities on the pressures are different.

The Lagrange variables are chosen so that in the new variables the sought free boundary corresponds to the line $\{y = 0\}$ where conditions (5.7), (5.8) take the form

$$P(-0,t) = P(+0,t), \tag{6.8}$$

$$\frac{\varphi_1(P)}{\mu_1}\frac{\partial P}{\partial y}(-0,t) = \frac{\varphi_2(P)}{\mu_2}\frac{\partial P}{\partial y}(+0,t). \tag{6.9}$$

Here the symbols $u(\pm 0, t)$ denote the limits

$$u(+0,t) = \lim_{\substack{y\to 0\\ y>0}} u(y,t), \quad u(-0,t) = \lim_{\substack{y\to 0\\ y<0}} u(y,t).$$

Let us clarify the structure of the domains $G_{1,T}$, $G_{2,T}$. The easiest case corresponds to problem (E_2). Since the density is bounded away from zero, then

$$G_{1,T} = \{(y,t) : -\infty < y < 0,\ 0 < t < T\},$$

$$G_{2,T} = \{(y,t) : 0 < y < \infty,\ 0 < t < T\}.$$

In problem (E_1), the boundary $x = x_i$ of $\Omega_{1,t}$, $i = 1, 2$, corresponds to the boundary

$$y = Y_i(t) \equiv \int_{h(t)}^{x_i} \rho(s,t)\,ds, \quad i = 1, 2, \tag{6.10}$$

of the domain $G_{i,T}$. The functions $Y_i(t)$, $i = 1, 2$, are easy to calculate through the data of problem (E_1). Indeed, calculating the derivative

$$\frac{dY_i}{dt} = \int_{h(t)}^{x_i} \frac{\partial \rho}{\partial t}\, ds - \rho(h, t)\, \frac{dh}{dt}$$

and applying the continuity equation and condition (3.3) on the free boundary (as in the derivation of (6.5)), we get

$$\frac{dY_i}{dt} = -\rho\,(x_i, t)\, v(x_i, t), \quad i = 1, 2.$$

The right-hand side of this equality is defined by the boundary conditions (5.4) and (5.6). Thus,

$$Y_i(t) = Y_i(0) - \int_0^t g_i(\tau)\, d\tau, \quad i = 1, 2, \tag{6.11}$$

and the domains $G_{i,T}$ take the form

$$G_{1,T} = \{(y, t) : \; Y_1(t) < y < 0, \; 0 < t < T\},$$

$$G_{2,T} = \{(y, t) : \; 0 < y < Y_2(t), \; 0 < t < T\}.$$

The constants $Y_i(0)$ are defined from formula (6.2) and the initial condition (5.10):

$$Y_i(0) = \int_{h(0)}^{x_i} \varphi_i(p_0(s))\, ds.$$

It remains to pose the boundary conditions for the pressure $P(y, t)$ at the boundaries $y = Y_i(t)$, $i = 1, 2$, and to define the initial pressure $P(y, 0)$. Having recalculated the derivative in x in the boundary conditions (5.4), (5.6), we get

$$\frac{\varphi_i^2(p)}{\mu} \frac{\partial P}{\partial y}(Y_i(t), t) = -g_i(t), \quad i = 1, 2. \tag{6.12}$$

At the instant $t = 0$ the function $X_0(y) = X(y, 0)$ is known, defining the inverse change of variables $y \to x$. Hence,

$$P(y, 0) = P_0(y), \quad y \in G(0), \tag{6.13}$$

where $P_0(y) \equiv p_0(X_0(y))$ is a known function,

$$G(0) = \{y : \; Y_1(0) < y < Y_2(0)\}.$$

Thus, the free boundary problem (E_1) in the Lagrange variables corresponds to the equivalent *problem* (L_1) of finding a function $P(y, t)$ satisfying equation (6.7) in the known domains $G_{1,T}$, $G_{2,T}$, the boundary conditions (6.8), (6.9) at the common boundary $\{y = 0\}$ of the domains $G_{1,T}$, $G_{2,T}$, the boundary conditions (6.12) at the exterior boundaries $\{y = Y_i(t)\}$, $i = 1, 2$, and the initial condition (6.13).

The analog of the Cauchy–Verigin problem (E_2) in the new variables is the *problem* (L_2) of finding a function $P(y, t)$ satisfying equation (6.7) in the domains

$$G_{i,T} = \left\{(y, t) : \ 0 < (-1)^i y < \infty, \ 0 < t < T\right\},$$

conditions (6.8), (6.9) at the contact surface $\{y = 0\}$, and the initial condition (6.13).

It is natural to restrict the study of problem (L_1) to the time interval $(0, T_0)$ where the domains

$$G_1(t) = \{Y_1(t) < y < 0\}, \quad G_2(t) = \{0 < y < Y_2(t)\}$$

do not disappear. That is, $T < T_0$, where T_0 is the minimal root of the equation

$$Y_1(t)Y_2(t) = 0.$$

This assertion has a clear physical meaning: the solutions of problem (L_1) (or (E_1)) are considered until the moment when one of the liquids is totally forced out by the other.

Problem (L_1) is well known in the theory of boundary value problems for parabolic equations and is referred to as the *diffraction problem* (see [97, Ch. 3, Sec. 13]). The character of nonlinearity of equation (6.7) and the boundary conditions (6.12) are such that one fails to refer directly to any of the results already known. The problem can, however, be solved fairly easily by means of the method developed in [97].

The main idea proposed in [97] for problems of this sort consists in replacing equation (6.7) in each of the domains $G_{i,T}$, $i = 1, 2$, and conditions (6.8), (6.9) at the contact boundary of these domains by the single equation with non-differentiable coefficients

$$\frac{\partial}{\partial t}\left(-\frac{1}{\varphi(P, y)}\right) = \frac{\partial}{\partial y}\left(\frac{\varphi(P, y)}{\mu(y)}\frac{\partial P}{\partial y}\right) \tag{6.14}$$

in the domain

$$G_T = \{(y, t) : \ Y_1(t) < y < Y_2(t), \ 0 < t < T\}.$$

Here

$$\varphi(P, y) = \varphi_i(P), \quad \mu(y) = \mu_i \quad \text{as } (y, t) \in G_{i,T}, \ i = 1, 2.$$

The solution $P(y, t)$ of equation (6.14) is to be understood as the generalized solution i.e., as a function having the integrable derivatives $\partial P/\partial t$, $\partial P/\partial y$ and satisfying the integral identity

$$\int_{G_T} \left\{\frac{\varphi_P}{\varphi^2}\frac{\partial P}{\partial t}\eta + \frac{\varphi}{\mu}\frac{\partial P}{\partial y}\frac{\partial \eta}{\partial y}\right\} dy\, dt = 0 \tag{6.15}$$

with an arbitrary function $\eta(y, t)$ continuously differentiable and compactly supported in G_T.

It is easy to show that if

$$P \in C^{1,0}\left(\overline{G_T}\right) \cap C^{2,1}\left(G_{1,T}\right) \cap C^{2,1}\left(G_{2,T}\right),$$

then equality (6.15) is equivalent to equation (6.7) in each domain $G_{i,T}$ with conditions (6.8), (6.9) on the line $\{y = 0\}$.

There is a variety of methods appropriate to solving equation (6.7) with the boundary conditions (6.12) and the initial condition (6.13). Let us make use of one of them: the approximation of the discontinuous functions $\varphi(P, y)$ and $\mu(y)$ by smooth functions $\varphi_\varepsilon(P, y)$, $\mu_\varepsilon(y)$ and the consequent extraction of a compact subsequence from the sequence of solutions $P_\varepsilon(y, t)$ of the equation

$$\frac{\partial}{\partial t}\left(-\frac{1}{\varphi_\varepsilon(P, y)}\right) = \frac{\partial}{\partial y}\left(\frac{\varphi_\varepsilon(P, y)}{\mu_\varepsilon(y)}\frac{\partial P}{\partial y}\right) \tag{6.16}$$

with conditions (6.12) and the initial condition

$$P(y, 0) = P_0^\varepsilon(y), \quad y \in G(0), \tag{6.17}$$

where the smooth functions $P_0^\varepsilon(y)$ approximate somehow the initial pressure $P_0(y)$ as $\varepsilon \to 0$.

Let us term problem (6.16), (6.12), (6.17) *problem* (L_1^ε). Analogously, the problem approximating (L_2) we will term *problem* (L_2^ε).

Having deduced a sufficient set of ε-uniform estimates for the family of solutions $\{P^\varepsilon(y, t)\}$ of problem (L_i^ε), we shall prove that the limit function $P(y, t)$ will be the sought classical solution of problem (L_i).

6.2 Solutions to approximate problems. The crucial point here is the choice of the functions $\varphi_\varepsilon(P, y)$, $\mu_\varepsilon(y)$, and $P_0^\varepsilon(y)$, approximating the functions $\varphi(P, y)$, $\mu(y)$, $P_0(y)$ and influencing the properties of solutions $P_1^\varepsilon(y, t)$ to problem (L_1^ε). It is easy to see that conditions (U.1)–(U.5) transform into the analogous conditions for $\varphi(P, y)$, $\mu(y)$, $\mu(y)$, $P_0(y)$, and $g_i(t)$:

(U.6) $\varphi \in C^3(-\infty, \infty)$ as a function of the variable P, uniformly in $y \in (-\infty, \infty)$;

$$\frac{\partial \varphi}{\partial P}(P, y) \geq \nu > 0;$$

there exists a constant p_∞ such that

$$\varphi(P, y) > 0 \quad \text{as } P > p_\infty \text{ uniformly in } y \in (-\infty, \infty).$$

(U.7)
$$g_1, g_2 \in C^{(2+\alpha)/2}[0, \infty), \quad 0 < \alpha \in< 1;$$

$$Y_1(0) < 0 < Y_2(0); \quad P_0(y) > p_\infty, \in G(0);$$

$$P_0 \in C^{3+\alpha}(G_1(0)) \cap C^{3+\alpha}\left(\overline{Q}\right);$$

If $|Y_i(0)| < \infty$ (the case of problem (L_1)), then $P_0 \in C^{3+\alpha}\left(\overline{Q}\right)$, where $Q \subset G_i(0)$ contains the point $Y_i(0)$ and is bounded away from the point $y = 0$ by a positive distance.

(U.8) In problem (L_1), at the points $y = Y_i(0)$ the first-order compatibility condition of the boundary condition (6.12) with the initial condition (6.13) is fulfilled.

(U.9) The functions

$$P_0(y) \quad \text{and} \quad V_0(y) = -\frac{\varphi(P_0(y), y)}{\mu(y)} \frac{dP_0}{dy}(y)$$

belong to the space $W_2^1(Q)$, where Q is a bounded domain included in $G(0)$ and containing the point $y = 0$.

Let us choose approximations P_0^ε, μ_ε, φ_ε so that

$$P_0^\varepsilon \in C^{3+\alpha}\left(\overline{G(0)}\right),$$

μ_ε and φ_ε are twice continuous differentiable in y (recall that the functions μ and φ are discontinuous), and for the approximating functions conditions (U.6)–(U.9) hold, besides, with $W_2^1(Q)$-norms of P_0^ε, V_0^ε bounded uniformly with respect to the approximation number ε.

Let $\tilde{\phi}(y) \in C^\infty(-\infty, \infty)$, $\tilde{\phi}(y) = 1$ for $y > 1$, and $0 < \tilde{\phi}(y) < 1$ for $|y| < 1$. The sequence of infinitely differentiable functions

$$\phi_\varepsilon(y) = \tilde{\phi}\left(\frac{y}{\varepsilon}\right)$$

approximates as $\varepsilon \to 0$ the function $\phi(y)$, where $\phi(y) = 0$ for $y < 0$, and $\phi(y) = 1$ for $y > 0$.

Set

$$\varphi_\varepsilon(P, y) = (1 - \phi_\varepsilon(y))\,\varphi_1(P) + \phi_\varepsilon(y)\varphi_2(P).$$

It is easy to see that $\varphi_\varepsilon(P, y)$ satisfies condition (U.6).

Lemma 6.1. *For all $\varepsilon > 0$ small enough there exist functions*

$$P_0^\varepsilon(y) \in C^{3+\alpha}\left(\overline{G(0)}\right) \quad \text{and} \quad \mu_\varepsilon(y) \in C^{2+\alpha}(-\infty, \infty),$$

which satisfy conditions (U.6)–(U.9), *along with the above-described function $\varphi_\varepsilon(P, y)$, such that $W_2^1(Q)$-norms of the functions P_0^ε and V_0^ε are bounded uniformly.*

Proof. When constructing the functions P_0^ε, μ_ε, we shall distinguish two cases

$$\text{a)} \quad \frac{dP_0}{dy}(\pm 0) \neq 0, \quad \text{b)} \quad \frac{dP_0}{dy}(\pm 0) = 0.$$

Let us observe that due to the compatibility conditions at the point $y = 0$ (condition (U.9)) and the strict positivity of the functions $\varphi(P, y)$, $\mu(y)$, the limit values

$$\frac{dP_0}{dy}(-0), \qquad \frac{dP_0}{dy}(+0)$$

are either both zero, or both nonzero.

In case a), let us consider an arbitrary sequence of functions $\{P_0^\varepsilon\}$ from $C^{3+\alpha}\left(\overline{G(0)}\right)$ coinciding with $P_0(y)$ for $|y| > \varepsilon$ and such that for $|y| < \varepsilon$

$$\frac{1}{2}\min\left(\left|\frac{dP_0}{dy}(-0)\right|, \left|\frac{dP_0}{dy}(+0)\right|\right) \leq \left|\frac{dP_0^\varepsilon}{dy}(+0)\right| \leq \left|\frac{dP_0^\varepsilon}{dy}(y)\right| \leq \frac{3}{2}\max\left(\left|\frac{dP_0}{dy}(-0)\right|, \left|\frac{dP_0}{dy}(+0)\right|\right). \quad (6.18)$$

Let us choose ε small enough to provide inequality (6.18) for the function $P_0(y)$ with $|y| < \varepsilon$.

Let $V_0^\varepsilon \in C^{2+\alpha}\left(\overline{G(0)}\right)$ converge to $V_0^\varepsilon(y)$ as $\varepsilon \to 0$ in the norm of W_2^1. Set

$$\mu_\varepsilon(y) = -V_0^\varepsilon(y)\left[\frac{dP_0^\varepsilon}{dy}\varphi_\varepsilon\left(P_0^\varepsilon(y), y\right)\right]^{-1}.$$

Because of the uniform convergence of P_0^ε, V_0^ε to P_0 and V_0, respectively, and due to property (6.18), $\mu_\varepsilon(y)$ satisfy the main condition

$$0 < \overline{\nu} \leq \mu_\varepsilon(y) \leq \overline{\nu}^{-1} = \text{const.}$$

All other requirements are fulfilled by the construction.

In the case b), let us set

$$P_0^\varepsilon(y) \equiv P_0(y), \qquad \mu_\varepsilon(y) = (1 - \phi_\varepsilon(y))\,\mu_1 + \phi_\varepsilon(y)\mu_2.$$

Then we have

$$V_0^\varepsilon(y) = \frac{\varphi_\varepsilon\left(P_0(y), y\right)}{\mu_\varepsilon(y)}\frac{dP_0}{dy}(y).$$

The function $V_0^\varepsilon(y)$ defined in this way can have a singularity only at the point $y = 0$. Hence, to prove the inclusion $V_0^\varepsilon \in W_2^1(Q)$ amounts to checking that $\left(dV_0^\varepsilon/dy\right)^2$ is integrable for $y > 0$ and for $y < 0$. It follows from the conditions of the lemma that $V_0 \in W_2^1(Q)$, and that $\mu(y)$, $\varphi(P_0, y)$ are differentiable in $\overline{Q_1} = Q \cap G_1(0)$. Hence,

$$\frac{dP_0}{dy} \in W_2^1(Q_1),$$

and making use of the equality b), for the function dP_0/dy we get the estimate

$$\left|\frac{dP_0}{dy}(\overline{y})\right| = \left|\int_0^{\overline{y}}\frac{d^2P_0}{dy^2}(y)\,dy\right| \sim \text{const. } \overline{y}^{-1/2}.$$

Moreover, due to the choice of μ_ε, φ_ε we have

$$\frac{d\mu_\varepsilon}{dy}(y) \sim \text{const.}\cdot\varepsilon^{-1}, \qquad \frac{\partial\varphi_\varepsilon}{\partial y} \sim \text{const.}\cdot\varepsilon^{-1}.$$

Therefore,

$$\int_0^\varepsilon \left|\frac{d^2V_0^\varepsilon}{dy}(y)\right| dy \sim \frac{\text{const.}}{\varepsilon^2}\int_0^\varepsilon \left|\frac{dP_0}{dy}(y)\right|^2 dy + \text{const.}\cdot\int_0^\varepsilon \left|\frac{d^2P_0}{dy^2}(y)\right|^2 dy$$
$$\sim \frac{\text{const.}}{\varepsilon^2}\int_0^\varepsilon y\,dy + \text{const.} \sim \text{const.}$$

The case $0 < y < \varepsilon$ is considered analogously. □

Lemma 6.2. *Let conditions* (U.6)–(U.8) *hold. Then for each* $\varepsilon > 0$ *small enough there exists a solution* $P^\varepsilon(y,t)$ *of problem* (L_1^ε) *belonging to space* $C^{3+\alpha,(3+\alpha)/2}\left(\overline{G_{T_\varepsilon}}\right)$, *where* $T_\varepsilon \in (0, T_0]$ *is defined by one of the following relations:*

$$\min_{(y,t)\in G_{T_\varepsilon}} \varphi_\varepsilon\left(P^\varepsilon(y,t), y\right) = \frac{1}{2}\rho_-, \tag{6.19}$$

$$\max_{(y,t)\in G_{T_\varepsilon}} \varphi_\varepsilon\left(P^\varepsilon(y,t), y\right) = \frac{3}{2}\rho_+, \tag{6.20}$$

$$\min_{0<t<T_\varepsilon} \{|Y_1(t)|, |Y_2(t)|\} = \delta, \tag{6.21}$$

$$\delta = \frac{1}{3}\min\{|Y_1|(0), |Y_2(0)|\},$$

$$\rho_- = \min_{y\in G(0)} \varphi\left(P_0(y), y\right), \quad \rho_+ = \max_{y\in G(0)} \varphi\left(P_0(y), y\right).$$

Proof. It follows from the standard references [97, 160] that "in the small" in time there exists a unique solution $P^\varepsilon(y,t)$ to problem (L_1^ε) belonging to the function class mentioned in the conditions of the lemma. Assume that we have constructed such a solution on an interval $[0, T_\varepsilon^1]$ but none of conditions (6.19)–(6.21) is true. As follows from the estimates on solutions for quasilinear second-order parabolic equations, [97], at the instant $t = T_\varepsilon^1$ the function $P^\varepsilon(y,t)$ possesses the same smoothness that it had at $t = 0$, and its norm is estimated through the norms of $P_0^\varepsilon(y)$ and $g_i(t)$ given in conditions (U.6)–(U.8), the constants bounding the density $\varphi_\varepsilon\left(P^\varepsilon(y,t), y\right)$ from below and from above, and the distances from the boundaries $\{y = Y_i(t)\}$, $i = 1, 2$, to the line $\{y = 0\}$. Hence, the solution can be extended to a larger interval, up to the instant when none of conditions (6.19)–(6.21) is true. □

Remark 6.1. If for all $\varepsilon > 0$ small enough

$$\left|\varphi_\varepsilon\left(P^\varepsilon, y\right)\right|_{t,G_{T_\varepsilon}}^{(\beta/2)} \le M \tag{6.22}$$

with constants M, β independent of the approximation number ε, then all intervals $(0, T_\varepsilon)$ contain the interval $(0, T_1)$, where

$$T_1 \le \min\left(T_0, \left(\frac{\rho_-}{4M}\right)^{2/\beta}\right). \tag{6.23}$$

Proof. By the definition of the Hölder norm

$$-Mt^{\beta/2} \leq \varphi_\varepsilon\left(P^\varepsilon(y,t), y\right) - \varphi_\varepsilon\left(P_0^\varepsilon(y), y\right) \leq Mt^{\beta/2},$$

or

$$\varphi_\varepsilon\left(P_0^\varepsilon(y), y\right) - Mt^{\beta/2} \leq \varphi_\varepsilon\left(P^\varepsilon(y,t), y\right) \leq \varphi_\varepsilon\left(P_0^\varepsilon(y), y\right) + Mt^{\beta/2}.$$

Since for all $\varepsilon > 0$ small

$$\varphi_\varepsilon\left(P_0^\varepsilon(y), y\right) \geq \min_{y\in G(0)} \varphi_\varepsilon\left(P_0^\varepsilon(y), y\right) \geq \frac{3}{4}\rho_-,$$

$$\varphi_\varepsilon\left(P_0^\varepsilon(y), y\right) \leq \max_{y\in G(0)} \varphi_\varepsilon\left(P_0^\varepsilon(y), y\right) \leq \frac{5}{4}\rho_+,$$

with the aid of the inequality $\rho_- \leq \rho_+$ for $t \leq T_1$ we get

$$\frac{1}{2}\rho_- \leq \varphi_\varepsilon\left(P^\varepsilon(y), y\right) \leq \frac{3}{2}\rho_+.$$

□

Lemma 6.3. *Let conditions* (U.6), (U.7), (U.9) *be true. Then for each* $\varepsilon > 0$ *small enough there exists a solution to problem* (L_2^ε) *belonging to the space* $C^{3+\alpha,(3+\alpha)/2}(R_T)$, *where* $R_T = \{(y,t) : |y| < \infty,\ 0 < t < T\}$ *and* $T > 0$ *is arbitrary.*

The proof is similar to the proof of Lemma 6.2 but, unlike problem (L_1^ε), in problem (L_2^ε) the maximum and minimum values of the density (pressure) are estimated through the initial data with the help of the maximum principle.

6.3 Uniform estimates for solutions of approximation problems. Making use of the independence in time of the coefficients of the approximating equations we deduce here the simplest integral estimates.

It follows from the local estimates for solutions of parabolic equations [97, Ch. 3,5], that given $\varepsilon < \delta/3$, in an arbitrary domain Q which belongs to G_T and has no intersection with the strip $\{|y| < \delta/3\}$, the $C^{3+\alpha,(3+\alpha)/2}\left(\overline{Q}\right)$-norms of the solutions $P^\varepsilon(y,t)$ of problems (L_1^ε), (L_2^ε) are uniformly bounded with respect to the approximation parameter ε. Let us recall that the crucial ingredients of these estimates are the smoothness of the coefficients of the corresponding equations and the boundedness of the extreme values of the solutions in a domain larger then Q. The smoothness of the coefficients is provided by the choice of $\varepsilon > 0$ small enough, while the boundedness of the solutions in problem $(L_1)^\varepsilon$ is guaranteed by estimates (6.19), (6.20). To be specific, the latter assertion is a byproduct of these estimates since the upper and lower boundedness of the density $\varphi_\varepsilon(P^\varepsilon, y)$ and the continuous dependence of the density on the pressure imply the boundedness of the pressure P^ε from below and from above.

Besides, if the domain Q does not belong to the strip $\{|y| < \varepsilon_1\}$, then for $\varepsilon < \varepsilon_1$ the norms $|P^\varepsilon|_Q^{(3+\alpha)}$ are uniformly bounded. Later on, this observation will make it possible to construct a diagonal process to extract from the sequence $\{P^\varepsilon\}$ a subsequence converging in the norm of the space $C^{3+\alpha,(3+\alpha)/2}(\overline{Q})$, where Q is an arbitrary domain from G_T not containing the line $\{y=0\}$.

Thus, in order to deduce the integral estimates for the families of solutions $\{P^\varepsilon(y,t)\}$ to problems $(L_1)^\varepsilon$, (L_2^ε), one can confine all considerations to the domain

$$Q_T^\lambda = \{(y,t):\ |y|<\lambda,\ 0<t<T\}$$

with $\lambda = 2\delta/3$, $T \le T_\varepsilon$, bearing in mind that all derivatives of the form $D_t^r D_y^s P^\varepsilon$ with $2r+s\le 3$ on the boundaries $\{y=\pm\lambda\}$ are uniformly bounded with respect to ε.

In what follows, we denote by M a constant depending only on the constants and norms mentioned in conditions (U.6)–(U.9), and the constants μ_1, μ_2.

Lemma 6.4. *Let*

$$V^\varepsilon(y,t) = -\frac{\varphi_\varepsilon(P^\varepsilon(y,t),t)}{\mu_\varepsilon(y)}\frac{\partial P^\varepsilon}{\partial y}(y,t),$$

where $P^\varepsilon(y,t)$ is a solution either of problem (L_1^ε), or of problem (L_2^ε). Then for all $\varepsilon>0$ small enough the following inequality holds:

$$\max_{t\in(0,T)}\int_{-\lambda}^{\lambda}\left|\frac{\partial V^\varepsilon}{\partial y}(y,t)\right|^2 dy + \int_{Q_T^\lambda}\left(\left|\frac{\partial V^\varepsilon}{\partial t}\right|^2 + \left|\frac{\partial^2 P^\varepsilon}{\partial t\partial y}\right|^2\right) dydt \le M. \qquad (6.24)$$

Proof. To simplify notation, let us omit the index "ε" wherever this does not cause confusion. Adopt the notation

$$a = \frac{\varphi^2}{\partial\varphi/\partial P}.$$

Then the equation for the pressure can be written in the form

$$\frac{\partial P}{\partial t} = -a\frac{\partial V}{\partial y}.$$

Differentiating this equation in y and making use of the equality

$$\frac{\partial^2 P}{\partial t\partial y} = -\frac{\mu}{\varphi}\frac{\partial V}{\partial t} - \mu V\frac{\partial V}{\partial y},$$

which appears after differentiation in time of the equality defining the function V, we get for V the parabolic equation

$$\frac{\mu}{\varphi}\frac{\partial V}{\partial t} = \frac{\partial}{\partial y}\left(a\frac{\partial V}{\partial y}\right) - \mu V\frac{\partial V}{\partial y}. \qquad (6.25)$$

It follows directly from this equation, the maximum principle, and the boundedness of the derivatives on the lines $\{y=\pm\lambda\}$, that

$$\max_{(y,t)\in Q_T^\lambda}|V(y,t)| \le M. \qquad (6.26)$$

The first energy estimate follows by multiplication of equation (6.25) by V and integration with respect to y in the limits $(-\lambda, \lambda)$ with $0 < t < T$. After straightforward transformations we arrive at the estimate

$$\frac{d}{dt}\int_{-\lambda}^{\lambda}\frac{\mu}{2\varphi}V^2\,dy+\int_{-\lambda}^{\lambda}a\left|\frac{\partial V}{\partial y}\right|^2 dy=-\frac{1}{2}\int_{-\lambda}^{\lambda}\mu V^2\frac{\partial V}{\partial y}\,dy+\left(aV\frac{\partial P}{\partial y}\right)\Big|_{y=-\lambda}^{y=\lambda}. \tag{6.27}$$

Here we made use of the equality

$$\frac{\mu}{\varphi}V\frac{\partial V}{\partial t}=\frac{1}{2}\frac{\partial}{\partial t}\left(\frac{\mu}{\varphi}V^2\right)-\frac{1}{2}V^2\frac{\partial V}{\partial y}\mu.$$

Observe that the function $a(P, y)$ is bounded from below by a positive constant uniformly in ε. Then (6.27) implies in the usual way the estimate

$$\int_{Q_T^\lambda}\left|\frac{\partial V}{\partial y}\right|^2 dy\,dt\le M. \tag{6.28}$$

The second main integral estimate relies on (6.28) and Gronwall's inequality. Let us multiply equation (6.25) by the derivative $\partial V/\partial t$ and integrate with respect to y. We will get

$$\int_{-\lambda}^{\lambda}\frac{\mu}{\varphi}\left|\frac{\partial V}{\partial t}\right|^2 dy+\frac{1}{2}\frac{d}{dt}\int_{-\lambda}^{\lambda}a\left|\frac{\partial V}{\partial y}\right|^2 dy=\left(a\frac{\partial V}{\partial y}\frac{\partial V}{\partial t}\right)\Big|_{y=-\lambda}^{y=\lambda}-\int_{-\lambda}^{\lambda}\mu V\frac{\partial V}{\partial y}\,dy$$
$$+\int_{-\lambda}^{\lambda}b\left|\frac{\partial V}{\partial y}\right|^3 dy. \tag{6.29}$$

Here the function

$$b(P, y)\equiv-\frac{1}{2}a\frac{\partial a}{\partial P}$$

is uniformly bounded in ε.

The first term of the right-hand side of (6.29) is bounded, while the second one can be estimated in the standard manner with the help of (6.26) and the Cauchy inequality with a small $\gamma > 0$:

$$\left|\int_{-\lambda}^{\lambda}\mu V\frac{\partial V}{\partial t}\frac{\partial V}{\partial y}\,dy\right|\le\gamma\int_{-\lambda}^{\lambda}\left|\frac{\partial V}{\partial t}\right|^2 dy+\frac{M}{\gamma}\int_{-\lambda}^{\lambda}\left|\frac{\partial V}{\partial y}\right|^2 dy.$$

The last term of the right-hand side of (6.29) is estimated with the help of the representation

$$w^3(\tilde{y}, t)=w^2(\tilde{y}, t)\left[w(-\lambda, t)+\int_{-\lambda}^{\lambda}\frac{\partial w}{\partial y}(y, t)\,dy\right]$$

and the inequality

$$\int|w|^3\,dy\le\gamma\int\left|\frac{\partial w}{\partial y}\right|^2 dy+\frac{M}{\gamma}\left(\int w^2\,dy\right)^2+M\int w^2\,dy$$

with $w = a\partial V/\partial y$. In this case

$$\frac{\partial w}{\partial y} = \frac{\partial}{\partial y}\left(a\frac{\partial V}{\partial y}\right) = \frac{\mu}{\varphi}\frac{\partial V}{\partial t} + \mu V\frac{\partial V}{\partial y}.$$

Thus,

$$\int_{-\lambda}^{\lambda} \frac{\mu}{\varphi}\left|\frac{\partial V}{\partial t}\right|^2 dy + \frac{1}{2}\frac{d}{dy}\int_{-\lambda}^{\lambda} a\left|\frac{\partial V}{\partial y}\right|^2 \leq \gamma M \int_{-\lambda}^{\lambda} \frac{\mu}{\varphi}\left|\frac{\partial V}{\partial t}\right|^2 dy$$
$$+ \frac{M}{\gamma}\left(\int_{-\lambda}^{\lambda} a\left|\frac{\partial V}{\partial y}\right|^2 dy\right)^2 + M\int_{-\lambda}^{\lambda} a\left|\frac{\partial V}{\partial y}\right|^2 dy. \quad (6.30)$$

Taking $\gamma > 0$ small enough and letting

$$z(t) = \int_{-\lambda}^{\lambda} a\left|\frac{\partial V}{\partial y}\right|^2 dy,$$

we arrive at Gronwall's inequality

$$\frac{dz}{dt} \leq \tilde{M}(t)z, \quad z(0) = z_0,$$

where the function

$$\tilde{M}(t) = M\left(1 + \int_{-\lambda}^{\lambda} a\left|\frac{\partial V}{\partial y}\right|^2 dy\right)$$

is integrable on the interval $(0, T)$ (estimate (6.28)). Hence,

$$\max_{0<t<T} z(t) \leq M.$$

Returning to relation (6.30) and integrating it with respect to time we get the estimate for the time derivative

$$\int_{Q_T^\lambda} \left|\frac{\partial V}{\partial t}\right|^2 dy\, dt \leq M.$$

To complete the proof, it remains to observe that

$$\frac{\partial^2 P}{\partial y \partial t} = -\frac{\mu}{\varphi}\frac{\partial V}{\partial y}.$$

□

Corollary 6.1. *For all small $\varepsilon > 0$ the function $V^\varepsilon(y, t)$ satisfies the Hölder condition*

$$|V^\varepsilon|_{G_T}^{(\beta)} \leq M \quad (6.31)$$

with an exponent $\beta > 0$ not depending on ε.

Proof. As was already mentioned, we could restrict our considerations by the domain Q^λ. The derivative in y of the function

$$W(y,t) = \int_0^y V^\varepsilon(\xi,t)\,d\xi$$

satisfies the Hölder condition in y

$$\left|\frac{\partial W}{\partial y}(y_2,t) - \frac{\partial W}{\partial y}(y_1,t)\right| = \left|V^\varepsilon(y_2,t) - V^\varepsilon(y_1,t)\right| = \left|\int_{y_1}^{y_2} \frac{\partial V^\varepsilon}{\partial y}(y,y)\,dy\right|$$
$$\leq |y_2 - y_1|^{1/2}\left(\int_{-\lambda}^{\lambda}\left|\frac{\partial V}{\partial y}(y,t)\right|^2 dy\right)^{1/2} \leq M|y_2 - y_1|^{1/2}.$$

On the other hand, the function $W(y,t)$ itself satisfies the Hölder condition in the variable t

$$|W(y,t_2) - W(y,t_1)| = \left|\int_{t_1}^{t_2}\int_0^y \frac{\partial V^\varepsilon}{\partial t}(\xi,t)\,d\xi\,dt\right|$$
$$\leq |t_2 - t_1|^{1/2}\left(\int_{Q_T^\lambda}\left|\frac{\partial V^\varepsilon}{\partial t}\right|^2 dy\,dt\right)^{1/2} \leq M|t_2 - t_1|^{1/2}.$$

Thus, one may have recourse to results already well-known [97, Ch. 2, Sec. 4], to conclude that the derivative in y of function $W(y,t)$ is Hölder continuous in t with some exponent β. □

Corollary 6.2. *For all small $\varepsilon > 0$ the functions $P^\varepsilon(y,t)$ are Hölder continuous*

$$|P^\varepsilon|_{G_T}^{(\beta)} \leq M. \tag{6.32}$$

Proof. As follows from the previous assertion, to verify Hölder continuity of the function $W(y,t)$ in all its arguments it suffices to check the boundedness of the function

$$\int_{Q_T^\lambda}\left|\frac{\partial W}{\partial t}\right|^2 dy\,dt + \max_{t\in(0,T)}\int_{-\lambda}^{\lambda}\left|\frac{\partial W}{\partial y}\right|^2 dy.$$

The functions $P^\varepsilon(y,t)$ possess stronger estimates, since the derivative $\partial P^\varepsilon/\partial y$ is proportional to the bounded function V^ε, and the derivative $\partial P^\varepsilon/\partial t$ is proportional to the function $\partial V^\varepsilon/\partial y$, square-integrable on the truncations:

$$\max_{Q_T^\lambda}\left|\frac{\partial P^\varepsilon}{\partial y}(y,t)\right| \leq M, \quad \max_{0<t<T}\int_{-\lambda}^{\lambda}\left|\frac{\partial P^\varepsilon}{\partial t}(y,t)\right|^2 dy \leq M.$$

□

Corollary 6.3. *For all $\varepsilon > 0$ small enough estimate* (6.22) *holds.*

The proof of this assertion follows from Corollary 6.2 and the smoothness of φ_ε with respect to the variable P:

$$\left|\varphi_\varepsilon(P^\varepsilon(y,t),y)-\varphi_\varepsilon(P^\varepsilon(y,t_1),y)\right| \le \max\left|\frac{\partial\varphi_\varepsilon}{\partial P}\right| \left|P^\varepsilon(y,t_2)-P^\varepsilon(y,t_1)\right|.$$

6.4 Passage to the limit. The estimates just derived for the families of solutions to problems (L_1^ε) and (L_2^ε) allow one to extract convergent subsequences and to perform the passage to the limit as $\varepsilon \to 0$.

First of all, let us observe that by Corollary 6.2 to Lemma 6.4, and Remark 6.1 to Lemma 6.2, all intervals $(0, T_\varepsilon)$ contain an interval $(0, T_1)$, where T_1 does not depend on ε. Without loss of generality, we may suppose that

$$T_1 = \inf_{\varepsilon>0} T_\varepsilon$$

and in the passage to the limit confine ourselves to the domain G_{T_1}.

The above-obtained estimates make it possible to extract a subsequence (to simplify the presentation, let us still use for it the previous notation $\{P^\varepsilon(y,t)\}$), such that in the domain G_{T_1} P^ε and V^ε converge uniformly to functions

$$P(y,t) \quad \text{and} \quad V(y,t) = -\frac{\varphi(P(y,t),y)}{\mu(y)}\frac{\partial P}{\partial y},$$

respectively. P^ε converges to P in the norm of $C^{3+\alpha,(3+\alpha)/2}(Q)$, where Q is an arbitrary domain from $\overline{G_T}$ not containing the line $\{y=0\}$.

The limit function $P(y,t)$ belongs to $C^{\beta,\beta/2}\left(\overline{G_{T_1}}\right)$ and, moreover, is bounded by

$$\int_{G_T}\left(\left(\frac{\partial V}{\partial t}\right)^2+\left(\frac{\partial^2 P}{\partial y\partial t}\right)^2\right)dy\,dt + \max_{t\in(0,T)}\int_{G(t)}\left|\frac{\partial V}{\partial y}\right|^2 dy \le M, \tag{6.33}$$

where $T \le T_1$.

It is easy to see that everywhere outside the line $\{y=0\}$ the function $P(y,t)$ satisfies equation (6.7) in the ordinary sense, initial and boundary conditions (6.13), (6.12), and the compatibility conditions (6.8), (6.9). The last assertion follows, for example, from the continuity of the functions P and V everywhere in the domain G_{T_1}.

Let us now consider the truncation $\{t = T_1\}$. Since $P(y, T_1)$ possesses the same properties as $P_0(y)$, the solution of problem (L_1) (or (L_2)) can be continued to some interval (T_1, T_2). This process can be continued up to the instant $t = T_0$ when one of the lateral boundaries of G_T intersects the axis $\{y=0\}$, or when the density $\varphi(y,t)$ either vanishes, or becomes infinite.

Lemma 6.5. *Let conditions* (U.6)–(U.8) *hold. Then on a small enough interval* $(0, T_*)$ *problem* (L_1) *has at least one classical solution P belonging to*

$$C^{\beta,\beta/2}\left(\overline{G_T}\right)\cap C^{3+\alpha,(3+\alpha)/2}\left(G_{1,T}\right)\cap C^{3+\alpha,(3+\alpha)/2}\left(G_{2,T}\right), \quad T<T_*.$$

The corresponding function $V(y,t)$ belongs to $C^{\beta,\beta/2}\left(G_{T_*}\right)$ and satisfies estimate (6.33). *The instant $T_* > 0$ is defined as the moment when the density either vanishes, or becomes infinite.*

Lemma 6.6. *Let conditions* (U.6), (U.7), (U.9) *be true. Then problem* (L_2) *has at least one classical solution*

$$P \in C^{\beta,\beta/2}\left(\overline{G_T}\right) \cap C^{3+\alpha,(3+\alpha)/2}\left(G_{1,T}\right) \cap C^{3+\alpha,(3+\alpha)/2}\left(G_{2,T}\right),$$

where $T > 0$ is an arbitrary number such that

$$V \in C^{\beta,\beta/2}\left(\overline{G_T}\right).$$

6.5 Existence of classical solutions to original problems. To prove the existence of classical solutions to problems (E_1) and (E_2) amounts to verifying that the inverse passage to the Euler coordinates is realized by smooth enough functions.

As was shown at the beginning of this section, if $x = X(y,t)$ is the transformation from Lagrangian to Euler coordinates, then

$$\frac{\partial X}{\partial t}(y,t) = \frac{1}{\mu}\frac{\partial P}{\partial x}(x,t) = -\frac{\varphi(P,y)}{\mu(y)}\frac{\partial P}{\partial y}(y,t).$$

Hence,

$$x = X(y,t) = X_0(y) - \int_0^t \frac{\varphi(P(y,\tau),y)}{\mu(y)}\frac{\partial P}{\partial y}(y,\tau)\,d\tau,$$

where $X_0(y)$ is a known smooth function.

The sought contact boundary is given by the equality

$$x = h(t) \equiv h_0 - \frac{1}{\mu_1}\int_0^t \varphi_1(P(0,\tau))\frac{\partial P}{\partial y}(-0,\tau)\,d\tau \tag{6.34}$$

and defines the domains

$$\Omega_{1,T} = \{(x,t) : \ X_1 < x < h(t),\ 0 < t < T\},$$

$$\Omega_{2,T} = \{(x,t) : \ h(t) < x < X_2,\ 0 < t < T\},$$

where the functions

$$p(x,t) \equiv P(y,t), \quad \rho(x,t) = \varphi(p(x,t),x), \quad v(x,t) = \frac{1}{\mu(x)}\frac{\partial P}{\partial x}(x,t)$$

satisfy in the ordinary sense the continuity equation, Darcy's law, and the boundary conditions as $x = X_i$, $i = 1,2$ (in the case of problem (E_1)). By virtue of the already demonstrated properties of the solutions to problems (L_1), (L_2), the function $h(t)$ is a member of $C^{1+\beta/2}[0,T]$, and the functions $\rho(x,t)$ and $v(x,t)$ are continuous in the closed domain $\overline{\Omega_T} = \overline{\Omega} \times [0,T]$, $\Omega = \{x : \ X_1 < x < X_2\}$. The last property provides the fulfillment of the required conditions at the contact line $\{x = h(t)\}$.

6.6 Uniqueness of solution to problem (E_1). If there exist two different solutions to problem (E_1) corresponding to the same input data, then in Lagrangian coordinates they correspond to two different solutions of problem (L_1) satisfying the same initial and boundary conditions (each of these solutions defines its own change of the independent variables). Problem (L_1) is an initial-boundary value problem for a nonlinear parabolic equation with discontinuous coefficients under the sign of the derivative and involves nonlinear boundary conditions on the lateral boundaries $\{y = Y_i(t)\}$, $i = 1, 2$. If all possible solutions are smooth enough (e.g. have first derivatives bounded in the time and in the space variable), then the uniqueness of a solution to (E_1) follows from standard methods, for example, by application of the first energy estimate to the integral identity for the difference of two possible solutions. In our situation, the demonstrable smoothness of solutions to problem (L_1) is somewhat lower but still appropriate to have recourse to the standard theory.

Let $(0, T_*)$ be the maximal existence interval for the solutions P_1 and P_2 of problem (L_1). This means that on the interval $(0, T)$ with $T < T_*$ the norms of P_1 and P_2 indicated in Lemma 6.4 are bounded by a constant depending only on T. (Perhaps some of these norms grow unboundedly as $T \to T_*$.) For the difference $\tilde{P} = P_1 - P_2$ we have

$$\int_{G(t)} \left\{ \frac{\partial}{\partial t}\left(A\tilde{P}\right)\eta + \left(B\frac{\partial \tilde{P}}{\partial y} + C\right)\frac{\partial \eta}{\partial y} \right\} dy = \sum_{i=1}^{2} (-1)^i \left(F_i \eta \tilde{P}\right)\Big|_{y=Y_1(t)}, \tag{6.35}$$

where

$$A = \frac{\varphi(P_2, y) - \varphi(P_1, y)}{(P_2 - P_1)\varphi(P_2, y)\varphi(P_1, y)} \int_0^1 \frac{\partial \varphi}{\partial P}(P_1 + \tau(P_2 - P_1), y)\, d\tau,$$

$$B = b(P_2, y) \equiv \frac{\varphi(P_2, y)}{\mu(y)}, \quad C = \frac{b(P_2, y) - b(P_1, y)}{P_2 - P_1} \frac{\partial P_1}{\partial y},$$

$$F_i = g_i(t) A|_{y=Y_i(t)}, \quad i = 1, 2,$$

and $\eta \in W_2^{1,0}(G_T)$.

What is the smoothness of the coefficients in identity (6.35)? As was shown above (Lemma 6.4 and the Corollaries),

$$\max_{0<t<T} \int_{G(t)} \left| \frac{\partial P_i}{\partial t} \right|^2 dy + \max_{G_T} \left| \frac{\partial P_i}{\partial y}(y, t) \right| + \int_{G_T} \left| \frac{\partial^2 P_i}{\partial y \partial t} \right|^2 dy\, dt \le M.$$

Hence,

$$|A, B, C|_{G_T}^{(0)} \le M.$$

Besides these estimates, we have to evaluate

$$\alpha(t) = \max_{y \in G(t)} \left| \frac{\partial A}{\partial t}(y, t) \right|.$$

Since

$$\alpha(t) \le M \max_{i=1,2} \max_{y \in G(t)} \left| \frac{\partial P_i}{\partial t}(y,t) \right|,$$

and for all $\lambda > 0$ small enough

$$\frac{\partial P_i}{\partial t}(y_0,t) = \frac{\partial P_i}{\partial t}(\lambda,t) + \int_\lambda^{y_0} \frac{\partial^2 P_i}{\partial y \partial t}(y,t)\,dy,$$

then

$$\alpha(t) \le M \left(1 + \max_{i=1,2} \left(\int_{G(t)} \left| \frac{\partial^2 P_i}{\partial y \partial t}(y,t) \right|^2 dy \right)^{1/2} \right)$$

and

$$\int_0^T \alpha(t)\,dt \le M. \tag{6.36}$$

Having chosen in identity (6.35) $\eta = \tilde{P}$, after some transformations we arrive at the equality

$$\frac{1}{2}\frac{d}{dt}\int_{G(t)} A\tilde{P}^2\,dy + \int_{G(t)} B \left|\frac{\partial \tilde{P}}{\partial y}\right|^2 dy = -\int_{G(t)} \left\{ C\tilde{P}\,\frac{\partial \tilde{P}}{\partial y} + \frac{1}{2}\frac{\partial A}{\partial t}\tilde{P}^2 \right\} dy$$
$$+\frac{1}{2}\sum_{i=1}^{2}(-1)^i \left(F_i\tilde{P}^2\right)\Big|_{y=Y_i(t)}. \tag{6.37}$$

To derive (6.37), we have used relations (6.11) wherever necessary to eliminate the derivatives of the form dY_i/dt, $i = 1, 2$.

Observe that the coefficients A, B are strictly positive:

$$0 < \min \frac{\partial \varphi}{\partial P}\,\frac{1}{\max \varphi^2} \le A \le \max \frac{\partial \varphi}{\partial P}\,\frac{1}{\min \varphi^2} < \infty,$$

$$0 < \frac{\min \varphi}{\max \mu} \le B \le \frac{\max \varphi}{\min \varphi^2} < \infty.$$

Then by the Cauchy inequality with $\gamma > 0$ small enough, we get from (6.37)

$$\frac{d}{dt}\int_{G(t)} A\tilde{P}^2\,dy + \int_{G(t)} B\left|\frac{\partial \tilde{P}}{\partial y}\right|^2 dy$$
$$\le \gamma \int_{G(t)} B\left|\frac{\partial \tilde{P}}{\partial y}\right|^2 dy + M(1+\alpha(t))\left(1+\frac{1}{\gamma}\right)\int_{G(t)} A\tilde{P}^2\,dy + M\sum_{i=1}^{2}\tilde{P}^2\left(Y_i(t),t\right).$$

Recall that M always stands for constants which depend only on norms of the functions included in the original equation, the initial and boundary conditions, and the constants in conditions (U.6)–(U.9).

In turn,

$$\tilde{P}^2(Y_i(t),t) \le M\left(1+\frac{1}{\gamma}\right)\int_{G(t)} A\tilde{P}^2 dy + \gamma\int_{G(t)} B\left|\frac{\partial\tilde{P}}{\partial y}\right|^2 dy.$$

Thus, choosing γ small enough and adopting the notation

$$z(t) = \int_{G(t)} A\tilde{P}^2(y,t)dy,$$

we obtain the inequality

$$\frac{dz}{dt} \le M\,(1+\lambda(t))\,z, \quad z(0) = z_0.$$

Since $\lambda(t) \in L_1(0,T)$ (estimate (6.36)), it follows by Gronwall's inequality that $z(t) \equiv 0$.

The uniqueness of a solution to problem (L_2) is proved in the same way.

6.7 Global in time solvability of problem (E_1). Here we present a few conditions which guarantee the existence of a classical solution on an arbitrary time interval.

The existence of a global in time solution to problem (E_1) follows from the global in time solvability of problem (L_1). It proves very difficult to answer the question of when the life span of the solution to problem (L_1) is defined only by the time of degeneracy of the domains $G_1(t)$ and $G_2(t)$. It comes down to the problem of estimating the maximum and minimum of a solution to a nonlinear parabolic equation with nonlinear boundary conditions. The complete study of this problem is beyond the scope of this book. Let us confine ourselves to the simplest case when the density lends itself to estimating without reference to a technique of any complexity.

Lemma 6.7. *In the conditions of Lemma* 6.5, *let* $g_1(t) \ge 0$, $g_2(t) \le 0$. *Then*

$$\rho(x,t) \equiv \varphi\,(P(y,t),y) \ge \rho_- = \min\left\{\min_{y\in G_1(0)}\varphi_1(P_0),\ \min_{y\in G_2(0)}\varphi_2(P_0)\right\}, \tag{6.38}$$

$$\rho(x,t) \le \frac{m(t)}{\min|X_i - h(t)|} + |X_2 - X_1|V^+ \max\left\{\frac{d\varphi_1}{dP}, \frac{d\varphi_2}{dP}\right\}, \tag{6.39}$$

where

$$m(t) = \int_0^t \{g_1(\tau) - g_2(\tau)\}\,d\tau + \int_\Omega \rho_0(x)\,dx,$$

$$V^+ = \max\left\{\frac{1}{\rho_-}|g_1|^{(0)}_{[0,\infty)}, \frac{1}{\rho_-}|g_2|^{(0)}_{[0,\infty)}, \max\frac{\varphi(P_0(y),y)}{\mu(y)}P_0(y)\right\}.$$

Proof. In the first case (estimate (6.38)) one merely has to show that, due to the monotone dependence of the density on the pressure, the latter attains its minimal value at the initial instant of motion:

$$P(y, y) \geq P_- = \min P_0(y).$$

Indeed: assuming the contrary, there exists a point $(\tilde{y}, \tilde{t}) \in \overline{G_{1,T}}$ where

$$\varphi_1(p(\tilde{y}, \tilde{t})) = \rho(\tilde{y}, \tilde{t}) < \rho_-,$$

whence $P(\tilde{y}, \tilde{t}) < P_-$, since

$$\min_{y \in G_i(0)} \varphi_i(P_0(y)) \geq \varphi_i(P_-) \quad \text{and} \quad \varphi_1(P(\tilde{y}, \tilde{t})) < \rho_- \leq \varphi_i(P_-).$$

Let $T_* > 0$ be the maximal time interval on which problem (L_1) has a solution, and let $0 < T < T_*$. In each domain $G_{i,T}$ the pressure $P(y, t)$ satisfies a second-order uniformly parabolic equation to which the maximum principle is applicable: the maximum and minimum values of the classical solutions to this equation can be attained only on the "parabolic" boundary of the problem domain.

Suppose, for instance, that the function P attains its minimum value at a point of the lateral boundary $\{y = Y_1(t)\}$, $t = \tilde{t} \leq T$. According to the parabolic version of Hopf's principle (the lemma about the sign of nontangential derivative [52, Ch. 2]), at this point the derivative in the direction of the inner normal $\partial P(Y_1(\tilde{t}), \tilde{t})/\partial y$ is strictly positive. On the other hand, by the conditions of the lemma and due to the boundary condition (6.12) this derivative is nonpositive. Hence, P cannot have a point of local minimum on the lateral boundary $\{y = Y_1(t)\}$. The opposite lateral boundary is treated likewise.

Assume now that the minimum value of P is attained on the common boundary $\{y = 0\}$ of the domains $G_{1,T}$, $G_{2,T}$ at the instant $\tilde{t} \leq T$. Evidently, this value is the minimum value of P in both the domains $G_{1,T}$ and $G_{2,T}$. Hence, one can use once again the property already utilized ([52, Ch. 2]) to get

$$\frac{\partial P}{\partial y}(0-, \tilde{t}) < 0, \qquad \frac{\partial P}{\partial y}(0+, \tilde{t}) > 0.$$

The last relations contradict condition (6.9) on the contact surface $\{y = 0\}$. Thus, the minimal value of P in the domain $\overline{G_T}$ coincides with P_- and everywhere in G_T the density ρ extends the value ρ_-.

To estimate the maximum of the density (or the pressure) we proceed in two steps.

At the first step we estimate the maximum modulus of $V(y, t)$, satisfying equation (6.25) in the domain G_T with $\varepsilon = 0$, and subject to the conditions

$$V(Y_i(t), t) = -\frac{\varphi_i(P)}{\mu_i} \frac{\partial P}{\partial y}(Y_i(t), t) = \frac{g_i(t)}{\varphi_i(P)}, \qquad i = 1, 2,$$

on the lateral boundaries of G_T. Due to inequality (6.38), the density $\varphi_i(P)$ is bounded from below. Hence, the modulus of $V(y,t)$ is bounded from above on the lateral boundaries $\{y = Y_i(t)\}$, $i = 1, 2$:

$$|V(Y_i(t), t)| \le \frac{1}{\rho_-}|g_i|^{(0)}_{[0,\infty)}, \quad i = 1, 2.$$

For equation (6.25) with $\varepsilon = 0$ the generalized maximum principle holds. This is easy to check by various methods. For instance, let us consider the regularized equation (6.25) with $\varepsilon > 0$ and the corresponding solution $V^\varepsilon(y,t)$, which is already shown to converge to $V(y,t)$ uniformly in ε as $\varepsilon \to 0$. For V^ε the ordinary maximum principle holds

$$\max_{(y,t)\in G_T} |V^\varepsilon| \le \max\left\{\frac{1}{\rho_-}|g_1|^{(0)}_{[0,\infty)}, \frac{1}{\rho_-}|g_2|^{(0)}_{[0,\infty)}, |V_0^\varepsilon|^{(0)}_{G(0)}\right\}.$$

Varying $\varepsilon > 0$ and choosing it as small as we please, we can always get the inequality

$$|V_0^\varepsilon|^{(0)}_{G(0)} \le |V_0|^{(0)}_{G(0)} + \gamma$$

with an arbitrary $\gamma > 0$. Clearly, this estimate will also hold for the limit function $V(y,t)$. Moreover, since γ is arbitrary, it may be taken merely to be zero.

At the second step, it will be convenient to revert to Euler's coordinates (x,t). In these coordinates the modulus of the function $v(x,t)$ given by the equality

$$V(y,t) \equiv v(x,t) = -\frac{1}{\mu(x)}\frac{\partial p}{\partial x}(x,t)$$

is bounded.

Integrate the continuity equation (5.1) over the domain Ω at $t > 0$; after simple transformations we arrive at the mass conservation law

$$\int_\Omega \rho(x,t)\,dx = \int_\Omega \rho_0(x)\,dx + \int_0^t (g_1(\tau) - g_2(\tau))\,d\tau \equiv m(t).$$

The density $\rho(x,t)$ is strictly positive in each of the domains $\Omega_1(t)$ and $\Omega_2(t)$. Hence,

$$\int_{\Omega_i(t)} \rho(x,t)\,dx \le \int_\Omega \rho(x,t) = m(t).$$

As has already been demonstrated, the density $\rho(x,t)$ is continuous in Ω_T. Therefore, the density is continuous in each of the closed domains $\overline{\Omega_{1,T}}$, $\overline{\Omega_{2,T}}$. Let us now make use of the mean value theorem. According to this theorem, each of $\Omega_{i,t}$ contains a continuous line $\{x = h_i(t)\}$, $0 < t < T$, $i = 1, 2$, such that along these lines

$$\rho(h_i(t), t) = \frac{1}{|X_i - h(t)|}\int_{\Omega_i(t)} \rho(x,t)\,dt \le \frac{m(t)}{\min_{i=1,2}|X_i - h(t)|}.$$

Thus, given an arbitrary point $x_0 \in \Omega_i(t)$,

$$\begin{aligned}\rho(x_0, t) &= \rho(h_i(t), t) + \int_{h_i(t)}^{x_0} \frac{\partial \rho}{\partial x}(x, t)\, dx \\ &\le \frac{m(t)}{\min_{i=1,2} |X_i - h(t)|} + |X_i - h(t)| \max_{x \in \Omega_i(t)} \left| \frac{\partial P}{\partial x}(x, t) \right| \\ &\le \frac{m(t)}{\min_{i=1,2} |X_i - h(t)|} + |X_2 - X_1| \left(\max \frac{d\varphi}{dp} \right) \max_{(x,t) \in \Omega_T} \left| \frac{\partial P}{\partial x} \right| .\end{aligned}$$

Since the derivative $\partial P / \partial x$ is bounded, so is the maximum of the density, provided the function

$$q(t) = |\ln (X_2 - h(t))\, (h(t) - X_1)|$$

is bounded. □

7 Proof of Theorem 5.3: Verigin's problem with the given pressure on the known boundaries

7.1 Equivalent problem (L_3). Existence of solution. As in the case of problem (E_1), we make use of Lagrangian variables. Everywhere in the domain G_T, except for the contact line $\{y = 0\}$, the pressure $P(y, t) \equiv p(x, t)$ satisfies the equation

$$\frac{\partial}{\partial t} \left(- \frac{1}{\varphi(P, y)} \right) = \frac{\partial}{\partial y} \left(\frac{\varphi(P, y)}{\mu(y)} \frac{\partial P}{\partial y} \right). \tag{7.1}$$

On the contact line $\{y = 0\}$ the compatibility conditions are fulfilled:

$$P(0-, t) = P(0+, t), \tag{7.2}$$

$$\frac{\varphi(P, 0-)}{\mu(0-)} \frac{\partial P(0-, t)}{\partial y} = \frac{\varphi(P, 0+)}{\mu(0+)} \frac{\partial P(0+, t)}{\partial y}. \tag{7.3}$$

Here $\varphi(P, y)$ and $\mu(y)$ are the same as in Section 6:

$$\varphi(P, y) = \varphi_1(P), \quad \mu(y) = \mu_1 \text{ for } y < 0,$$

$$\varphi(P, y) = \varphi_2(P), \quad \mu(y) = \mu_2 \text{ for } y > 0.$$

The domain G_T is the image of Ω_T after the passage to Lagrangian coordinates:

$$G_T = \{(y, t) : \ Y_1(t) < y < Y_2(t), \ 0 < t < T\},$$

$$G_{1,T} = G_T \cap \{y > 0\}.$$

However, in contrast to problem (E_1), in problem (E_3) the domain G_T is unknown. That is, the reformulation also leads to a free boundary problem. A natural question is:

how necessary is this change of independent variables? To our opinion, the equivalent problem posed in the Lagrange coordinates is easier to investigate. Despite additional difficulties caused by the change of variables (the images of the boundaries $\{x = X_i\}$, $i = 1, 2$ are also the sought functions), this problem is reduced to the well-studied Stefan problem. Namely, at the free boundary $\{y = Y_i(t)\}$, $i = 1, 2$, we have the conditions

$$P(Y_i(t), t) = P_i(t), \tag{7.4}$$

$$\frac{dY_i}{dt}(t) = \frac{\varphi_i(P_i(t))}{\mu_i} \frac{\partial P}{\partial y}(Y_i(t), t). \tag{7.5}$$

The last condition is analogous to Stefan's condition and is derived in the same way as relation (6.11) of Section 6.

At the initial instant

$$P(y, 0) = P_0(y), \quad Y_i(0) = Y_i^o, \quad i = 1, 2, \tag{7.6}$$

where $P_0(y)$, Y_i^0 are known functions.

Problem (7.1)–(7.6) of finding functions Y_1, $Y_2(t)$, $P(y, t)$ we will term *problem* (L_3). The proof of existence of a classical solution to problem (L_3) repeats almost literally the analogous proof for problem (L_1). The difference consists in the necessity to seek for functions $Y_i(t)$. That is why we have to perform the complementary step – to make use of the fixed point theorem. Namely, let us fix a set $\mathcal{M}$ of vector functions $R(t) = \{R_1(t), R_2(t)\}$. Given a function $R(t)$, let us construct the domain

$$G_T^R = \{(y, t): \ R_1(t) < y < R_2(t), \ 0 < t < T\}.$$

In the domain G_T^R, one has to solve problem (7.1)–(7.4), (7.6) to define the function $\tilde{P}(y, t)$. The proof of solvability of this problem is identical to that of problem (L_1).

Next, we consider the functions

$$Z_i(t) = Y_i^0 + \int_0^t \frac{\varphi_i(P_i(\tau))}{\mu_i} \frac{\partial \tilde{P}}{\partial y}(R_i(\tau), \tau)\, d\tau, \tag{7.7}$$

defining the operator

$$Z = \Phi(R) \equiv \{Z_1(t), Z_2(t)\}$$

on the set $\mathcal{M}$. The fixed points of this operator $Y = \{Y_1(t), Y_2(t)\}$ and the solution $P(y, t)$ of the auxiliary problem corresponding to these points yield a solution to the original problem (L_3).

Thus, let us take for $\mathcal{M}$ the set of vector-valued functions from $C^{(2+\alpha)/2}[0, T]$ such that

$$R_i(0) = Y_i^0, \quad \left|R_i(t) - Y_i^0\right| \le \delta, \quad i = 1, 2, \tag{7.8}$$

$$\frac{dR_i}{dt}(0) = q_i^0 = \frac{\varphi_i(P_i(0))}{\mu_i} \frac{dP_0}{dt}(Y_i^0), \quad i = 1, 2, \tag{7.9}$$

$$|R_i|_{[0,T]}^{((2+\alpha)/2)} \le N, \quad i = 1, 2. \tag{7.10}$$

Here $3\delta = \min\left\{\left|Y_1^0\right|\right\}$, $N > N_0$ is an arbitrary number,

$$N_0 = \max\left\{\left|Y_1^0\right|, \left|Y_2^0\right|\right\} + \max\left\{\left|q_1^0\right|, \left|q_2^0\right|\right\}.$$

The solvability of the auxiliary problem is shown in the following assertion.

Lemma 7.1. *Let conditions* (U.1)–(U.3), (U.5) *hold, and let* $R \in \mathcal{M}$ *be an arbitrary vector function defining the domain* G_T^R. *Given an arbitrary interval* $(0, T)$, *there exists a unique solution of problem* (7.1)–(7.4), (7.6) *such that*

$$\tilde{P} \in C^{\beta,\beta/2}\left(\overline{G_T^R}\right) \cap C^{2+\alpha,(2+\alpha)/2}\left(\overline{Q}\right),$$

where $Q \subset G_T^R$ *is an arbitrary subdomain bounded away from the line* $\{y = 0\}$ *by a positive distance.*

It follows from the properties of the space $C^{2+\alpha,(2+\alpha)/2}\left(\overline{Q}\right)$ that the derivative $\partial\tilde{P}/\partial y$ belongs to the space $C^{(1+\alpha)/2}[0, T]$ on the line $y = R_i(t)$. Hence the function $Z_i(t)$ defined by (7.7) belongs to $C^{(3+\alpha)/2}[0, T]$ and the operator $\Phi(R)$ is a compact operator on $\mathcal{M}$. Moreover, Φ is completely continuous on $\mathcal{M}$. The proof of the continuity of Φ is standard (the continuous dependence of the solutions of a parabolic equation on its coefficients) and therefore is omitted here.

To apply Schauder's fixed point theorem amounts to showing that Φ maps $\mathcal{M}$ into itself. We shall state this fact for a small enough time interval $[0, T]$.

Let us confine consideration to the case $T < 1$. Then in a neighborhood of the line $\{y = R_i(t)\}$, $i = 1, 2$, the norm of the solution $\tilde{P}$ in the space $C^{2+\alpha,(2+\alpha)/2}$ is bounded by a constant M_0 depending only on N and the problem data. We have:

$$|Z_i|_{[0,T]}^{(0)} \leq \max\left|Y_i^0\right| + T\,M,$$

$$\left|\frac{dZ_i}{dt}\right|_{[0,T]}^{(0)} \leq \max\left|q_i^0\right| + T^{(1+\alpha)/2}M_0,$$

$$\left|\frac{dZ_i}{dt}(t+\tau) - \frac{dZ_i}{dt}(t)\right| \leq M_0\tau^{(1+\alpha)/2} \leq M_0T^{1/2}\tau^{\alpha/2},$$

$$\left|Z_i(t) - Y_i^0\right| \leq T\,M_0.$$

Hence,

$$|Z|_{[0,T]}^{((2+\alpha)/2)} \leq N_0 + 3\,T^{1/2}M_0 < M, \qquad \left|Z_i(t) - Y_i^0\right| < \delta,$$

if $T > 0$ is small enough.

Thus, $\Phi(R)$ maps $\mathcal{M}$ into itself and, by Schauder's theorem, problem (L_3) has at least one solution on a small enough time interval. It is easy to see that away from the contact line $\{y = 0\}$ the smoothness of the solution obtained is defined only by the problem data:

$$P \in C^{3+\alpha,(3+\alpha)/2}\left(\overline{Q}\right),$$

where $Q \subset G_T$ is an arbitrary domain bounded away from the line $\{y = 0\}$ by a positive distance.

By virtue of the restrictions imposed on the data of problem (E$_3$), and due to the maximum principle for the pressure in problem (L$_3$), the density $\varphi(P, y)$ is bounded from above and from below by positive constants. Thus, the solution of problem (L$_3$) can be continued up to the time instant T_* when either $Y_1(T_*)Y_2(T_*) = 0$ (the degeneracy of one of domains $G_{1,T}$ or $G_{2,T}$), or one of the norms

$$|Y_i|_{[0,T]}^{((2+\alpha)/2)} = q_i(T)$$

becomes infinite as $T \to T_*$.

Let us show that the existence time of a solution to problem (L$_3$) is defined, in fact, only by the time of degeneracy of the domains $G_{i,T}$. Indeed, let

$$|Y_1(t)| > 2\delta_1, \quad Y_2(t) > 2\delta_1 = \text{const.} > 0 \quad \text{for} \quad 0 < t < T.$$

In the plane of the original Euler coordinates, the strip

$$S_1 = \{Y_1(t) < y < -\delta_1,\ 0 < t < T\}$$

corresponds to the strip

$$\Pi_1 = \{(x, t) :\ X_1 < x < h_1(t),\ 0 < t < T\},$$

where $\{x = h_1(t)\}$ is the image of the line $\{y = -\delta_1\}$ under the mapping $(y, t) \mapsto (x, t)$.

Since the density is bounded, the line $\{x = h_1(t)\}$ is separated from the free boundary $\{x = h(t)\}$:

$$\delta_1 = \int_{h(t)}^{h_1(t)} \rho(x, t)\, dx \le |h(t) - h_1(t)| \max \rho(x, t).$$

Therefore, we can make use of local estimates for solutions of parabolic equations near the boundary $\{x = X_1\}$: the $C^{3+\alpha,(3+\alpha)/2}(\overline{\Pi_1})$–norm of the solution $p(x, t)$ is bounded in the strip Π_1 by a constant depending only on the problem data and the constant δ_1.

Recalling the rule of the change of variables

$$\frac{\partial P}{\partial y}(Y_1(t), t) = \varphi_1(P_1(t))\frac{\partial P}{\partial x}(X_1, t),$$

we conclude that $q_1(t)$ is bounded by a constant depending only on δ_1. The boundary $\{x = X_2\}$ is considered in the same manner.

Lemma 7.2. *Let conditions* (U.1)–(U.3), (U.5) *be fulfilled. Then on a small enough time interval* $(0, T_*)$ *problem* (L$_3$) *has at least one solution*

$$P \in C^{\beta,\beta/2}(\overline{G_T}) \cap C^{3+\beta,(3+\beta)/2}(\overline{Q}), \quad Y_i \in C^{(3+\beta)/2}[0, T_*],$$

$i = 1, 2$, *where* Q *is an arbitrary domain separated from the line* $\{y = 0\}$ *by a positive distance.*

Also, the function V *defined in Section 6 belongs to the space* $C^{\beta,\beta/2}\left(\overline{G_T}\right)$ *and satisfies estimate* (6.33).

The instant $T_* > 0$ *is defined as the moment when one of the domains* $G_{i,T}$ *degenerates:*

$$\lim_{T\to T_*} Y_1(T)Y_2(T) = 0.$$

7.2 Uniqueness of solution to problem (L_3). As in problem (E_1), the uniqueness of a solution to problem (E_3) is a byproduct of the uniqueness result for its Lagrangian counterpart (L_3). However, to pose an initial-boundary value problem for the difference of two possible solutions some additional transformations are required. This is because different solutions correspond to different domains G_T.

Suppose that $\min\{|Y_1(t)|, Y_2(t)\} > 0$ on the interval $(0, T)$ and consider the new independent variables (ξ, t) introduced by the rules

$$t = t, \quad \xi = \frac{y}{|Y_i(t)|} \text{ for } (y,t) \in G_{i,T}, \quad i = 1, 2. \tag{7.11}$$

In this change of independent variables the domain $G_{i,T}$ transforms into the domain

$$Q_i = \left\{(\xi, t) : 0 < (-1)^i\xi < 1,\ 0 < t < T\right\}.$$

In each domain Q_i, $i = 1, 2$, the function $u(\xi, t) \equiv P(y, t)$ satisfies the equation

$$|Y_i(t)|\,\frac{\partial}{\partial t}\left(-\frac{1}{\varphi_i(u)}\right) = -\frac{\partial}{\partial \xi}\left(\frac{1}{\varphi_i(u)}\right)\xi : \frac{dY_i}{dt} + \frac{\partial}{\partial \xi}\left[\frac{\varphi_i(u)}{\mu_i|Y_i(t)|}\,\frac{\partial u}{\partial \xi}\right]$$

and the conditions

$$u = P_i(t), \quad |Y_i|\frac{dY_i}{dt} = \frac{\varphi_i(P_i(t))}{\mu_i}\,\frac{\partial u}{\partial \xi}$$

on the lateral boundaries $\{\xi = (-1)^i\}$, $i = 1, 2$.

On the line $\{\xi = 0\}$, where the coefficients become discontinuous, the following conditions hold:

$$u(0-, t) = u(0+, t), \quad \frac{\varphi_1(u)}{\mu_1|Y_1(t)|}\,\frac{\partial u}{\partial \xi}(0-, t) = \frac{\varphi_2(u)}{\mu_2|Y_2(t)|}\,\frac{\partial u}{\partial \xi}(0+, t).$$

One may consider now two possible solutions

$$\{u^{(1)}, Y_1^{(1)}, Y_2^{(1)}\}, \quad \{u^{(2)}, Y_1^{(2)}, Y_2^{(2)}\}$$

and deduce an initial-boundary value problem for the difference $\tilde{u} = u^{(1)} - u^{(2)}$, $\tilde{Y}_i = Y_i^{(1)} - Y_i^{(2)}$, $i = 1, 2$. We have:

$$a_{1i}\,\frac{\partial \tilde{u}}{\partial t} - \frac{\partial}{\partial \xi}\left[a_{2i}\,\frac{\partial \tilde{u}}{\partial \xi} + a_{3i}\tilde{u} + a_{4i}\tilde{Y}_i\right] + a_{5i}\,\frac{\partial \tilde{u}}{\partial \xi} + a_{6i}\tilde{u} = b_{1i}\,\frac{d\tilde{Y}_i}{dt} + b_{2i}\tilde{Y}_i \tag{7.12}$$

$$\text{in } Q_i, \; i = 1, 2,$$

$$\tilde{u} = 0, \quad b_{3i} \frac{\partial \tilde{u}}{\partial \xi} = \frac{d\tilde{Y}_i}{dt} + b_{4i}\tilde{Y}_i \quad \text{as } \xi = (-1)^i, \tag{7.13}$$

$$\tilde{u}(0-,t) = \tilde{u}(0+,t), \tag{7.14}$$

$$a_{21}\frac{\partial \tilde{u}}{\partial \xi}(0-,t) + a_{31}\tilde{u} + a_{41}\tilde{Y}_1 = a_{22}\frac{\partial \tilde{u}}{\partial \xi}(0+,t) + a_{32}\tilde{u} + a_{42}\tilde{Y}_2 \quad \text{as } \xi = 0, \tag{7.15}$$

$$\tilde{u}(\xi,0) = 0, \quad \tilde{Y}_i(0) = 0. \tag{7.16}$$

In (7.12)–(7.16) we used the notation

$$a_{1i} = |Y_1^{(1)}|\frac{\varphi_i(u^{(1)})}{\varphi_i^2(u^{(1)})}, \quad a_{2i} = \frac{\varphi_i(u^{(1)})}{\mu_i|Y_i^{(1)}|},$$

$$a_{3i} = \frac{\varphi_i\left(u^{(1)}\right) - \varphi_i\left(u^{(2)}\right)}{\left(u^{(1)} - u^{(2)}\right)\mu_i|Y_i^{(1)}|}\frac{\partial u^{(2)}}{\partial \xi}, \quad a_{4i} = -\frac{\varphi_i\left(u^{(2)}\right)}{\mu_i|Y_i^{(1)}||Y_i^{(2)}|},$$

$$a_{5i} = -\frac{\varphi_i\left(u^{(1)}\right)}{\varphi_i^2\left(u^{(1)}\right)}\xi\frac{dY^{(1)}}{dt},$$

$$a_{6i} = \frac{a_i\left(u^{(1)}\right) - a_i\left(u^{(2)}\right)}{u^{(1)} - u^{(2)}}\left(Y_i^2\frac{\partial u^{(2)}}{\partial t} - \xi\frac{dY_i}{dt}\right), \quad a_i(u) = \frac{\varphi_i(u)}{\varphi_i^2(u)},$$

$$b_{1i} = \xi\, a_i(u^{(1)})\frac{\partial u^{(2)}}{\partial \xi}, \quad b_{2i} = -a_i(u^{(1)})\frac{\partial u^{(2)}}{\partial t},$$

$$b_{3i} = \frac{\varphi_i\left(P_i(t)\right)}{\mu_i|Y_i^{(1)(t)}|}, \quad b_{4i} = \frac{\varphi_i\left(P_i(t)\right)}{\mu_i|Y_i^{(1)}||Y_i^{(2)}|}\frac{\partial u^{(2)}}{\partial \xi}((-1)^i,t).$$

Since all coefficients and their derivatives in ξ are Hölder continuous everywhere outside the line $\{\xi = 0\}$, the equation is uniformly parabolic in each domain Q_1 and Q_2:

$$a_{1i},\ a_{2i} \geq a_0 = \text{const.} > 0;$$

The coefficients a_{ji}, b_{ki}, $j = 1,\ldots,5$, $k = 1,\ldots,4$, $i = 1,2$, are uniformly bounded, and the functions a_{6i}, $\partial a_{1i}/\partial t$ are bounded in the norm

$$\int_0^T \left\{\max_\xi |a_{6i}(\xi,t)| + \max_\xi \left|\frac{\partial a_{1i}}{\partial t}\right|\right\} dt, \quad i = 1,2.$$

Moreover,

$$\max_{t\in(0,T)} \int_{-1}^{1} |b_{2i}(\xi,t)|^2 d\xi < \infty, \quad i = 1,2.$$

First, let us estimate the L_2-norm of the difference $\tilde{u}$ through the norm

$$\|\tilde{Y}\|^{(1)}_{2,[0,T]} = \left(\sum_{i=1}^{2}\int_0 \left(\tilde{Y}_i^2(t) + \left|\frac{d\tilde{Y}_i}{dt}(t)\right|^2\right) dt\right)^{1/2}.$$

For this we use all the conditions (7.12)–(7.16) except for the second condition of (7.13). This can be done in a routine way by multiplication of equation (7.12) by $\tilde{u}$ with the consequent integration of the result with respect to ξ in the limits $(-1, 0)$ and $(0, 1)$, and summing the results of integration. Integrating by parts and using conditions (7.14), (7.15), and then making use of the Cauchy inequality with a small $\varepsilon > 0$, after some obvious transformations we arrive at the inequality

$$\frac{d}{dt}\int_{-1}^{1} \tilde{a}_1\, \tilde{u}^2\, d\xi \leq A(t)\int_{-1}^{1} \tilde{a}_1 \tilde{u}^2\, d\xi + B\sum_{i=1}^{2}\left\{\tilde{Y}_i^2(t) + \left|\frac{d\tilde{Y}_i}{dt}(t)\right|^2\right\},$$

where $\tilde{a}_1 = a_{11}$ for $\xi < 0$ and $\tilde{a}_1 = a_{12}$ for $\xi > 0$, B is a positive constant, and $A(t)$ is a function summable on the interval $(0, T)$.

It follows from Gronwall's inequality that

$$\max_{0<t<T}\left(\int_{-1}^{1} \tilde{u}^2(\xi, t)\, d\xi\right)^{1/2} \leq M \left\|\tilde{Y}\right\|_{2,[0,T]}^{(1)}. \tag{7.17}$$

(Recall that elsewhere M denotes a constant which can be calculated only through the problem data.)

Let

$$\Pi_i = \left\{(\xi, t) : \frac{1}{2} < (-1)^i \xi < 1,\ i = 1, 2\right\}.$$

(All other estimates can be derived in the domains Π_i that adjoin the corresponding lateral boundaries $\{\xi = (-1)^i\}$.) Viewing equation (7.12) as a non-homogeneous equation with the right-hand side

$$f_i = b_{1i}\frac{d\tilde{Y}_i}{dt} + b_{2i}\tilde{Y}_i, \quad i = 1, 2,$$

and evoking local estimates for the solutions of linear parabolic equations [97, Ch. 4, Sec. 10], we get

$$|u|_{\Pi_i}^{(2+\alpha)} \leq M\left(|\tilde{Y}_i|_{[0,T]}^{((2+\alpha)/2)} + \|u\|_{2,Q_i}\right), \tag{7.18}$$

$$\|u\|_{2,\Pi_i}^{(2)} \leq M\left(T^{1/2}|\tilde{Y}_i|_{[0,T]}^{(1)} + \|u\|_{2,Q_i}\right). \tag{7.19}$$

The second condition in (7.13) can be viewed as a linear differential equation for the function $\tilde{Y}_i(t)$, $i = 1, 2$. We have:

$$|\tilde{Y}|_{[0,T]}^{((3+\alpha)/2)} \leq M\left|\frac{\partial \tilde{u}}{\partial \xi}\right|_{\Pi_i}^{(1+\alpha)} \leq M\,|\tilde{u}|_{\Pi_i}^{(2+\alpha)}, \tag{7.20}$$

$$|\tilde{Y}_i|_{[0,T]}^{(0)} \leq M\,T\,|\tilde{u}|_{\Pi_i}^{(1)}, \quad i = 1, 2. \tag{7.21}$$

It follows from estimate (7.17) that

$$\|u\|_{2,Q_i} \leq M\,T \sum_{i=1}^{2} |\tilde{Y}_i|^{(1)}_{[0,T]}. \tag{7.22}$$

Sequentially applying estimates (7.20). (7.18), and (7.22), we get

$$\sum_{i=1}^{2} |\tilde{Y}_i|^{((3+\alpha)/2)}_{[0,T]} \leq \sum_{i=1}^{2} |\tilde{Y}_i|^{(2+\alpha)/2}_{[0,T]}. \tag{7.23}$$

On the other hand, the interpolation inequality ([97, Ch. 2, Sec. 3])

$$|\tilde{Y}|^{((2+\alpha)/2)}_{[0,T]} \leq T^{1/4} |\tilde{Y}_i|^{((3+\alpha)/2)}_{[0,T]} + C\,T^{-(2+\alpha)/4} |\tilde{Y}_i|^{(0)}_{[0,T]},$$

estimate (7.21), the analogous inequality for the function $\tilde{u}$

$$|\tilde{u}|^{(1)}_{\Pi_i} \leq T^{(1+\alpha)/2} |\tilde{u}|^{(2+\alpha)}_{\Pi_i} + C\,T^{-1/2} |\tilde{u}|^{(0)}_{\Pi_i},$$

and estimates (7.18), (7.22) provide the estimate

$$\sum_{i=1}^{2} |\tilde{Y}_i|^{((2+\alpha)/2)}_{[0,T]} \leq M \left(T^{1/4} \sum_{i=1}^{2} |\tilde{Y}_i|^{((3+\alpha)/2)}_{[0,T]} + T^{\alpha/4} \sum_{i=1}^{2} |\tilde{u}|^{(0)}_{\Pi_i} \right).$$

To estimate the extreme values of the solution, let us have recourse once again to the interpolation inequalities

$$|\tilde{u}|^{(0)}_{\Pi_i} \leq T^{1/2} |\tilde{u}|^{(2)}_{2,\Pi_i} + C\,T^{-3/4} \|\tilde{u}\|_{2,\Pi_i},$$

and estimates (7.19), (7.22):

$$\sum_{i=1}^{2} |\tilde{u}|^{(0)}_{\Pi_i} \leq M\,T^{1/4} \sum_{i=1}^{2} |\tilde{Y}_i|^{(1)}_{[0,T]}.$$

Thus,

$$\begin{aligned} \sum_{i=1}^{2} |\tilde{Y}_i|^{((2+\alpha)/2)}_{[0,T]} &\leq M \left(T^{1/4} \sum_{i=1}^{2} |\tilde{Y}_i|^{((3+\alpha)/2)}_{[0,T]} + T^{(1-\alpha)/4} \sum_{i=1}^{2} |\tilde{Y}_i|^{(1)}_{[0,T]} \right) \\ &\leq M\,T^{(1-\alpha)/4} \sum_{i=1}^{2} |\tilde{Y}_i|^{((3+\alpha)/2)}_{[0,T]}. \end{aligned}$$

Returning to estimate (7.23), we finally get:

$$\sum_{i=1}^{2} |\tilde{Y}_i|^{((3+\alpha)/2)}_{[0,T]} \leq M\,T^{(1-\alpha)/4} \sum_{i=1}^{2} |\tilde{Y}_i|^{((3+\alpha)/2)}_{[0,T]}.$$

If the time interval $[0, T]$ is small enough, the last inequality is possible only if

$$\tilde{Y}_1(t) \equiv \tilde{Y}_2(t) \equiv 0.$$

If this process is continued, the interval $[0, T]$ is exhausted within a finite number of steps.

Relying on the existence and uniqueness results for problem (L_3), we conclude similarly to the previous section that the assertion of Theorem 5.3 is true.

Chapter II

Equivalence transformations of evolution equations

8 Main ideas. A historical survey

This chapter deals with the class of one-dimensional evolution equations of divergence form. One of the simplest representatives of this class is the equation

$$u_t = [\varphi(u, u_x)]_x \,. \tag{8.1}$$

A particular case of (8.1) is the nonlinear heat conduction equation

$$u_t = [k(u)u_x]_x \,. \tag{8.2}$$

The equation

$$[S(v)]_t = [C(v, v_x)]_x \,, \tag{8.3}$$

often occurring in applications, can be reduced to (8.1) by the change of the sought function $u = S(v)$.

In 1982, practically simultaneously, W. Strampp [165] and J. G. Kingston, C. Rogers [85] published papers in which they pointed out a nonlocal transformation conserving the family of equations (8.1). Let us note that W. Strampp was dealing with equation (8.3), equivalent to (8.1), while the authors of [85] treated a divergence equation of arbitrary order with two independent variables. The above-mentioned transformation consists in passing in (8.1) to the new independent variable ξ, replacing x, and the new sought function ω, replacing $u(x, t)$, by the formulas

$$\begin{cases} \xi = \int_0^x u(y,t)dy + \int_0^t \varphi\,[u(0,s), u_x(0,s)]\,ds, \\ \omega(\xi, t) = [u(x,t)]^{-1} \,. \end{cases} \tag{8.4}$$

The function ω obeys the equation

$$\omega_t = \left[-\omega\varphi\left(\omega^{-1}, -\omega^{-3}\omega_\xi\right)\right]_\xi \,. \tag{8.5}$$

The authors of [165, 85] interpreted (8.4) as a Bäcklund transformation. (The reader is referred to monograph [143] devoted to the theory and applications of Bäcklund transformations; see also [72, Sec. 14].) It is proven that the transformation (8.4) may be given another interpretation [137]: having viewed equation (8.1) as expressing the continuity law of an imaginary motion of a liquid with density u, this transformation corresponds to passing in this equation to Lagrangian coordinates. As a matter of fact, in application to the one-dimensional evolution equations both approaches are equivalent. However, one can also pass to Lagrangian coordinates in multi-dimensional equations of divergence form [109, 65], but apparently there are no natural generalizations of Bäcklund transformations in the multi-dimensional case.

Let us term the change of variables (8.4) an *L-transformation*. This transformation establishes the correspondence between equation (8.1) and equation (8.5) of the same class by the rule

$$\varphi(u,p) \mapsto \psi(u,p) \equiv -u\varphi\left(u^{-1}, -u^{-3}p\right). \tag{8.6}$$

It is essential that the variable x does not enter equation (8.1) in explicit form. In Section 9 we construct a nonlocal transformation which makes invariant another class of second-order equations, namely, $u_t = [\lambda(x, u_x)]_x$. Analogs of L-transformations for higher-order equations and some systems of equations are found in paper [136] and the already-cited paper [85]; these results are presented in Section 14.

Relation (8.6) implies the equality

$$-u\psi(u^{-1}, -u^{-3}p) = \varphi(u,p)$$

which means that the L-transformation is *involutive*, i.e. $L^2 = I$. It may happen, though, that equation (8.5), having been subject to an L-transformation, coincides with the original. These equations we shall term *L-invariant*. A comparison of (8.1) and (8.5) gives the conditions of L -invariance

$$\varphi(u,p) = -u\varphi(u^{-1}, -u^{-3}p), \tag{8.7}$$

which can be viewed as a functional equation for the function φ. Classes of particular solutions of (8.7) containing arbitrary functions of one variable were discovered in [31, 136]. The general solution of equation (8.7) for positive u is presented in [135]. Section 12 is devoted to Lagrange-invariant equations.

Having applied an L-transformation to equation (8.1), we obtain equation (8.5) which admits a Lie group larger than does (8.1). For instance, equation (8.2)

$$u_t = \left[u^{-2}\exp\left(u^{-1}\right)u_x\right]_x$$

possesses a three-parameter group. An L-transformation transforms it into the equation

$$\omega_t = \left[\exp(\omega)\omega_\xi\right]_\xi,$$

which possesses a four-parameter group. It is natural to term the property of extension of the group admitted by equation (8.1) the *hidden symmetry property* of this equation

[136] caused by nonlocal transformations. Examples of explicit solutions to equations invariant with respect to nonlocal symmetries are constructed in [137, 2, 3]. In [1] I. Sh. Akhatov, R. K. Gazizov, and N. H. Ibragimov developed the theory of nonlocal symmetries of evolution equations and applied it to the group analysis of the gas dynamics equations. Examples of equations with hidden symmetries and their nonlocal-invariat solutions are given in Section 10.

An important class of equations with hidden symmetries is constituted by those equations which become linear under transformations of the type (8.4). The first example of such equations

$$u = \left(u^{-2}u_x\right)_x \tag{8.8}$$

was found by M. Storm [164] in 1951. Later on, equation (8.8) attracted a number of investigators. Section 15 is especially devoted to this equation. Starting from a linear equation of the form (8.1) and combining the L-transformation with the Florin–Hopf–Cole transformation and with pointwise transformations, one may obtain new classes of linearizable equations. The questions of linearization of similar (and more general) equations were studied in papers by A. Munier, J. R. Burgan, J. Gutierrez, E. Fialkow, M. R. Feix [112], G. Rosen [146, 145], A. S. Fokas, B. Fuchssteiner [48], A. S. Fokas, Y. C. Yortsos [49], V. A. Dorodnitsyn, S. R. Svirshchevskii [44], V. V. Pukhnachov [137]. Some of the results of these papers are presented in Section 11. With the help of transformations analogous to (8.4) one even can linearize some higher-order equations [85, 137] — see Section 14.

Until now we were mainly talking about "autonomous" equations (that is, not containing any of the independent variables in explicit form) of the type (8.1) and their higher-order analogs. The equivalence transformations studied in this chapter can be easily extended to the case when the right-hand side of an equation, solved with respect to u_t, depends on t. As for any explicit dependence on x, to conserve the class of divergence equations with one independent space variable via the passage to Lagrangian coordinates, this dependence has to be of a special form. Several examples of such equations are considered in Section 13. One of them appears in the analysis of solutions of the heat conduction equation with the gradient nonlinearity $u_t = \operatorname{div}[k(|\nabla u|)\nabla u]$, which is invariant with respect to rotations. Another one is the multi-dimensional analog of equation (8.8). Those solutions of this equation which possess cylindrical or spherical symmetries obey the one-dimensional equation

$$u_t = r^{-\nu}\left(r^{\nu}u^{-2}u_r\right)_r, \quad \nu = 1, 2,$$

for which an L-transformation exists.

One of the topics of Section 13 is transforming to Lagrangian coordinates the nonlinear Schrödinger equation for fields with cylindrical or spherical symmetries. This equation occurs in nonlinear optics, in plasma physics, and has the form

$$2i\frac{\partial E}{\partial z} + \frac{1}{r^{\nu}}\left(r^{\nu}\frac{\partial E}{\partial r}\right) + E|E|^2 = 0. \tag{8.9}$$

Here E is a complex-valued function and z plays the role of the evolution variable. This transformation was found in 1977 by L. M. Degtyaryov, V. V. Krylov [39]. The authors of this paper used the mass Lagrangian coordinates to construct an effective algorithm for numerical treatment of an initial-boundary value problem for equation (8.9). Paper [39] is also remarkable for the fact that it gives, apparently, the first example of application of Lagrangian coordinates in an "abstract" situation when the equation under study does not possess any direct hydrodynamical interpretation.

The remainder of this section contains a survey of works where Lagrangian coordinates were used to study the Cauchy problem and initial-boundary value problems for parabolic equations of the type (8.1) and, particularly, to study free boundary problems. It is worth noting here that some features of transformation (8.4) force us to confine our considerations to the sign-definite (positive, as a rule) solutions of the corresponding problems. Also, the positive solutions of parabolic equations are of both much theoretical and practical interest because the sought function in equations (8.1), (8.2) usually represents a positive magnitude: the temperature, the concentration, the density, a function of distribution, the density of probability, etc.

For instance, the equation

$$u_t = \left(u^m\right)_{xx}, \quad x \in \mathbb{R},\ t > 0, \tag{8.10}$$

covers the process of heat propagation (of the passive solvent propagation) in a nonlinear medium with the coefficient of heat conductivity (diffusion) being a power function of the temperature (the concentration), or the process of isentropic filtration of the polytropic gas in a porous medium. In the latter case, the nonnegative function $u(x,t)$ is identified with the gas density and $m-1$ with the polytropic exponent. Let us consider for equation (8.10) the Cauchy problem

$$u = u_0(x), \quad x \in \mathbb{R},\ t = 0, \tag{8.11}$$

where the nonnegative function $u_0 \in L^1(\mathbb{R})$ is compactly supported. A characteristic feature of the solutions of problem (8.10), (8.11) is that for $m > 1$ the support of the solution is also finite for each $t > 0$. This is a consequence of the degeneracy of equation (8.10) at the level $u = 0$.

There is a voluminous literature devoted to the Cauchy problem for degenerate parabolic equations represented by equation (8.10) (see Section 17 and the references therein). In 1980 J. G. Berryman [20] proposed a new approach to the study of problem (8.10)–(8.11) based on the passage to the mass Lagrangian coordinate ξ. The passage to Lagrangian coordinates leads to the initial-boundary value problem

$$\begin{cases} \left(\theta^{-1/m}\right)_t = \theta_{\xi\xi} & 0 < \xi < 1,\ t > 0, \\ \theta(0,t) = \theta(1,t) = 0, & t \geq 0, \\ \theta(\xi,0) = \theta_0(\xi), & 0 \leq \xi \leq 1. \end{cases}$$

Such a reformulation of the problem is convenient for the study of the asymptotic behavior of the support of the solutions (unknown on the plane (x,t)), for investigation of stability of the solutions. Also, the reduction of the original problem to an initial-boundary value problem with the time independent problem domain gives certain advantages in the numerical treatment. (For problem (8.10)–(8.11) a related algorithm was first constructed and realized in [65].)

The idea of using Lagrangian coordinates as a tool to analyze the Cauchy problem for degenerate parabolic equations of the type (8.1)–(8.3) turned out to be very fruitful. Chapter 3 is entirely devoted to its development. Earlier, in Chapter 1, the effectiveness of the method was demonstrated in application to the problem of advancement of the surface of contact discontinuity (Verigin's problem). There the crucial point was that the interface between two liquids (a priori unknown) described a vertical line in the plane of Lagrangian coordinates. The contents of Chapter 1 is an amplified version of paper [105] by A. M. Meirmanov. The results of this paper were announced earlier in note [106]. These publications were preceded by paper [108] by the same author where a free boundary problem for equations of nonstationary filtration was considered. To be precise, it was concerned with the motion of a heavy compressible fluid in a vertical well in a porous soil. Apparently, paper [108] is the first example of analysis of qualitative properties of a solution to a free boundary problem for a parabolic equation with the use of Lagrangian coordinates.

Another possibility of employing Lagrangian coordinates in applied problems is to approximate the coefficients of the original nonlinear equation and the boundary conditions so that the approximate problem would become linearizable via the passage to Lagrangian coordinates. This question was discussed in papers [49, 144] where the one-dimensional problem of two-phase filtration of fluids in a porous medium was considered, and in paper [142] in application to the problem of precipitation of sediments in a vertical column. The sought function in the first problem is the saturation $S(x, t)$, and in the second one the ratio of the volume concentrations of the fluid and the admixture. Under very special (although realistic) dependences of the phase permabilities on the saturations in the first case, and the coefficients of concentrations and the hydraulic conductivities on the ratio of concentrations in the second case, in both cases the function S can be shown to satisfy the equation

$$\frac{\partial S}{\partial t} = \frac{\partial}{\partial x}\left[\frac{1}{(\beta S+\gamma)^2}\frac{\partial S}{\partial x}\right] + \left[\lambda - \frac{\omega\beta}{(\beta S+\gamma)^2}\right]\frac{\partial S}{\partial x}.$$

Here β, γ, λ, and ω are constants. The last equation is linearized by means of an L-transformation. Moreover, it turns out that the natural boundary conditions in both problems also become linear in the ξ, t plane. As a result, we obtain the explicitly solvable models of one-dimensional two-phase filtration. It was this idea that laid the groundwork for construction of explicit solutions to the Cauchy–Verigin problem presented at the end of Section 16.

Concluding this section, let us mention papers [137, 135, 141, 153] where the one-dimensional one-phase Stefan problem was studied with the help of Lagrangian coordinates. (The statement of the one-phase Stefan problem is given in Section 4.) Assuming that the heat flux is given at the known boundary, or that all the lateral boundary of the problem domain is free, in the plane of Lagrangian coordinates we get a fixed boundary problem for equation (8.2). Proceeding in this way, we construct, for example, a functional-arbitrary family of explicit solutions of Stefan's problem for equation (8.8). This result was announced in note [135] and is presented in complete detail in Section 16.

9 Reciprocity transformations of second-order equations

All considerations of this section are local. It is assumed that all functions we are dealing with are sufficiently smooth functions of their arguments and that all solutions of the equations under study are classical.

Let us start our considerations with the equation

$$u_t = [\varphi(u, u_x)]_x \,. \tag{9.1}$$

This equation has the form of a conservation law with density u. Also, it does not contain explicitly the independent variables x, t. Set

$$\rho(x,t) = u, \quad V(x,t) = -u^{-1}\varphi(u, u_x)\,. \tag{9.2}$$

Then (9.1) takes the form

$$\rho_t + (\rho V)_x = 0. \tag{9.3}$$

If ρ is identified with the density and V with the velocity of some one-dimensional motion of a continuous medium, then (9.3) becomes the continuity equation of this motion.

Let us pass in (9.3) to the new independent variables ξ (Lagrangian coordinate), $t = \tau$, and introduce the new sought functions $\rho = \gamma(\xi, \tau)$, $x = X(\xi, \tau)$. The function X determines the position of the "liquid particle" at the moment τ, and ξ characterizes its initial location. If the function $V(x,t)$ is known, then to define the dependence of X on (ξ, τ) one has to solve the trajectory equation

$$X_\tau = V(X, \tau) \tag{9.4}$$

with the initial condition $X(\xi, 0) = X_0(\xi)$. Let us subject X_0 to the relations

$$X_0(0) = 0, \quad \rho(X_0, 0)\,(X_0)_\xi = 1\,.$$

The latter means that ξ is the mass Lagrangian coordinate: the value of ξ equals the liquid mass per square unit included between the planes $x = 0$ and $x = X_0$ at the moment $\tau = 0$.

Now observe that by virtue of (9.4) the following relations hold

$$\rho_t + V\rho_x = \gamma_\tau, \quad V_x = X_{\xi\tau}/X_\xi,$$

where the left-hand sides are considered as composite functions of ξ and τ. These equalities enable one to rewrite equation (9.3) in the form

$$\left(\gamma X_\xi\right)_\tau = 0.$$

Taking into account the fact that ξ is the mass Lagrangian coordinate and, hence, $\gamma X_\xi = 1$ as $\tau = 0$, we obtain, having integrated the preceding equality in τ:

$$\gamma X_\xi = 1 \tag{9.5}$$

independently of τ. In particular, this implies a sufficient condition for the invertibility of the coordinate transformation performed: $\gamma \neq 0$ in a domain D on the plane (ξ, τ), or the equivalent condition $u \neq 0$ in the domain Ω being the image of D under the mapping $x = X(\xi, \tau)$, $t = \tau$. If the solution u of equation (9.1) is bounded from above or from below, then the equivalence transformation $\tilde{u} = au + b$ with appropriate constants a and b (which preserves the class of these equations), reduces the problem to the case of a positive function $\tilde{u}$. We remark that in those applications where equation (9.1) arises, $u(x, t)$ usually stands for the temperature, the concentration, or the density, which must be nonnegative, i.e., a function bounded from below.

Let us write down the second of equalities (9.2) in the new variables and then make use of relations (9.5) and $u_x = \gamma_\xi / X_\xi$ to eliminate γ. Then $V[X(\xi, \tau), \tau]$ becomes a function of the arguments X_ξ, $X_{\xi\xi}$. Substituting the relation for V into (9.4), we arrive at the equation

$$X_\tau = -X_\xi \varphi\left(\frac{1}{X_\xi}, -\frac{X_{\xi\xi}}{X_\xi^3}\right). \tag{9.6}$$

Differentiating it in ξ and taking the deformation $\omega = X_\xi$ for the new sought function, we get the equation for ω

$$\omega_\tau = \left[-\omega\varphi\left(\omega^{-1}, -\omega^{-3}\omega_\xi\right)\right]_\xi. \tag{9.7}$$

We see that the passage to Lagrangian coordinates (we term it "L-transformation") establishes the correspondence between equation (9.1) and equation (9.7) of the same form by the rule

$$\varphi(u, p) \mapsto \psi(u, p) = -u\varphi\left(u^{-1}, -u^{-3}p\right).$$

An important property of the L-transformation is that it is *involutive*. Indeed, since

$$-u\psi(u^{-1}, -u^{-3}p) = \varphi(u, p),$$

the L-transformation transforms equation (9.7) into (9.1). This gives every reason to term (9.1) and (9.7) *reciprocal* with respect to L-transformation. Applying it twice, the L-transformation transforms equation (9.1) into itself.

If $\varphi_p > 0$ in a neighborhood of a point (u_0, p_0) of the plane (u, p), then $\psi_p > 0$ in a neighborhood of the point $(u_0^{-1}, -u_0^{-3}p_0)$. Thus, the L-transformation transforms a parabolic equation into a parabolic one. By (9.5), the positive solution of equation (9.1) corresponds to a positive solution of (9.7).

Let us indicate the L-transformation explicitly. Using the equalities $u(x, t) = \gamma(\xi, \tau)$, $t = \tau$, and (9.5), we have $u(x, t) = 1/\omega(\xi, t)$. It remains to define the function $x = X(\xi, t)$. It is uniquely defined by a solution of the system of compatible equations

$$X_\xi = \omega, \quad X_t = -\omega\varphi\left(\omega^{-1}, -\omega^{-3}\omega_\xi\right),$$

with the condition $X(0,0) = 0$. (The latter of these equations is a direct consequence of (9.6).) Equation (9.7) expresses the compatibility condition for this system. Having integrated it, we get the following representation for the solution u of equation (9.1) in terms of the solution ω of equation (9.7)

$$\begin{aligned} u(x,t) &= [\omega(\xi,t)]^{-1}, \\ x &= \int_0^\xi \omega(\eta,t)d\eta - \int_0^t \omega(0,s)\varphi\left[1/\omega(0,s), -\omega_\xi(0,s)/\omega^3(0,s)\right]ds. \end{aligned} \tag{9.8}$$

Let us take into consideration a function Σ(x,t) such that

$$X[\Sigma(x,t),t] = x.$$

Proceeding from (9.5), (9.6), it is easy to derive the equalities $\Sigma_x = u$, $\Sigma_t = \varphi(u, u_x)$, where u is a solution of (9.1). The compatibility condition for these two equations for the function Σ is automatically fulfilled. Taking into account that $\omega[\Sigma(x,t),t] = 1/u(x,t)$ and $\Sigma(0,0) = 0$, we get the parametric representation for a solution of equation (9.7) in terms of the function ω:

$$\begin{cases} \omega(\xi,t) = [u(x,t)]^{-1}, \\ \xi = \int_0^x u(y,t)dy + \int_0^t \varphi[u(0,s), u_x(0,s)]\,ds. \end{cases} \tag{9.9}$$

Formulas (9.8) are an explicit realization of the L-transformation of equation (9.1). On the other hand, relations (9.8) may be interpreted as performing the passage to Lagrangian coordinates (x,t) in equation (9.7), where (ξ,τ) are now viewed as the Euler variables.

It is to be noted that the definition of Lagrangian coordinates is arbitrary up to the single-variable function $X_0(\xi)$. Although, if $X_0(\xi)$ is chosen not as above, in particular, if the mass Lagrangian coordinate ξ is replaced by the geometrical Lagrangian coordinate η, $X_0(\eta) = \eta$ (as was proposed in [65]), then the L-transformation does not preserve the class of equations (9.1): the variable ξ appears in the transformed equation in an explicit form.

Let us consider now the equation

$$v_t = [\phi(x, v_x)]_x. \tag{9.10}$$

This equation already admits a direct dependence of ϕ on x but the sought function v appears here only under the sign of the derivative. In what follows, we construct a nonlocal transformation that preserves the class of such equations. This construction is based on a simple fact: the passage in (9.10) to the new space variable v and the choice of x for the new sought function

$$x = u(v,t), \tag{9.11}$$

i.e. von Mises's transformation, transforms (9.10) to an equation of the form (9.1)

$$u_t = [-\phi\,(u,\, 1/u_v)]_v\,. \tag{9.12}$$

To equation (9.12) the L-transformation is applicable:

$$\begin{aligned} \omega(\xi, t) &= [u(v, t)]^{-1}\,, \\ \xi &= \int_0^v u(y, t)dy - \int_0^t \phi\,[u(0, s),\, 1/u_v(0, s)]\,ds, \end{aligned} \tag{9.13}$$

whence the equation for ω:

$$\omega_t = \left[\omega\phi\left(\omega^{-1}, -\omega^3/\omega_\xi\right)\right]_\xi\,. \tag{9.14}$$

Finally, let us perform von Mises's change of variables in equation (9.14):

$$\xi = \xi(\omega, t). \tag{9.15}$$

The new sought function obeys the equation

$$\xi_t = \left[-\omega\phi\left(\omega^{-1}, -\omega^3\xi_\omega\right)\right]_\omega\,. \tag{9.16}$$

The possibility of performing von Mises's transformation is guaranteed by the conditions $v_x \neq 0$, $\omega_\xi \neq 0$ which are always assumed in the corresponding domains of the planes (x, t), (ξ, t). For the regularity of the L-transformation and of its inversion the equality $\omega \neq 0$ is needed (or its counterpart $x \neq 0$). If these inequalities are fulfilled, the composition of von Mises's transformation (9.11), L-transformation (9.13), and von Mises's transformation (9.15) converts equation (9.10) into equation (9.16) of the same class. Each of these transformations is invertible and we may term this composition of transformations, as well as its inversion, the *reciprocity transformation* of equations (9.10) and (9.16).

The constructions described can be easily generalized to the equations

$$s(u)u_t = [C(u, u_x)]_x\,, \quad r(x)v_t = [D(x, v_x)]_x\,,$$

where $s(u) \neq 0$, $r(x) \neq 0$. (The first of them was considered in [165].) Indeed, the substitution $\tilde{u} = S(u)$, where S is the primitive of s, transforms the first one into equation (9.1) for the function $\tilde{u}$. Having performed in the second equation von Mises's transformation (9.11), for the function $\underline{u} = R\,[u(v, t)]$ (R is the primitive of r) we get an equation of the type (9.12).

To conclude this section, let us remark that all these considerations can be applied to the case of divergence equations of the type (9.1), (9.10) with the functions φ, ψ explicitly depending on the variable t.

10 Hidden symmetry of evolution equations

In this section we shall be interested in an equation of the form

$$u_t = [\varphi(u, u_x)]_x \tag{10.1}$$

or in its special case, the equation

$$u_t = [k(u)u_x]_x , \tag{10.2}$$

from the point of view of their group properties.

Let us consider first the nonlinear heat conduction equation (10.2). (When saying "nonlinear heat conductivity" we mean that the coefficient of heat conductivity $k(u) \neq$ const.) It is reasonable to pose for this equation, including an "arbitrary" element $k(u)$, the problem of group classification: to identify from all equations of the type (10.2) those whose admissible Lie groups are larger than for the equation of the general form. Note that the transformation of type $E = E(a, b, c)$

$$\tilde{u} = au + b, \ \tilde{k} = ck, \quad a, b, c = \text{const.}, \ ac \neq 0, \tag{10.3}$$

transforms equation (10.2) into an equivalent one but with another function k. Equations connected by pointwise equivalence transformations possess Lie groups similar to $E(a, b, c)$ and in the group classifications are considered to be indistinguishable.

The problem of the group classification of equation (10.2) was posed and solved in 1959 by L. V. Ovsyannikov [123] (see also monograph [121] where the below-utilized terminology of the group analysis of differential equations is introduced). Let us formulate the main result of paper [123]. If $k \neq$ const., $k \neq e^u$, $k \neq u^m$, then the full group admitted by equations (10.2) is three-parameter. The basis of the corresponding Lie algebra is generated by the infinitesimal operator

$$X_1 = \partial/\partial x, \quad X_2 = \partial/\partial t, \quad X_3 = 2t\partial/\partial t + x\partial/\partial x. \tag{10.4}$$

If $k = e^u$, then the group extends to a four-parameter group on account of the operator

$$X_4 = t\partial/\partial t - \partial/\partial u. \tag{10.5}$$

If $k = u^m$, $m \neq -4/3$, then the full group of equation (10.2) is also four-parameter. The basis of the Lie algebra here is constituted by the operators (10.4) and

$$X_5 = mx\partial/\partial x + 2u\partial/\partial u. \tag{10.6}$$

Finally, in the exceptional case $k = u^{-4/3}$ equation (10.2) admits the five-parameter group of pointwise transformations generated by operators (10.4), (10.6) with $m = -4/3$, and the additional operator

$$X_6 = -x^2\partial/\partial t + 3xu\partial/\partial u. \tag{10.7}$$

The foregoing constructions are based on the following observation: besides pointwise transformations (10.3), equation (10.2) and the general equation (10.1) admit the nonlocal equivalence transformation which is performed via the passage to Lagrangian coordinates (L-transformation). The explicit form of L-transformation of equation (10.1) is given by formulas (9.9) and its specification for the case $\varphi(u, p) = k(u)p$ is the following:

$$\begin{aligned} \xi &= \int_0^x u(y,t)\,dy + \int_0^t k\,[u(0,s)]\,u_x(0,s)\,ds, \\ \tau &= t, \quad \omega(\xi,\tau) = [u(x,t)]^{-1}; \end{aligned} \tag{10.8}$$

it is assumed here that $u \neq 0$. Since transformation (9.9) is not pointwise, then the Lie groups which are admitted by the original and transformed equations need not be isomorphic. This feature allows one to construct equations (10.1) for which the corresponding equation (9.7) possesses a larger Lie group. An example of such a situation is furnished by the nonlinear heat conduction equation under an appropriate choice of function $k(u)$. Thus, the equation

$$u_t = \left[u^{-2}\exp\left(u^{-1}u_x\right)_x\right] \tag{10.9}$$

admits a three-parameter Lie group [123]. Having subjected it to L-transformation, we get the equation

$$\omega_\tau = \left(e^{\omega}\omega_\xi\right)_\xi, \tag{10.10}$$

which possesses the four-parameter full group.

The group of pointwise transformations of the equation

$$u_t = \left(u^{-2/3}u_x\right)_x \tag{10.11}$$

will be three-parameter. As the result of L-transformation of the latter equation we obtain the equation

$$\omega_t = \left(\omega^{-4/3}\omega_\xi\right)_\xi, \tag{10.12}$$

whose full group is the largest (five-parameter) among all nonlinear equations of type (10.2). Having rewritten in the new notation the additional operator of the Lie algebra of this group, we arrive at the corresponding finite transformation T_α

$$\tilde{\tau} = \tau, \quad \tilde{\xi} = (1+\alpha\xi)^{-1}, \quad \tilde{\omega} = (1+\alpha\xi)^3\omega \tag{10.13}$$

with a parameter $\alpha \in \mathbb{R}$, that conserves equation (10.12). There are no finite transformations admitted by equation (10.11) in the space (x, t, u) which correspond to transformation (10.13).

Let us denote by U the manifold in $\mathbb{R}^3$ determined by a solution $u = u(x,t)$ of equation (10.11), and by Ω the corresponding manifold $\omega = \omega(\xi, \tau)$. The mapping of U into Ω realized by the change of variables (10.8) is denoted by $\Omega = LU$, and mapping (10.13) by $\tilde{\Omega} = T_\alpha \Omega$. Let us define the family of transformations S_α by the formula

$$S_\alpha U = L^{-1}\tilde{\Omega} \equiv L^{-1} T_\alpha L U.$$

(Due to (10.8), the invertibility of the L-transformation is guaranteed by the condition $u \neq 0$.) It follows from (10.13) and the definition of S_α that

$$\forall \alpha, \beta \in \mathbb{R} \quad S_\alpha S_\beta U = S_{\alpha+\beta} U; \quad \text{besides, } S_0 U = U.$$

Thus, the set of transformations S_α form a one-parameter group which, obviously, is not a Lie group.

The transformation S_α renders each solution of equation (10.11) again a solution of this equation. This property can be termed the *hidden symmetry* of equation (10.11). An analogous property of hidden symmetry is also intrinsic for equation (10.9). Along with (10.13) one should make use of the transformation

$$\tilde{\xi} = \xi, \quad \tilde{\tau} = \tau \exp(\alpha), \quad \tilde{\omega} = \omega - \alpha,$$

generated by operator (10.5) and preserving equation (10.9), reciprocal to (10.10).

Note that both equations (10.9) and (10.10) are not isolated by the groups they admit and even do not have a nontrivial Lie–Bäcklund group [44]. The discovered invariance of these equations with respect to the nonlocal transformations S_α is of another nature.

The term "hidden symmetry" (as well as the encompassing ones "nonlocal" and "non-pointwise" symmetries) were recently given a certain fame due to different methods of finding such nonlocal symmetries of evolution equations that cannot be found within the confines of the classical Lie approach (see, for instance, [137, 136, 3, 1, 174, 173, 120] and the references therein). To the best of our knowledge, the very first example of this sort was discovered by V. A. Fok [50] in 1935. It was established that after Fourier transformation the stationary Schrödinger equation for the hydrogen atom admitted the group of rotations in four-dimensional space. Later on, the hidden symmetry property of a number of linear equations of mathematical physics, particularly Maxwell's equation, was studied in works by V. I. Fushchich, A. G. Nikitin and their collaborators (see [57, 56, 55] and the bibliography therein). The methods used in [50, 57, 56, 55] to find hidden symmetries are specific for linear equations.

As far as nonlinear equations of the type (10.1) are concerned, besides (10.2), a fairly easy analysis may be given here of the hidden symmetry properties of equations of the class

$$v_t = [K(v_x)]_x \,, \tag{10.14}$$

describing the shear flow of a viscoplastic medium and processes of non-Newtonian filtration. It is to be noted that in the cases when $K(u)$ is the primitive of the function $k(u)$ from equation (10.2), there is a simple connection between solutions of the latter equation and equation (10.14): $u = v_x$. In this case the equation

$$v_t = k(v_x) v_{xx} \tag{10.15}$$

is an equivalent form of (10.14).

On the other hand, the substitution $v = w_x$ connects equation (10.14) to the equation

$$w_t = K\left(w_{xx}\right), \tag{10.16}$$

which, generally speaking, is neither of divergence form nor quasilinear. Thus, we have the sequence of three equations (10.2), (10.14), and (10.16) connected one to another by the transformations $u = v_x = w_{xx}$. In the series of works [3, 120, 86] the group properties of equations of this sequence were studied and nonlocal symmetries were discovered admitted by its members for special choices of $K(u)$.

Let us return to equation (10.15), equivalent to (10.14). In the generic case, this equation admits a four-parameter Lie group. If $k = \exp(v_x)$ and $k = v_x^m$, $m \neq -2$, then the full group of equation (10.15) is five-parameter. For $m = 0$ this equation is linear, in the case $m = -2$ it can be linearized with the help of von Mises transformation (cf. Section 9). The problem of the group classification of the equation under discussion is solved in [2]. Besides the cases already pointed out, there is the exceptional dependence

$$k = \left(1 + v_x^2\right)^{-1} \exp\left(\nu \arctan v_x\right), \quad \nu \geq 0, \tag{10.17}$$

when the full group extends to a five-parameter one. The additional operator of the Lie algebra of this group has the form

$$Y_5 = -\nu t \partial_t - v \partial_x + x \partial_v. \tag{10.18}$$

(It is curious to note that equation (10.15) in the complex domain, and with k given by formula (10.17), can be transformed into the equation with power nonlinearity $k = v_x^{m-1}$ [168].) As is shown in [3], in the sole case of (10.17) the transformations of the group generated by operator (10.18) and preserving equation (10.15) do not correspond to any pointwise transformation admitted by the two other equations (10.2) and (10.16) of the three-term sequence originating from the given function k. Since the transition from v to w is nonlocal, $v = w_x$, it follows, in particular, that the equations

$$\begin{aligned} w_t &= \arctan\left(w_{xx}\right), \\ w_t &= \nu^{-1} \exp\left(\nu \arctan\left(w_{xx}\right)\right), \quad \nu > 0, \end{aligned}$$

also possess hidden symmetries. These equations are implied by (10.16) after the specialization of k of the form (10.17) with the equality

$$\frac{dK(u)}{du} = k(u).$$

The case when the Lie group admitted by equation (10.2) can be extended in comparison with that of the corresponding equation (10.15) is exhausted by the dependence

$$k(u) = u^{-4/3}.$$

Another possibility when a nontrivial operator (e.g. not reducible to the appearance of the operator $\partial/\partial w$ in the Lie algebra) extending the full Lie group appears in the passage from (10.15) to equation (10.16) can be realized only in the cases $k = u^{-4/3}$ and $k = u^{-2/3}$. Here equation (10.16) is equivalent to the equations

$$w_t = w_{xx}^{-1/3}, \tag{10.19}$$

$$w_t = w_{xx}^{1/3}, \tag{10.20}$$

respectively. Each of the equations (10.19), (10.20) admits a seven-parameter group of pointwise transformations [3]. Equations (10.19), (10.20) are also remarkable because the dimension of the Lie algebras they admit is the maximal among all equations not equivalent to the linear equation $u_t = F(t, x, u, u_x, u_{xx})$. This assertion is a very partial implication of the results of B. A. Magadeev [101], who solved the problem of the group classification of the equations

$$u_t = F\left(t, x, u, \frac{\partial}{\partial x}, \dots, \frac{\partial^n}{\partial x^n}\right)$$

with respect to contact transformations. Moreover, it follows from [143, 72, 173] that equation (10.20) can be transformed to (10.19) by means of contact transformations. Not having the possibility to go deeply into the theory of contact transformations we refer the reader to monographs [143, 72, 173] and the survey paper [161] presenting the modern state of the subject.

So, we have at our disposal a few equations displaying the hidden symmetry property. Let us demonstrate how one can profit by this property to construct exact solutions of equations not invariant in the classical sense. For example, let us consider equation (10.12). It admits the infinitesimal operator

$$\Sigma = \partial_\tau - \xi^2 \partial_\xi + 3\xi\omega\partial_\omega.$$

One may take for the basis of invariants of this operator

$$I_1 = \xi^{-1} - \tau, \quad I_2 = \xi^3 \omega.$$

According to the techniques of [121], one has to find a solution of equation (10.12) invariant with respect to operator Σ in the form

$$\omega = \xi^{-3} f\left(\xi^{-1} - \tau\right). \tag{10.21}$$

The ordinary differential equation for f, obtained by the substitution of (10.21) into (10.12), can be integrated and then reduced to the form

$$\frac{df}{d\xi} = -f^{4/3}(f + C), \tag{10.22}$$

where C is the constant of integration. For the sake of definiteness we suppose that $C > 0$. All nonnegative solutions of equation (10.22) can be obtained by a translation along the axis ξ from the single solution defined for $\xi > 0$ and satisfying the condition

$f \to \infty$ as $\xi \to 0$. Without loss of generality, we may therefore assume that $f(\xi)$ is defined by the relation

$$\xi = \int_f^\infty \frac{dg}{g^{4/3}(g+C)}.$$

The constructed solution $\omega = \omega_*(\xi, \tau)$ of equation (10.12) is defined in the domain

$$\Delta = \{\xi, \tau : \xi > 0, \tau > 0, \xi\tau > 0\}.$$

The function ω_* is positive in this domain and grows unboundedly near that part of the boundary where $\xi\tau = 1$. The last feature does not allow one to use the explicit formulas (10.8) and to get the solution $u = u_*(x,t)$ of equation (10.11), reciprocal to (10.12). The required representation for the solution u_* is derived by the replacement of the loop of integration in the second line of formulas (9.8) by the curve connecting the points (ξ, t), $(\xi, 0)$, and $(\infty, 0)$, so that

$$x = -\int_\xi^\infty \omega(\eta, 0)\, d\eta + \int_0^t \omega^{-4/3}(\xi, s)\omega_\xi(\xi, s)\, ds.$$

Substituting $\omega = \omega_*(\xi, t)$ here and into the first line of formulas (9.8), after simple transformations we find the parametric representation for the solution u_* of equation (10.11):

$$\begin{aligned} u_* &= \xi^3 \left[f\left(\xi^{-1} - t\right)\right]^{-1}, \\ x &= -\int_0^{\xi^{-1}} \xi f(\xi)\, d\xi + \int_{\xi^{-1}-t}^{\xi^{-1}} \left\{\xi^{-1}[f(\xi) + C] - 3f^{-1/3}(\xi)\right\} d\xi. \end{aligned} \tag{10.23}$$

The integrals in (10.23) converge as $\xi \to 0$ and $\xi \to \infty$ for each $t \in [0, \xi^{-1})$. This follows from the asymptotic dependences

$$f = \left(\tfrac{3}{4\xi}\right)^{3/4} \left[1 + O\left(\xi^{3/4}\right)\right] \quad \text{as } \xi \to 0,$$

$$f = \left(\tfrac{3}{C\xi}\right)^{3} \left[1 + O\left(\xi^{-3}\right)\right] \quad \text{as } \xi \to \infty.$$

In particular, as $\xi \to 0$, we have

$$\lim_{\xi \to} \int_{\xi^{-1}-t}^{\xi^{-1}} \left[C\xi^{-1} - 3f^{-1/3}(\xi)\right] d\xi = \frac{Ct^2}{2}.$$

Thereby, the image of the ray $\{\xi = 0, \tau > 0\}$ under the mapping $(\xi, t) \mapsto (x, t)$ via the L-transformation generated by equation (10.12) will be the parabola l_1 given by the equation $x = x_1(t) \equiv -A + Ct^2/2$, $t > 0$ with

$$A = \int_0^\infty \xi f(\xi)\, d\xi.$$

Passing to the limit in the first line of (10.23), we get

$$u_* = \left(\frac{C}{3}\right)^3 \quad \text{as } x = x_1(t),\, t > 0. \tag{10.24}$$

Letting $t = 0$ in (10.23), we see that this transformation transforms the ray $\{\xi > 0, \tau = 0\}$ into the segment $l_0 = \{-A < x < 0, t = 0\}$. If $\psi(x)$ is defined by the relation

$$x = -\int_0^{\psi} \xi f(\xi)\, d\xi \quad \text{for } x \in (-A, 0],$$

then, due to (10.23), the values of u_* at the points of this segment can be calculated by the formula

$$u_*(x,0) \equiv u_0(x) = [\psi(x)]^{-3}\{f[\psi(x)]\}^{-1} \quad \text{as } -A < x < 0. \tag{10.25}$$

It follows that $u_0 \to (C/3)^3$ as $x \to -A$, i.e., at the point $x = -A$, $t = 0$ the compatibility condition for initial and boundary data (10.24), (10.25) is fulfilled. Also, u_0 grows unboundedly when $x \to -0$. Moreover,

$$u_0 = (48)^{3/5}(-5x)^{-9/5}\left[1 + O\left(|x|^{3/5}\right)\right] \quad \text{as } |x| \to -0.$$

Now set $\xi = t^{-1}$ in (10.22). The equation

$$x = x_2(t) \equiv Ct^2 + \int_0^t \left[(t-\xi)f(\xi) - 3f^{-1/3}(\xi)\right] d\xi$$

defines on the plane (x,t), $t > 0$ a curve emanating from the point $(0,t)$. Using equation (10.22) and the asymptotics of the function f as $\xi \to \infty$, we find $x_2'(0) = 0$, $x_2'' = C$ for all $t > 0$. Thereby, $x_2 = Ct^2/2$, i.e. the curve l_2 is also a parabola. Also, $u_*(x,t) \to 0$ as $x \to x_2(t) - 0$, $t > 0$. Thus, l_2 is a line of degeneracy of equation (10.11). It corresponds to that part of the boundary $\xi\tau = 1$ of the domain Δ where ω_* is defined. Hence, under the mapping $(\xi,\tau) \mapsto (x,t)$ the image D of the domain Δ is the domain bounded by the parabolas l_1, l_2, and the segment l_0.

Equation (10.11) belongs to the class of so-called *fast diffusion* equations: if u is interpreted as the concentration of a substance and the equation itself is considered as a nonlinear diffusion equation, then the diffusion coefficient $u^{-2/3} \to \infty$ as $u \to 0$. More information on the fast diffusion equation can be found in [151, Ch. 2]; see also the references to that chapter.

Returning to the function u_*, we may interpret it as a solution of equation (10.11) in the domain D satisfying conditions (10.24), (10.25), and the condition

$$u_* \to 0 \quad \text{as } x \to x_2(t) - 0,\, t > 0. \tag{10.26}$$

It is easy to show that along with (10.26) the following relations hold:

$$\partial_t u_* \to 0, \quad \partial_x u_* \to 0, \quad \partial^2_{xx} u_* \to 0 \quad \text{as } x \to x_2(t) - 0,\, t > 0.$$

A natural desire would be to continue u_* by zero into the domain $\{x > x_2(t),\ t > 0\}$, so as to make the continued function (let us denote it by U_*) a weak solution of equation (10.11) in the domain $D_+ = \{(x,t) : x > x_1(t), t > 0\}$. However, this is impossible for the following reason: the diffusion flux $-u_*^{-2/3}\partial u_*/\partial x$ does not tend to zero as $x \to x_2(t) - 0$. Making use of the relations

$$\lim_{x\to x_2-0} u_*^{-2/3}\frac{\partial u_*}{\partial x} = -\lim_{\xi\to t^{-1}+0} \omega_*^{-7/3}\frac{\partial \omega_*}{\partial \xi}$$

and referring to (10.21)–(10.23), we find:

$$u_*^{-2/3}(u_*)_x \to -t^{-2} \quad \text{as } x \to x_2(t) - 0, t > 0. \tag{10.27}$$

Since the solution u_* of equation (10.11) satisfies relation (10.27) at the line of degeneracy, in D_+ the function U_* obeys the equation

$$u_t = \left[u^{-2/3}u_x - t^{-2}H(x - x_2(t))\right]_x, \tag{10.28}$$

where $H(x)$ is the Heaviside function.

Equations similar to (10.28) have so far not been given a detailed mathematical treatment. In 1990, in a private communication J. L. Vazquez pointed out that one could interpret as solutions of equations of the type (10.28) the solutions of the fast diffusion equations with discontinuities of the diffusion flux at the line (or the surface) of degeneracy.

Besides the above-outlined features, the solutions of the initial-boundary value problem (10.11), (10.24)–(10.26) possess a number of additional interesting properties. The function $u_0(x)$ in the initial condition (10.25) has a strong singularity at the point $x = 0$ — this function even does not belong to class $L_1(-A, 0)$. On the other hand, for each $t > 0$ the function u_* turns out to be smooth up to the line of degeneracy. In particular, this implies that "the mass"

$$M(t) = \int_{x_1(t)}^{x_2(t)} u_*(x,t)\,dx$$

associated with the solution of problem (10.11), (10.24)–(10.26) is finite for each $t > 0$ while $M \to \infty$ as $t \to 0$.

As was already mentioned, having continued u_* by zero into the domain $D_+ \setminus D$, we go out of the class of generalized solutions to equation (10.11). There is another way to continue a solution of this equation across the line of degeneracy; however, the continued function fails to remain nonnegative. Let us make in equation (10.11) the substitution $u = z^3$. The function z is a solution of the equation

$$z^2 z_t = z_{xx} \quad \text{in } D. \tag{10.29}$$

According to (10.26), (10.27), at l_2 this function satisfies the conditions

$$z = 0, \quad z_x = -\frac{t^{-2}}{3} \quad \text{as } x = \tfrac{Ct^2}{2}, t > 0. \tag{10.30}$$

By the Cauchy–Kovalevskaya theorem, there exists a solution z_* of the Cauchy problem (10.29), (10.30) analytic in a neighborhood of the curve l_2, $t > 0$. The function z_*^3 coincides with u_* in D_+ and is the analytic continuation of the solution u_* of equation (10.11) across the line of degeneracy for $x > x_2$, $t > 0$ ($x - x_2$ is assumed small enough). By (10.30), it is clear that z_* (and along with it the continuation of function u_*) assumes negative values for small positive values of $x - x_2(t)$.

Solution (10.23) is an infrequent example of the exact solution to the fast diffusion equation being not a self-similar one. Moreover, this solution does not belong at all to the set of solutions of equation (10.11) invariant with respect to the group of pointwise transformations. (Let us recall that the generators of this group are the operators X_1, X_2, X_3 (10.4), and X_5 (10.6).) To prove this assertion one has to subject the function $\Phi = u - u_*(x, t)$ to the operator

$$X = \sum_{j=1}^{3} \alpha_j X_j + \alpha_5 X_5,$$

set $u = u_*$ in the expression for $X\Phi$, and equate the result to zero. If the last equality is fulfilled identically in x, t only for $\alpha_j = \alpha_5 = 0$, $j = 1, 2, 3$, the desired assertion follows.

From (10.4)–(10.6) we get:

$$\begin{aligned} X\Phi|_{u=u_*} = & -\left[\alpha_1 + \left(\alpha_3 - \frac{2}{3}\alpha_5\right) x\right] \frac{\partial u_*}{\partial x} \\ & - (\alpha_2 + 2\alpha_3 t) \frac{\partial}{\partial x}\left(u_*^{-2/3} \frac{\partial u_*}{\partial x}\right) - 2\alpha_5 u_*. \end{aligned} \tag{10.31}$$

(Here $\partial u_*/\partial t$ is replaced by its expression by virtue of equation (10.11).) Substituting here u_* which is implicitly defined by formulas (10.23), passing from the derivatives in x to the derivatives in ξ, and then having recourse to equation (10.22), we transform the invariance condition $X\Phi = 0$ with $u = u_*$ to the form

$$\begin{aligned} & \left[\alpha_1 + \left(\alpha_3 - \frac{2}{3}\alpha_5\right) x\right] \left[\xi^4 f^{-5/3}(f + C) - 3\xi^5 f^{-2}\right] \\ & \quad - (\alpha_2 + 2\alpha_3 t) \left[\xi^4 C f^{-5/3} (f + C) + 6\xi^4 f^{-2} (f + C) - 9\xi^5 f^{-7/3}\right] \\ & \quad - 2\alpha_5 \xi^3 f^{-1} = 0. \end{aligned}$$

Let us multiply equality (10.31) by $\xi^{-1} f$ and pass to the limit as $\xi \to 0$. The result of the operation is the relation

$$-(\alpha_2 + 2\alpha_3 t) \frac{C^4}{9} = 0,$$

which has to be satisfied identically in t. If $C \neq 0$, this is possible only with $\alpha_2 = \alpha_3 = 0$. Using the equalities obtained, we once again make recourse to (10.31), multiply this relation by ξ, and then let $\xi \to 0$. Then we have the relation

$$2\left[-\alpha_1 + \frac{2}{3}\alpha_5\left(-A + \frac{Ct^2}{2}\right)\right]\left(\frac{C}{3}\right)^6 = 0,$$

whence $\alpha_1 = \alpha_5 = 0$, if $C \neq 0$. Thus, for $C \neq 0$ relation (10.31) is fulfilled identically in ξ and t if $\alpha_j = \alpha_5 = 0$ $(j = 1, 2, 3)$, which proves the assertion.

The case $C = 0$ is specific. Here the solution of equation (10.22) with the condition $f \to \infty$ as $\xi \to 0$ has the form $f = (3\xi/4)^{3/4}$ and the invariance condition holds if $\alpha_1 = \alpha_2 = 0$, $\alpha_3 = 2$, $\alpha_5 = -9/2$. So, for $C = 0$ solution (10.23) renders a solution invariant with respect to the operator

$$5x\partial_x + 4t\partial_t - 9u\partial_u$$

e.g. a self-similar solution. The parametric representation for this solution is of the following form:

$$u_* = \left(\frac{3}{4}\right)^{-3/4} \xi^3 \left(\xi^{-1} - t\right)^{3/4}$$

$$x = -4\left(\frac{3}{4}\right)^{3/4} \left(\xi^{-1} - t\right)^{1/4} \left(\frac{9}{5}\xi^{-1} - t\right).$$

As shown above, equation (10.23) is not invariant in the ordinary sense, generally speaking. However, it is invariant with respect to a nonlocal transformation admitted by equation (10.11). This gives a reason to term the solution u_* of this equation the *nonlocal-invariant* solution.

The possibility of employing the L-transformation to seek nonlocal-invariant solutions was pointed out already in [137]. The next example is taken from [151]. Let us consider equation (10.10). It admits the operator $\tau\partial/\partial\tau - \partial/\partial\omega$. One of the solutions invariant with respect to this operator is

$$\omega = \ln\left[(T - t)^{-1}\frac{\xi^2}{2}\right],$$

where $T > 0$ is a constant. Applying to it the L-transformation (9.8) we find a solution of equation (10.9), reciprocal to (10.10),

$$u(x,t) = \left\{\ln\left[\frac{1}{2}(T - t)^{-1}\psi^2(x,t)\right]\right\}. \qquad (10.32)$$

Here the function ψ is implicitly given by the equality

$$x = \psi\left\{\ln\left[(T - t)^{-1}\psi\right] - 1\right\}.$$

One can show that solution (10.32) is not a solution of equation (10.9) invariant with respect to pointwise transformations.

Another possibility of obtaining nonlocal-invariant solutions is concerned with consideration of the three-term sequence (10.2), (10.14), (10.16). Papers [3, 120] point out a method of finding exact solutions of an equation of this sequence, relying on classical-invariant solutions of the remainders.

The class of equations displaying the hidden symmetry property can be amplified by combining nonlocal and pointwise transformations, say, the von Mises ones. Proceeding in this direction one can discover equations which explicitly contain the space variable. Thus, the equation

$$\left(r^{N-1}u\right)_t = \left(r^{N-1}u^{-2}u_r\right)_r$$

with an integer $N \geq 2$ possesses nonlocal symmetries. This equation describes the radially symmetric heat propagation processes in the solid H_2 and in a number of simple metals. (See Sections 13, 15.)

In concluding the section, let us mention one other method to obtain equations with hidden symmetries. We will demonstrate it by the example of the nonlinear heat conduction equation (10.2). It is based on the following simple observation: the composition LE of the pointwise equivalence transformation (10.3) and the L-transformation preserves the class of equations (10.2), but, generally speaking, $LE \neq EL$.

Consider the equation

$$u_t = \left[\frac{u^m}{(u+1)^{m+2}}u_x\right]_x, \quad m = \text{const.}, \tag{10.33}$$

$m \neq 0, -2$. Let us apply to equation (10.32) transformation (10.3) with $a = c = 1$, $b = -1$ and then pass in the transformed equation to Lagrangian coordinates. The resulting equation

$$\omega_\tau = \left[(1-\omega)^m \omega_\xi\right]_\xi$$

by means of the pointwise transformation $\omega = 1 - \phi$ transforms into the equation

$$\phi_\tau = \left(\phi^m \phi_\xi\right)_\xi . \tag{10.34}$$

The last equation for $m \neq -4/3$ admits the four-parameter group, while the full group of equation (10.33) with $m \neq -2$ is three-parameter. (If $m = 0$, this equation can be linearized by means of L-transformation.) Thus, equation (10.33), equivalent to (10.34), possesses the hidden symmetry property since, evidently, there are no transformations of the type (10.3) transforming one of these equations into the other.

In the family (10.33) one may isolate the equation

$$u_t = \left[\frac{u_x}{u^{4/3}}(1+u)^{-2/3}\right]_x$$

equivalent to equation (10.34) with the exceptional value of the exponent $m = -4/3$ [123]. Besides the three-parameter Lie group it admits a two-parameter group of nonlocal transformations.

The results presented in this section are published in [133].

11 Linearization by means of Lagrangian coordinates

There is already a wide literature devoted to the problems of linearization of differential equations, i.e. of finding transformations which render the original equation a linear one. Besides the above-cited books [143, 72, 121] and papers [112, 146, 145, 48, 49, 44], these questions are intensively discussed in monograph [5] and survey works [35, 111]. The material of this section is of illustrative character: relying on equations of the class (8.1) we demonstrate the capabilities of L-transformation as one of the linearization methods.

Set in equation (8.1) $\varphi(u, p) = p$. Then it becomes the linear heat conduction equation

$$u_t = u_{xx}. \tag{11.1}$$

By (9.7), the equation reciprocal to (11.1) (and, hence, linearizable) is the equation

$$\omega_\tau = \left(\omega^{-2}\omega_\xi\right)_\xi. \tag{11.2}$$

The fact of linearization of (11.2) was first established in [164]. Had one considered instead of (11.1) the general linear equation of the form (8.1) with $\varphi = \alpha u + \beta p + \gamma$ (α, β, γ are constants) it would not lead to any substantial amplification of the family of linearizable equations. In this case the equation reciprocal to (8.1) is

$$\omega_\tau = \left(\beta\omega^{-2}\omega_\xi - \gamma\omega\right)_\xi,$$

which can be transformed into (11.2) by a pointwise transformation. It is curious to observe that after the substitution of φ of this form into equation (8.1) the constant γ disappears, but remains in the transformed equation. Meanwhile, the transformed equation does not contain the constant α.

Now let $\varphi = u_x - u^2/2$. In this case equation (8.1) is the famous Burgers' equation

$$u_t = u_{xx} - uu_x. \tag{11.3}$$

It is well known that the substitution

$$u = 2\frac{\phi_x}{\phi} \tag{11.4}$$

transforms (11.3) into a linear equation for the function ϕ. Transformation (11.4) was discovered in 1948 by V. A. Florin [47] and later on was rediscovered by E. Hopf [68], J. G. Cole [37]. Passing in (11.3) to Lagrangian coordinates by formulas (8.4), we obtain a new example of linearizable equation:

$$\omega_\tau = \left(\omega^{-2}\omega_\xi + \frac{\omega^{-1}}{2}\right)_\xi. \tag{11.5}$$

Assume that D, K, and v are constants, also $D > 0$, $K > 0$, $v \neq 0$. Let us perform in (1-.5) the change of variables

$$\xi = \frac{2vy}{D}, \qquad \omega = \frac{D^{1/2}(1 + K^{-1}C)}{2v}.$$

The new sought function $C(y, \tau)$ obeys the equation

$$\frac{\partial C}{\partial \tau} = \left(D\frac{\partial^2}{\partial y^2} - v\frac{\partial}{\partial y}\right)\left(\frac{C}{1 + K^{-1}C}\right). \tag{11.6}$$

Equation (11.6) appeared in [145] as a model of the admixture diffusive-convective transfer by a liquid filtrating through a porous medium with constant velocity v; D is the diffusion coefficient. A part of the substance might absorb on the porous surface. It was assumed that the absorption-dissorption processes were instantaneous, and that the equilibrium values of the surface and volume concentrations of the substance were connected by the finite relation (the Langmuir–Hinshelwood equation). If $C(y, \tau)$ is the summary admixture concentration, then $C/(1 + K^{-1}C)$ is the "free" concentration of the substance involved in the transport process. Since equation (11.6) linearizes and, finally, reduces to (11.1), one has possibility to obtain a parametric representation for a solution of the Cauchy problem for equation (11.6) [145].

Let us return to equation (11.5) and perform the substitution

$$\xi = 2\beta\mu\tilde{x} - \beta\lambda\tilde{t}, \quad \tau = (2\beta\mu)^2\tilde{t}, \quad \omega = \beta S + \gamma,$$

where β, γ, λ, μ are constants; we always assume that $\beta\mu \neq 0$. The equation for $S(\tilde{x}, \tilde{t})$ coincides with (8.12), whence the linearizability of the latter. (If $\beta\mu = 0$, then equation (8.12) is either linear, or can be transformed to the linear equation (11.2) by a pointwise transformation.) Equation (8.12) has wide physical applications (see Section 8). Having a possibility to linearize it, one can effectively solve a number of initial-boundary value problems for this equation [49, 144, 142].

Let us now consider an equation of a form more general than (8.1). Namely,

$$u_t = [\Phi(x, t, u, u_x)]_x\,. \tag{11.7}$$

The explicit dependence of Φ on x implies that the class of equations (11.7) is not preserved after the passage to Lagrangian coordinate. (Conversely, the presence of the variable t in the arguments of the function Φ does not affect the invariance of this class of equations with respect to L-transformations — see Section 9.) It may happen, however, that an equation of the type (9.6) for the function $x = X(\xi, \tau)$, obtained at one of the intermediate stages of the passage to Lagrangian coordinates, will be linear or linearizable. Then equation (11.7) will be linearizable as well.

To illustrate this possibility, let us present two examples. Let in the equation

$$u_t = \left[u^{-2}u_x + q(t)xu\right]_x \tag{11.8}$$

$q(t)$ be an arbitrary function of t. Following the algorithm of Section 9, we get as an analog of equation (9.6) the linear equation

$$X_\tau = X_{\xi\xi} - q(\tau)X.$$

(The linearizability of equation (11.8) was discovered in [35, 166].) The next example is furnished by the equation

$$u_t = \left(u^{-2}u_x + x\right)_x. \tag{11.9}$$

Here the function $X(\xi, \tau)$ satisfies Burgers' equation (11.3), whence the linearizability of (11.9), [44].

Having started from linear equation (11.7), one can obtain another family of linearizable equations,

$$u_t = [a(x,t)u_x + b(x,t)u + c(x,t)]_x,$$

where a, b, c are arbitrary functions of x and t [35]. In this case the corresponding equation for the function X has the form

$$X_\tau = a(X,\tau)X_\xi^{-2}X_{\xi\xi} - b(X,\tau) - c(X,\tau)X_\xi.$$

Some other examples of second-order linearizable equations containing the space variable in explicit form can be found in Section 13 and in [112].

An additional possibility of extending the class of linearizable equations appears as the result of the superposition of L-transformation with pointwise transformations. Thus, if in equation (8.3) $C = K(v)v_x$ and the functions S and K are connected by some differential relation [164, 112, 146], then the substitution $\omega = S(v)$ reduces this equation to (11.2) and, hence, this equation can be linearized (see Section 15 for the detailes). One more example of this sort is considered in [48]. This paper deals with the equation

$$s_t = \psi s_{xx} + \left(\frac{\psi\psi''}{\psi'} - \frac{\psi'}{2}\right)s_x^2 - \frac{1}{2}\psi s_x. \tag{11.10}$$

Here ψ is an arbitrary function of s, the prime denotes differentiation in s. Equation (11.10) possesses the following remarkable property: after the substitution $s = \eta(\tilde{s})$, the function $\tilde{s}$ satisfies, independently of η, again equation (11.10) with $\psi(s)$ replaced by $\tilde{\psi}(\tilde{s}) = \psi\left[\eta(\tilde{s})\right]$. Letting $\eta(\tilde{s}) = \psi^{-1}(\tilde{s}^2)$, we transform (11.10) into the equation

$$\tilde{s}_t = \tilde{s}^2\left(\tilde{s}_{xx} - \frac{\tilde{s}_x}{2}\right). \tag{11.11}$$

The last equation transforms into the linear equation (11.5) by means of the transformation $\tilde{s} = 1/\omega$.

In conclusion of the section, let us remark that the L-transformation is applicable for linearization of certain higher-order equations (see [85, 137]). Examples of such equations are given in Section 14.

12 Lagrange-invariant equations

Let us term *Lagrange-invariant* (or simply *L-invariant*) an equation that transforms into itself under L-transformation. In this section, we confine ourselves to consideration of L-invariant equations of the type (8.1). In Section 14 we indicate also a class of third-order Lagrange-invariant equations of the form $u_t = [\Phi(u, u_x, u_{xx})]_x$.

The following problem was first posed in [31]: to find L-invariant equations in the encompassing class of one-dimensional evolution equations of divergence form. This paper dealt with the class of heat conduction equations (8.2) arbitrary up to a single-variable function $k(u)$. A direct check shows that equation (8.2) is L-invariant if and only if k satisfies the functional equation

$$k(u) = u^{-2}k(u^{-1}). \tag{12.1}$$

The general solution of this equation has the form $k = u^{-1}\lambda(u)$, where the function λ is subject to the condition $\lambda(u) = \lambda(u^{-1})$ and is arbitrary in all other respects [31].

Note that due to (12.1) the L-invariant equation (8.2) is necessarily degenerate at the values $u = 0$, $u = \pm\infty$ or, at least, at one of these values. On the other hand, if we are interested in the solutions of equation (12.1) in $C^l[a, a^{-1}]$, l a natural number and $a \in (0, 1)$, then we have to set $\lambda \in C^l[a, a^{-1}]$ and, moreover, to claim that all odd-order derivatives of the function $\mu(v) = \lambda(e^v)$ of order not exceeding l vanish at the point $v = 0$. It is easy to exhibit examples of such functions: they are $\lambda_1 = f_1(u)f_1(u^{-1})$, $\lambda_2 = f_2(u) + f_2(u^{-1})$ (here $f_1, f_2 \in C^l[a, a^{-1}]$ are arbitrary), and also an arbitrary function of the arguments $\lambda_1(u)$ and $\lambda_2(u)$. The simplest representative of the class of L-invariant equations (8.2) is the equation

$$u_t = \left(u^{-1}u_x\right)_x, \tag{12.2}$$

occurring in plasma physics.

If equation (8.2) is L-invariant, then as the result of L-transformation every of its solution yields once again a solution of (8.2). The following example shows that the original and the transformed equations can be strongly distinguished. Equation (12.2) admits the solution

$$u_1 = \frac{2t}{x^2 + 4t^2}. \tag{12.3}$$

By (8.4), L-transformation of solution (12.3) has the form

$$u_2 = \frac{2t}{(\cos x)^2}. \tag{12.4}$$

These two solutions of equation (12.2) are invariant with respect to some groups of dilatations admitted by this equation. These solutions are substantially different because there are no Lie groups admitted by equation (12.2), transforming (12.3) into (12.4). (All such transformations are exhausted by the translations in x and t, some dilatations in the space x, t, u, and their superpositions.) The correspondence between (12.3) and (12.4) is established by the nonlocal L-transformation.

Let us consider now a more general equation of the class (8.1). As was shown in Section 8, equation (8.1) is L-invariant if and only if equality (8.7) holds. For the sake of convenience, we present and re-number it once again:

$$\varphi(u, p) = -u\varphi(u^{-1}, -u^{-3}p). \tag{12.5}$$

Relation (12.5) is a functional equation for the function of two variables $\varphi(u, p)$. In what follows, we restrict ourselves to seek the solutions of equation (12.5) defined in one of two open half-planes of the plane u, p: $u > 0$ or $u < 0$. A partial justification of this restriction is the fact that given some t fixed, the mapping $\xi = \xi(x, t)$, defined by the first of formulas (8.4), is a diffeomorphism if and only if the smooth function $u(x, t)$ keeps the sign.

Thus, let in equation (12.5) $u > 0$. Let us perform in this equation the substitution

$$\varphi(u, p) = u^{1/2}F(u, u^{-3/2}p). \tag{12.6}$$

Let a continuous function $F(u, p)$ be defined in the half-plane $u > 0$ of the plane (u, p), satisfy the equation

$$F(u, q) = -F(u^{-1}, -q), \tag{12.7}$$

and be arbitrary in all other respects. Then the function φ given by (12.6) is a solution of the functional equation (12.5) [135]. The proof of this assertion consists in the direct substitution of (12.6) into (12.5) with the use of property (12.7). Equality (12.7) may be given a more symmetric form by means of introduction of the function $G(v, q) = F(e^v, q)$; then (12.7) implies the relation

$$G(v, q) = -G(-v, -q) \tag{12.8}$$

(as was observed by H. F. Weinberger).

Thus, the class of L-invariant equations (8.1) is very broad: we can take the function φ arbitrary in the strip $0 < u < 1$, $p \in \mathbb{R}$, and continue it to the half-plane $u > 1$, $p \in \mathbb{R}$ with the help of equalities (12.6), (12.7). (It should be taken into account that the smoothness of φ in this strip still does not guarantee the smoothness of the continued function at the points of the line $u = 1$.) Let us mention in passing that the required smoothness conditions can be obtained from the representation $\varphi(e^v, e^{3v/2}q) = e^{v/2}G(v, q)$ and relation (12.8). Let us point out two partial solutions of equation (12.5): the solution linear in p

$$\varphi_1 = u^{-1}f(u)p + u^{1/2}g(u),$$

and the "self-similar" solution [137]

$$\varphi_2 = u^{-1}p\psi\left(u^{-3/2}p\right).$$

Here $\psi(s)$ is an arbitrary even function; for $0 < u < 1$ the values of the functions f and g satisfy the relations $f(u^{-1}) = f(u)$, $g(u^{-1}) = -g(u)$. Besides (12.2), one of the simplest L-invariant equations of the form (8.1) is the equation

$$u_t = \left(|u_x|^{-2/3}u_x\right)_x, \tag{12.9}$$

which appears in considerations of the rectilinear-trajectory motions of non-Newtonian fluids (i.e. pseudo-plastic).

Transformation (8.4) defines an automorphism on the set of positive solutions of equations of the type (8.1). One may profit from this property to obtain from an "igniting" solution of equation (8.1) a multiparameter family of solutions in the case when this equation admits a non-trivial Lie group (larger than the group of translations in x and t). This method of reproduction of solutions to equation (8.1) is relative to the method of obtaining a family of solutions of an evolution equation by means of the differential substitutions developed by S. V. Khabirov [83, 82]. We will demonstrate the idea of this method by the example of equation (12.9), [133].

Equation (12.9) admits the Lie group G_5 generated by the operators

$$X_1 = \partial_t, \quad X_2 = \partial_x, \quad X_3 = \partial_u, \quad X_4 = \tfrac{2}{3}t\partial_t + u\partial_u,$$

$$X_5 = \tfrac{2}{3}t\partial_t + x\partial_x - u\partial_u.$$

Let us consider one of the solutions of equation (12.9) invariant with respect to the dilatation transformation with operator X_5,

$$u_0 = \frac{1}{t^{3/2}} h\left(\frac{x}{t^{3/2}}\right). \tag{12.10}$$

The function $h(\xi)$ satisfies the ordinary differential equation

$$-\frac{3}{2}\left(\xi h' + h\right) = \left(|h|^{-2/3} h'\right)'$$

(the prime denotes differentiation in ξ). We are interested in the positive even solutions of the latter equation which are regular for all ξ and decrease as $\xi \to \infty$. Such solutions form the one-parameter family

$$h = \left[\frac{27}{16}\xi^4 + \frac{4}{3\sqrt{3}}\left(\frac{I}{C_0}\right)^2\right]^{-1/2}$$

with the parameter C_0. The constant I is defined by the equality

$$I = \int_{\mathbb{R}} \frac{dz}{\sqrt{z^4+1}} \equiv \frac{1}{\sqrt{\pi}}\left[\Gamma\left(\frac{1}{4}\right)\right]^2.$$

This parametrization is convenient because C_0 coincides with the value of the integral which is preserved for every solution of equation (12.9), namely,

$$\int_{\mathbb{R}} u_0(x,t)\,dx = C_0.$$

This makes it possible to interpret (12.10) as a solution of the Cauchy problem

$$u(x,0) = C_0\delta(x)$$

for equation (12.9).

Having subjected solution (12.10) to L-transformation (8.4) and taking into account the evenness of $h(\xi)$, we get a new solution of equation (12.9) represented in the form

$$u_1 = t^{3/2}\left\{h\left[H^{-1}(x)\right]\right\}^{-1}, \tag{12.11}$$

where H^{-1} is the inverse of the primitive H of the function h; $H(0) = 0$. Solution (12.11), as well as (12.10), is a self-similar solution (e.g. a solution invariant with respect to the dilatation transformation) of equation (12.9), but with another operator X_4. Solution (12.11) possesses some pathological properties: being positive as $t \to 0$, it vanishes at $t = 0$; the function $\varphi(x) = \{h[H^{-1}(x)]\}^{-1}$ is defined only for $|x| < C_0/2$ and grows unboundedly as $x \to \pm C_0/2$; $\varphi = 2C_0I^{-1}(C_0/2 \mp x)^{-2} + o(1)$ when $x \to \pm C_0/2$.

Let us now make use of the invariance of (12.9) with respect to the operator X_4. This means that along with the function u_0 the equation mentioned also has as a solution the function $u_0 + C_1$, where C_1 is an arbitrary constant. In what follows, we suppose $C_1 > 0$. Let us apply to the function $u_0 + C_1$ the L-transformation and denote the result by $u_2 = \omega(\xi, t)$. Also, let us take into account that due to the evenness of $h(\xi)$ the second integral on the right-hand side of the first of relations (8.4) vanishes.

Replacing ξ and x in the final expression (which is possible because of L-invariance of (12.9)), we get the parametric representation for the new solution $u_2(x, t)$ of equation (12.9):

$$x = H\left(\xi/t^{3/2}\right) + C_1\xi, \quad u_2 = t^{3/2}/\left(h\left(\xi/t^{3/2}\right) + C_1t^{3/2}\right). \tag{12.12}$$

In contrast to solutions (12.10) and (12.11), this solution already depends on two arbitrary constants C_0 and C_1.

The mapping $\mathbb{R} \mapsto \mathbb{R}$, given by the first formula of (12.12) for arbitrary fixed t, is diffeomorphic for $t > 0$. If $t \to 0+$, this mapping tends to a discontinuous one: $x = C_1\xi + C_0/2$ for $\xi > 0$, $x = C_1\xi - C_0/2$ for $\xi < 0$. The function $u_2(x, t)$ is a classical solution of equation (12.9) for all $t > 0$. Letting $t \to 0$ in the second equality of (12.12) and taking into account that $h(0) > 0$, $h(\xi) = O\left(\xi^{-2}\right)$ as $\xi \to \pm\infty$, we find that this solution satisfies the boundary condition

$$u_2(x, 0) = \begin{cases} 1/C_1 & \text{as } x \neq 0, \\ 0 & \text{as } x = 0. \end{cases} \tag{12.13}$$

Thus, having started from the "igniting" self-similar solution (12.10), we have constructed a solution to the Cauchy problem for equation (12.9) with discontinuous initial data (12.13). The solution u_2 not only fails to be self-similar but is even not invariant with respect to any of the Lie groups admitted by equation (12.9). It is curious to note that solution (12.12) contains a parameter C_0 not involved into the initial conditions. This means non-uniqueness of solution of the Cauchy problem (12.9), (12.13).

Subjecting the solution $u_2 + C_2$ ($C_2 > 0$ is a constant) to transformation (8.4), we get a solution u_3 of equation (12.9), containing three arbitrary parameters. This method of reproduction of solutions to (12.9) is not unique. The point is that this equation possesses a hidden symmetry. One of the nonlocal transformations conserving it is generated by the extension of the Lie group under passage from (12.9) to the equation for the function ω such that $\omega_x = u$. The last equation,

$$\omega_t = |\omega_{xx}|^{-2/3}\omega_{xx},$$

admits the operator $Z = \omega\partial/\partial x$, which does not correspond to any of the operators of the Lie group G_5 [120, 3]. Another nonlocal transformation is discovered if one passes from equation (12.9) to the equation for the function $v = u_x$. The equation for v has the form

$$v_t = \left(|v|^{-2/3}v_x\right)_x$$

and belongs to the set of equations with hidden symmetries considered in Section 10. Combining L-transformation with the above-described nonlocal transformations and pointwise transformations of the Lie group G_5, we get additional possibilities of multiplying the number of explicit solutions of equation (12.9).

These results are contained in paper [133].

13 Equations with spherical and cylindrical symmetries

13.1 Gradient-nonlinear equations. In this section, we consider the equation that describes the process of heat propagation in a medium with constant density and heat conductivity, and with the heat conductivity coefficient depending nonlinearly on the temperature gradient ∇T,

$$T_t = \nabla \cdot [k(|\nabla T|)\nabla T]. \tag{13.1}$$

Here $\nabla \equiv (\partial/\partial x_1, \ldots, \partial/\partial x_n)$, n is the spatial dimension; the function k assumes non-negative values. Equation (13.1) is invariant with respect to rotations in $\mathbb{R}^n$, which makes it possible to seek its spherically (or radially) symmetrical solutions of the form $T = \theta(r,t)$, where $r = (x_1^2 + \cdots + x_n^2)^{1/2}$. The function θ obeys the equation

$$r^{n-1}\theta_t = \left[r^{n-1}k(\theta_r)\theta_r\right]_r. \tag{13.2}$$

A typical example of the gradient–nonlinear equation is

$$r^{n-1}\theta_t = \left(r^{n-1}|\theta_r|^l\theta_r\right)_r. \tag{13.3}$$

Equation (13.3) is distinguished from the other equations of the type (13.2) by the following remarkable property: if $l \neq -2$, $l \neq -2(n+1)^{-1}$, there exists a nonlocal transformation which transforms (13.3) into an equation of the same form but with another value of the exponent n. Let us prove this assertion. We shall assume that

$\theta_r \neq 0$ in a neighborhood of some point (r_0, t_0) of the half-plane $r > 0$ where our calculations are performed (obviously, the variable r can assume only non-negative values).

Let us pass in equation (13.3) to the new independent variables (θ, t), having taken r for the new sought function:

$$r = w(\xi, t). \tag{13.4}$$

Since $\theta_r = 1/w_\theta$, $\theta_t = -w_t/w_\theta$, equation (13.3) transforms into

$$w^{n-1}w_t = \left(-w^{n-1}|w_\theta|^{-l}w_\theta^{-1}\right)_\theta .$$

Let us perform in this equation the change of variables

$$\theta = x, \quad w = [u(x,t)]^{1/n} . \tag{13.5}$$

The new sought function u satisfies the equation of type (8.1)

$$u_t = \left[-n^{l+2}u^{(l+2)(n-1)/n}|u_x|^{-l}u_x^{-1}\right]_x . \tag{13.6}$$

Moreover, it follows from (13.4), (13.5), and the inequality $r > 0$ that $u > 0$ in a neighborhood of the point (x_0, t_0) — the image of the point (r_0, t_0). The latter provides the possibility to perform in equation (13.6) the L-transformation (8.4) where the lower limits of integration are replaced by x_0 and t_0, respectively. By virtue of (8.5) and (13.6), the equation for the new sought function $w(\xi, t)$ has the form

$$w_t = \left[-n^{l+2}w^{(2ln+l+2n+2)/n}|w_\xi|^{-l}w_\xi^{-1}\right]_\xi .$$

Let us observe that due to (13.4), (13.5), (8.4), the inequality $\theta_r \neq 0$ yields $w_\xi \neq 0$. This permits one to replace in the equation for w the space variable by the sought function. Letting

$$\xi = v(w, t), \tag{13.7}$$

we get the following equation for v:

$$v_t = \left[n^{l+2}w^{(2ln+l+2n+2)/n}|v_w|^l v_w\right]_w . \tag{13.8}$$

To reduce (13.8) to the needed form it suffices to perform the change of variables

$$w = C\tilde{r}^m, \quad v(w,t) = \tilde{\theta}(\tilde{r}, t) \tag{13.9}$$

with appropriate constants m and C. This proves possible if $l \neq -2$ and $l \neq -2(n+1)^{-1}$. Indeed, let us set

$$m = -\frac{(l+2)n}{ln+l+2} \tag{13.10}$$

and choose the normalizing factor

$$C = \left|\frac{(l+2)n}{l+2}\right|^{(l+2)n/ln+l+2} .$$

Then it follows from (13.8), (13.9) that the function $\tilde{\theta}(\tilde{r}, t)$ is a solution of the equation

$$\tilde{r}^{m-1}\tilde{\theta}_t = \left(\tilde{r}^{m-1}|\tilde{\theta}_{\tilde{r}}|^l\tilde{\theta}_{\tilde{r}}\right)_{\tilde{r}}. \tag{13.11}$$

Thus, the invertible chain of transformations (13.4), (13.5), (8.4), (13.7), (13.9) realizes the reciprocity transformation of equations (13.3) and (13.11). In addition, the exponents m and n are connected by relation (13.10). In particular, if $l = -4(n+2)^{-1}$, then the original and transformed equations coincide. In this case the mentioned reciprocity transformation defines a nonlocal automorphism on the set of solutions of equation (13.3). In this way, this equation proves similar to L-invariant equations considered in Section 12. Another interesting case corresponds to the value $l = -2(n+1)(2n+1)^{-1}$. Here $m = 1$, meaning the equivalence of equations (13.3) and (13.11), of which the former is satisfied by the radially symmetric solutions of the equation with gradient nonlinearity, and the latter describes solutions with plane waves.

Let us consider now the exceptional values of the exponent l: $l = -2$ and $l = -2(n+1)^{-1}$. In the first case $m = 0$ and the transformation (13.9) becomes degenerate. However, the substitution

$$r = (nx)^{1/n}, \quad \tilde{t} = -t, \quad \theta_r = u(x, \tilde{t})$$

reduces equation (13.2) with $l = -2$ to the form

$$u_{\tilde{t}} = \left(u^{-2}u_x\right)_x.$$

The latter equation transforms into the linear equation, $\omega_t = \omega_{\xi\xi}$, by transformation (8.4) where one should replace t by $\tilde{t}$ [164]. If $l = -2(n+1)^{-1}$, then the equation for defining the exponent m becomes unsolvable. (In the general case it leads to equality (13.10).) In this case, equation (13.3) itself becomes

$$r^{n-1}\theta_t = \left[r^{n-1}|\theta|^{-2/(n+1)}\theta_r\right]_r. \tag{13.12}$$

In comparison with the general equation (13.3), equation (13.12) possesses an additional symmetry which allows one to reduce it to an equation of simpler form. Namely, letting

$$r = e^y, \quad \theta = e^{-ny}\zeta(y, t),$$

we get the following equation for the function ζ:

$$\zeta_t = \left[|\zeta_y - n\zeta|^{-2/(n+1)}\left(\zeta_y - n\zeta\right)\right]_y. \tag{13.13}$$

Equation (13.13) does not contain the new space variable y explicitly. In particular, this provides the possibility to construct solutions of (13.13) of the "travelling-wave" type, $\zeta = f(y - Ct)$, where $C = \text{const}$. They correspond to the solutions of (13.12) of the form $\theta = r^{-n}f(\ln r - Ct)$. Later on, we will restrict ourselves to the case $C > 0$ which corresponds to radial heat waves propagating to infinity as t grows.

Having integrated in η the equation for the function f yielded by (13.13), we get the equation

$$-Cf - b = |f' - nf|^{-2/(n+1)} \left(f' - nf\right), \tag{13.14}$$

where $b = \text{const.}$ and the prime denotes differentiation in η. It is easy to show that for $b \geq 0$ equation (13.14) has a family of solutions defined for all η and such that $f \to f_-$ as $\eta \to -\infty$, $f \to f_+$ as $\eta \to \infty$. The case $b = 0$ is especially interesting: here $f_- = 0$ and the corresponding solution of equation (13.12) is bounded as $r \to 0$. This solution can be directly presented as

$$\theta = C^{(n+1)/2} n^{(n-1)/2} \left(r^{2n/(n-1)} + e^{2nCt/(n-1)}\right)^{-(n-1)/2}. \tag{13.15}$$

As opposed to most explicit (for instance, self-similar) solutions of degenerate parabolic equations, solution (13.15) is defined for all $t \in \mathbb{R}$. One may thus interpret it as a solution of "the Cauchy problem with data at $t = -\infty$" for equation (13.12)

$$\theta \to \theta_{-\infty}(r) = C^{(n+1)/2} n^{(n-1)/2} r^{-n} \quad \text{as} \quad t \to -\infty.$$

The "initial" function $\theta_{-\infty}(r)$ has a singularity at the point $r = 0$. This function depends on r so that the heat flux Q across the sphere $x_1^2 + \cdots + x_N^2 = r^2$ does not depend on r. The heat flux is equal to the product of the quantity $-|\theta_r|^{-2/(n-1)}\theta_r$ and the surface square of this sphere $\Omega_n r^{n-1}$, where Ω_n is the surface square of the unit sphere in $\mathbb{R}^n$. Each value $Q > 0$ corresponds to the unique value of parameter C which the family of solutions (13.15) depends on,

$$C = n\left(\frac{\Omega_n}{Q}\right)^{2/(n-1)}.$$

In conclusion of this section, let us consider the linear equations (13.3), i.e. the equations with $l = 0$. The equation $r^2\theta_t = \left(r^2\theta_r\right)_r$ is the one the spherically symmetric solutions of the linear heat conduction equations obey in the three-dimensional case. It is known that by means of the substitution $\theta = V/r$ this equation can be reduced to the equation with the constant coefficients $V_t = V_{rr}$. At the same time, such a transformation does not exist for the equation $r\theta_t = (r\theta_r)_r$ describing cylindrically symmetrical solutions. Extending the class of the equivalence transformations, we can establish the correspondence between the solutions of the linear equation

$$r^{n-1}\theta = \left(r^{n-1}\theta\right)_r \tag{13.16}$$

and solutions of an analogous equation but with another value of the exponent n. It amounts to specifying the arguments presented at the beginning of this section to the case $l = 0$. Letting in (13.10) $l = 0$, we get $m = -n$. The substitution of the values $m = -n$, $l = 0$ into (13.11) yields the equation

$$\tilde{r}^{-n}\tilde{\theta}_t = \left(\tilde{r}^{-n-1}\tilde{\theta}_{\tilde{r}}\right)_{\tilde{r}}. \tag{13.17}$$

Let us formulate the final result. Assume that a solution θ of equation (13.16) satisfies the condition $\theta_r \neq 0$ in a neighborhood of the point (r_0, t_0) (recall that the variable r in this equation is nonnegative). Then the superposition of transformations (13.4), (13.5), (8.4), (13.7), (13.9) puts into one-to-one correspondence the functions $\theta(r,t)$ and $\tilde{\theta}(\tilde{r},t)$ of which the latter is a solution of equation (13.17), defined in a neighborhood of the point (r_0, t_0) of the half-plane $\tilde{r} > 0$. Also, in this neighborhood $\tilde{\theta}_{\tilde{r}} \neq 0$. In particular, for $n = 1$ we get the reciprocity transformation of the equation

$$\tilde{r}^{-2}\tilde{\theta}_t = \left(\tilde{r}^{-2}\tilde{\theta}_{\tilde{r}}\right)_{\tilde{r}}$$

and the equation with the constant coefficients $\theta_t = \theta_{rr}$.

Equation (13.16) with $n = 1, 2, 3$ explicitly describes the one-dimensional processes of heat propagation with plane, cylindrical, or spherical geometries, respectively. The physical sense of the solutions of equation (13.17) with the same values of n is not so clear. Nonetheless, it can be used for approximate descriptions of the heat conduction process close to the one-dimensional one. For instance, this situation appears in "thin" axisymmetric bars, i.e. in the domains bounded by two planes $x_1 = a$, $x_1 = b$ $(0 < a < b \leq \infty)$, and the lateral surface of rotation described by the equation $x_2^2 + x_3^2 = \varepsilon x_1^{-n-1}$ with $\varepsilon > 0$; here $n > -1$ need not be an integer. Suppose that the lateral surface of the bar is heat insulated. Then for small ε and fixed a and b the distribution of the temperature T in the bar will depend weakly on the variables x_2, x_3, and under an appropriate choice of the scales the dependence of T on x_1 and t is well approximated by solutions of the equation $S(x_1)T_t = \left[S(x_1)T_{x_1}\right]_{x_1}$, where S is the square of the cross-section of the bar. Accepting the notation $x_1 = \tilde{r}$, $T(x_1, t) = \tilde{\theta}(\tilde{r}, t)$, and taking into account the formula $S = \pi\varepsilon x_1^{-n-1}$, we arrive at equation (13.17) for the function $\tilde{\theta}$.

The results of this section were published in paper [134].

13.2 Equation of heat conduction in solid hydrogen. The process of heat conduction in the solid hydrogen is described by the equation

$$\rho c_p T_t = \nabla(k \nabla T), \tag{13.18}$$

where $T \geq 0$ is the absolute temperature, ρ is the density, c_p is the heat capacity under constant pressure, and k is the heat conductivity coefficient. If the body is homogeneous and the dependence of the heat flux vector on the temperature gradient is linear, then the nonnegative magnitudes ρ, c_p, k may depend on the temperature T. As was noticed in [146], for solid hydrogen in the range $T < 14°\ K$ the dependences mentioned can be approximated by the relations

$$\rho = \text{const.}, \quad c_p = A\,T^3, \quad k = B\,T^3\left(C + T^4\right)^{-2}, \tag{13.19}$$

with A, B, C positive constants. Let us introduce the new sought function u by the relation

$$u = \left(\rho\frac{A}{B}\right)^{1/2} (C + T)^4 .$$

It follows from (13.18), (13.19) that this function satisfies the equation

$$u_t = \nabla\left(u^{-2}\nabla u\right). \tag{13.20}$$

Following the tradition originated by [146], equation (13.20) is usually referred to as the heat conduction equation in solid hydrogen. For the sake of fairness it is to be noted, however, that this equation was studied yet earlier in paper [164] where it appeared in another physical context. It was paper [164] where the transformation linearizing equation (13.20) with the single space variable was found (this fact was already mentioned in Section 8).

In this subsection, we shall consider the radially symmetric solutions of equation (13.20), that is, the solutions of the form $u = v(r,t)$, with the usual notation $r = (x_1^2 + \cdots + x_n^2)^{1/2}$. The function v satisfies the equation

$$v_t = \left(r^{n-1}v^{-2}v_r\right)_r. \tag{13.21}$$

In what follows, n is an arbitrary natural number, although the most interesting cases are $n = 1, 2, 3$. Following [137], where equation (13.21) with $n = 3$ was considered, and [151], we shall show that an analog of L-transformation reduces this equation to "autonomous" form (when the space variable is not explicitly contained in the transformed equation). Also, we will clarify here a special role not only of the value $n = 1$ but of the value $n = 2$, too.

Letting in (13.21) $v = \rho$, $-v^{-3}v_r = V$, we get

$$r^{n-1}\rho_t + \left(r^{n-1}\rho V\right)_r = 0.$$

The latter equation is the continuity equation of the radially symmetric flow in the n-dimensional space of the liquid having density ρ and velocity V. Our aim is to obtain from this equation a new equation for the magnitude r considered now as a function of t and the mass Lagrangian coordinate ξ : $r = R(\xi, t)$. The passage to Lagrangian coordinates is performed in the same way as in Section 9, however now the integral of the continuity equation has the form

$$\gamma R^{n-1}R_\xi = 1, \tag{13.22}$$

where we denote $\gamma(\xi, t) = v\left[R(\xi, t), t\right]$. The trajectory equation $r_t = V$, the definition of V in terms of the function v, the equalities $v_r = \gamma_\xi / R_\xi$ and (13.22), imply the following equation for the function R:

$$R_t = R^{n-1}\left(R^{n-1}R_\xi\right)_\xi. \tag{13.23}$$

Now we distinguish three cases: $n = 1, n = 2$, and $n \geq 3$.

If $n = 1$, equation (13.22) becomes linear. For $n = 2$ the substitution $R = e^\psi$ reduces (13.23) to the form

$$\psi_t = \left(e^{2\psi}\psi_\xi\right)_\xi,$$

e.g. to the exponent-nonlinear heat conduction equation. Finally, in the case $n \geq 3$ the transformation reducing (13.23) to divergence form is $R = \omega^{-1/(n-2)}$. The function ω satisfies the equation

$$\omega_t = \left[\omega^{-2(n-1)/(n-2)}\omega_\xi\right]_\xi .$$

The nonnegative values of R (and these are the only solutions we are interested in) correspond to nonnegative ω. Following the procedure described in Section 9 and performing in the equation for ω the inverse passage to the Euler coordinates $\tilde{x}$, t, and the new sought function $\tilde{u}(\tilde{x}, t) = 1/\omega(\xi, t)$, we arrive at the equation

$$\tilde{u}_t = \left[\tilde{u}^{2/(n-2)}\tilde{u}_{\tilde{x}}\right]_{\tilde{x}} . \tag{13.24}$$

Equation (13.24) arises in a number of applications (see Chapter III, which is mainly devoted to the study of this equation and its analogs). For example, if $\tilde{u}$ stands for the density of a polytropic gas filling a porous medium, then (13.24) covers the isentropic filtration process of this gas ($2/(n-2)$ is the polytropic exponent). Letting $n = 3$, we come to the conclusion: equation (13.21) which describes the spherically symmetric heat propagation process in solid hydrogen is equivalent to the gas filtration equation (13.24) in plane geometry with the polytropic exponent equal to two.

13.3 The nonlinear Schrödinger equation. The object of consideration of this subsection is the nonlinear (more exactly, cubic) Schrödinger equation. In the notation of [39] it has the form

$$2i\frac{\partial E}{\partial z} + \frac{1}{r^\nu}\frac{\partial}{\partial r}\left(r^\nu\frac{\partial E}{\partial r}\right) + |E|^2 E = 0. \tag{13.25}$$

Here E is a complex-valued function, i is the imaginary unit; z plays the role of the space variable. Equation (13.25), where the parameter ν takes one of the values 0, 1, 2, is widely used in nonlinear optics and plasma physics.

Equation (13.25) is considered in the half-strip $0 < r < R$, $z > 0$, and under the additional restrictions

$$\frac{\partial E}{\partial r}(0, z) = 0, \quad E(R, z) = 0 \quad \text{as } z \geq 0, \tag{13.26}$$

$$E(r, 0) = E_0(r) \quad \text{as } 0 \leq r \leq R. \tag{13.27}$$

The limit variant of problem (13.25)–(13.27) corresponding to $R = \infty$ is also of certain interest. Here the second condition of (13.26) is replaced by the condition $E \to 0$ as $r \to \infty$; the function E_0 is assumed to be rapidly decreasing as $r \to \infty$. The first of these conditions allows one to continue the function $E(r, z)$ evenly with respect to r in the domain $r < 0$, $z > 0$, thus obtaining the Cauchy problem for the nonlinear Schrödinger equation (13.25).

If $\nu = 0$, the above-formulated Cauchy problem can be effectively studied by the method of the inverse scattering problem [176]. This method is inapplicable in the cases $\nu = 1, 2$, and in the study of the initial-boundary value problems.

A characteristic feature of the discussed problems for equation (13.25) is that their solutions display singularities: the function E becomes infinite at some point (at the focus). The occurrence of singularities corresponds to physical phenomena such as self-focusing of the wave beams in nonlinear media or collapsing of Langmuir waves in the no-collision plasma (the corresponding references are given in [39]).

The analytic methods allow one to establish the very fact of occurrence of singularities and to clarify their self-similar character. Any more detailed study of the behavior of the function E near the focus lends itself only to numerical methods. However, their construction with the use of any of the grids fixed in the Euler coordinates r, z is surely inefficient. In paper [39] a method for numerical treatment of problem (13.25)–(13.27) was proposed based on the passage to the so-called energy Lagrangian coordinates. Here the grid, fixed in Lagrangian coordinates, is equivalent to a sliding grid in Euler coordinates, which automatically reconstructs in the process of solution.

It is to be noted that equation (13.25) is not in divergence form, therefore it is not evident how to perform the transformations of the type described in Section 9. This possibility is realized by the passage from the original sought function E, having the sense of the complex amplitude of the electric field, to two new real-valued functions — the energy density ρ and the group velocity of oscillations Φ — so that

$$E = \rho^{1/2} e^{i\Phi}. \tag{13.28}$$

For the sake of definiteness, later on we consider the case $\nu = 1$ corresponding to the cylindrically symmetric wave field.

The substitution of (13.28) into equation (13.25) leads to the system of equations for the functions ρ and Φ:

$$\rho_z + \rho_r \Phi_r + \rho \Phi_{rr} + \frac{\rho}{r} \Phi_r = 0,$$

$$\Phi_z + \frac{1}{2}\Phi_r^2 - \frac{\rho_{rr}}{4\rho} + \frac{\rho_r^2}{8\rho^2} - \frac{\rho_r}{4r\rho} - \frac{\rho}{2} = 0.$$

Differentiating the last equation in r and introducing the notation $z = t$, $v = \Phi_r$, we obtain the system of equations for the functions ρ and v:

$$\partial_t \rho + \frac{1}{r} \partial_r (r\rho v) = 0, \tag{13.29}$$

$$\partial_t v + v \partial_r v + \frac{1}{\rho} \partial_r p + g = 0, \tag{13.30}$$

where

$$p = -\frac{\rho}{4r} \partial_r \left(\frac{r}{\rho} \partial_r \rho \right) - \frac{\rho^2}{4}, \tag{13.31}$$

$$g = \frac{1}{4r\rho^2} (\partial_r \rho)^2. \tag{13.32}$$

Equations (13.29) and (13.30) formally coincide with the equations of continuity and momentum of the one-dimensional radially symmetrical motion of a gas with density ρ and velocity v. Also, p has the meaning of the pressure and g of the acceleration of the mass forces. However, if in the standard model of isentropic gas motion the pressure and the density are connected via a functional dependence, here the differential relation (13.31) occurs. This connection is typical for relaxing media. Moreover, due to (13.32), the acceleration of the exterior forces depends on the density gradient, which is also not typical for gas dynamics.

The hydrodynamical form of presentation of equation (13.25) allows one to perform the passage to Lagrangian coordinates without any difficulty.

Let us introduce the mass Lagrangian coordinate ξ connected with the Euler coordinate r by the formula

$$\xi = \int_0^r \rho(s,t)\,ds \tag{13.33}$$

and denote $\rho(r,t) = \gamma(\xi,t)$, $v(r,t) = V(\xi,t)$. Here r is already considered as a function of ξ and t. Passing in system (13.29)–(13.30) to the new variables and adding to it the trajectory equation (since the function $r(\xi,t)$ is now a sought function, too) we get:

$$\partial_t r = V, \tag{13.34}$$

$$\partial_t (1/\gamma) = \partial_\xi (rV), \tag{13.35}$$

$$\partial_t V + r\partial_\xi P + G = 0, \tag{13.36}$$

$$P = -\frac{\gamma^2}{4}\left[\partial_\xi\left(r^2\partial_\xi\gamma\right) + 1\right], \tag{13.37}$$

$$G = \frac{r}{4}\left(\partial_\xi\gamma\right)^2. \tag{13.38}$$

Before formulating the initial and boundary conditions for system (13.34)–(13.38), let us remark that for each solution of equation (13.25) the following quantities are conserved

$$I_1 = \int_0^R |E|^2 r\,dr, \tag{13.39}$$

$$I_2 = \int_0^R \left(\left|\frac{\partial E}{\partial r}\right|^2 - \frac{1}{4}|E|^4\right) r\,dr. \tag{13.40}$$

It follows from (13.28) and (13.39) that on the ray $r = R$, $t > 0$ of the plane of Euler coordinates the value of the Lagrangian coordinate ξ is constant and equal to $I_1 = l$. Thereby we define the domain where system (13.34)–(13.38) is to be solved: $0 < \xi < l$, $t > 0$. The substitution of (13.28) into (13.40) and the consequent passage in the integrand to the variable ξ imply that

$$I_2 = \int_0^l \left[V^2 + \frac{r^2}{4}\left(\partial_\xi\gamma\right)^2 - \frac{\gamma}{2}\right] d\xi \tag{13.41}$$

is constant. Moreover, (13.34), (13.35) imply the conservation condition for the mass of "the liquid volume":

$$\gamma r r_\xi = 1. \tag{13.42}$$

The first boundary condition in (13.26) leads to the relations

$$\partial_\xi \gamma(0,t) = 0, \quad V(0,t) = 0 \quad \text{as } t \geq 0. \tag{13.43}$$

The second condition of (13.26) directly implies only that $\gamma(l,t) = 0$. To obtain a condition for the function V, let us integrate equation (13.35) in ξ from 0 to l. To calculate the integral on the left-hand side, let us then make use of equalities (13.42) and $r(l,t) = R = \text{const}$. They show that the integral equals zero, whence $V(l,t) = 0$. So, we have:

$$\gamma(l,t) = 0, \quad V(l,t) = 0 \quad \text{as } t \geq 0. \tag{13.44}$$

To conditions (13.43), (13.44) the initial conditions have to be added:

$$r(\xi,0) = r_0(\xi), \quad \gamma(\xi,0) = \gamma_0(\xi), \quad V(\xi,0) = V_0(\xi) \quad \text{as } 0 \leq \xi \leq l. \tag{13.45}$$

The functions r_0, γ_0, V_0 can be easily expressed through the function E_0 from the initial condition (13.27) of the original problem. For example, $r_0(\xi)$ is defined as the solution of the Cauchy problem

$$|E_0(r_0)|^2 r_0 dE_0/d\xi = 1, \quad r_0(0) = 0.$$

Now we are in position to give an equivalent formulation for problem (13.25)–(13.27) in Lagrangian coordinates: to find a solution r, γ, V of system (13.34)–(13.38) in the domain $0 < \xi < l$, $t > 0$ satisfying the boundary conditions (13.43), (13.44), and the initial conditions (13.45). Note that by (13.34), (13.42), and the first equality of (13.45) one can exclude from this system the functions r, γ, and thus obtain the complete integro-differential equation for the function V; we will not stop at this point, however.

In [39], problem (13.34)–(13.38), (13.43)–(13.45) was treated numerically with the initial function $E_0 = A\exp(-Br^2)$, where A and B were variable positive parameters. The calculations were performed on the grid uniform in ξ, with the knot number $N = 150$; the time step τ decreased inversely proportionally to the modulus of the electric field strength at the beam axis, $\tau = \tau_0\left[\gamma(0,t)\right]^{-1/2}$.

The computations of the self-focusing process were carried on until the moment when the energy density at the beam axis exceeded its initial value in four exponents. These computations confirmed the self-similar character of the solutions near the focus. The correctness of the computational procedure is also verified by the fact that the values of the difference analogs of (13.41) and the left-hand side of (13.42) remained constant in the first six exponents.

The method of [39] was generalized in [90] to the problem of light focusing in a nonlinear optical medium with absorption, and to the problem of light defocusing. In the first case the left-hand side of (13.25) acquires the additional term $ih\left(|E|^2\right)E$, and in the second case the nonlinear term in (13.25) is replaced by $-f(|E|^2)E$. Here $f(s)$, $h(s)$ are real-valued functions assuming nonnegative values. These generalizations are not simply formal: thus, if the absorption is taken into account, the integrals I_1, I_2 are not conserved for the solutions of the analog of problem (13.25)–(13.27). The results of [90] prove the effectiveness of numerical treatment of the above-mentioned problems of nonlinear optics via the method based on the passage to the hydrodynamical variables and the consequent introduction of Lagrangian coordinates.

14 Equivalence transformations for higher-order equations and systems of equations

14.1 The Kingston–Rogers theorem. In the preceding sections of this chapter we were considering L-transformations of the divergence equations which had second order with respect to the space variable and first order with respect to time. Beginning with paper [164], such equations were the object of application of transformations of the type (8.4). In particular, it was pointed out that transformation (8.4) preserved the class of equations (8.1). This gives a reason to term the L-transformation the *reciprocity transformation*.

In paper [85], 1982, J. G. Kingston and C. Rogers made an important observation: for the fulfillment of the reciprocity transformation in an equation with two independent variables only two properties were essential, those of the divergence form of the equation and its autonomy with respect to one of the independent variables (which may always be thought of as the space variable). The second variable, usually associated with time, transforms identically. The order of the equation may be arbitrary. Now we present the main result of the above-mentioned paper. We keep the authors' notation and terminology.

Theorem 14.1 ([85]). *The conservation law*

$$\frac{\partial}{\partial t}\left[T\left(\frac{\partial}{\partial x};\frac{\partial}{\partial t};u\right)\right]+\frac{\partial}{\partial x}\left[F\left(\frac{\partial}{\partial x};\frac{\partial}{\partial t};u\right)\right]=0 \tag{14.1}$$

transforms into the mutually associated conservation law

$$\frac{\partial}{\partial t'}\left[T'\left(\frac{\partial}{\partial x'};\frac{\partial}{\partial t'};u'\right)\right]+\frac{\partial}{\partial x}\left[F'\left(\frac{\partial}{\partial x'};\frac{\partial}{\partial t'};u'\right)\right]=0 \tag{14.2}$$

by the reciprocity transformation

$$dx' = T dx - F dt, \quad dt' = dt \tag{14.3}$$

$$T'\left(\frac{\partial}{\partial x'};\frac{\partial}{\partial t'};u'\right) = \frac{1}{T\left(\frac{\partial}{\partial x};\frac{\partial}{\partial t};u\right)} \tag{14.4}$$

$$F'\left(\frac{\partial}{\partial x'};\frac{\partial}{\partial t'};u'\right)=-\frac{F\left(\frac{\partial}{\partial x};\frac{\partial}{\partial t};u\right)}{T\left(\frac{\partial}{\partial x};\frac{\partial}{\partial t};u\right)}. \tag{14.5}$$

Proof. The conservation law (14.2) expresses the fact that the form $T'dx' - F'dt$ is a total differential. Replacing there dx', dt' by formulas (14.3) and using equalities (14.4), (14.5), we find

$$T'dx' - F'dt' = TT'dx - \left(T'F + F'\right)dt = dx,$$

as desired.

Let us show that transformation (14.3) is involutive. Applying it for the second time, we get

$$dx'' = T'dx' - F'dt' = dx, \quad dt'' = dt' = dt,$$

$$T''\left(\frac{\partial}{\partial x''};\frac{\partial}{\partial t''};u\right)=\frac{1}{T'(\frac{\partial}{\partial x'};\frac{\partial}{\partial t'};u)}=T\left(\frac{\partial}{\partial x};\frac{\partial}{\partial t};u\right),$$

$$F''\left(\frac{\partial}{\partial x''};\frac{\partial}{\partial t''};u\right)=\frac{F\left(\frac{\partial}{\partial x};\frac{\partial}{\partial t};u\right)}{T\left(\frac{\partial}{\partial x};\frac{\partial}{\partial t};u\right)T'\left(\frac{\partial}{\partial x'};\frac{\partial}{\partial t'};u\right)}=F\left(\frac{\partial}{\partial x};\frac{\partial}{\partial t};u\right).$$

This means that equation (14.1), having been transformed twice, coincides with the original. □

Viewing formally equation (14.1) as the continuity equation of the one-dimensional motion of some continuum, one can identify T with the density and F with the flux of mass, so that F/T becomes the velocity of this medium. There arises the following problem: how could one define the density T' and the flux F' for the transformed equation (14.2) in terms of the given functions T and F? A method to solve this problem is given in [85].

It follows from (14.5)–(14.6) that the operations of differentiation in the original and new variables are expressed by the formulas

$$\partial_x = \frac{1}{T'}\partial_{x'}, \quad \partial_t = \frac{F'}{T'}\partial_{x'} + \partial_{t'}.$$

Substituting them into (14.4), (14.5), we have

$$T'\left(\frac{\partial}{\partial x'};\frac{\partial}{\partial t'};u\right)=\frac{1}{T\left(D',\partial',u\right)}, \tag{14.6}$$

$$F'\left(\frac{\partial}{\partial x'};\frac{\partial}{\partial t'};u\right)=\frac{F\left(D',\partial',u\right)}{T\left(D',\partial',u\right)}, \tag{14.7}$$

where

$$D' = \frac{1}{T'\left(\frac{\partial}{\partial x'};\frac{\partial}{\partial t'};u\right)}\frac{\partial}{\partial x'}, \tag{14.8}$$

$$\partial' = \frac{F'\left(\frac{\partial}{\partial x'};\frac{\partial}{\partial t'};u\right)}{T'\left(\frac{\partial}{\partial x'};\frac{\partial}{\partial t'};u\right)}\frac{\partial}{\partial x'}+\frac{\partial}{\partial t'}. \tag{14.9}$$

Relations (14.6), (14.7) constitute a system of two functional equations for defining the pair of functions T' and F'. The question of solvability of this system is not trivial and in the general form is not yet solved. In [85], two situations were pointed out for which this system admitted explicit solutions. We confine ourselves to the consideration of one of these two cases

$$T = u, \quad F = E\left(u, \frac{\partial u}{\partial x}, \dots, \frac{\partial^n}{\partial x^n}\right). \tag{14.10}$$

Then relations (14.6)–(14.9) show

$$T' \equiv u' = \frac{1}{u}, \tag{14.11}$$

$$\begin{aligned} F'\left(\frac{\partial}{\partial x'}; \frac{\partial}{\partial t'}; u\right) &= -\frac{1}{u} E\left(D'^{(0)}u, D'^{(1)}u, \dots, D'^{(n)}u\right) \\ &= -u'E\left(D'^{(0)}\left(\frac{1}{u'}\right), D'^{(1)}\left(\frac{1}{u'}\right), \dots, D'^{(n)}\left(\frac{1}{u'}\right)\right). \end{aligned}$$

According to (14.8), we use here the notation $D' = u\partial_{x'}$. The expression $D'^{(k)}$ denotes the operator D' applied k times.

So, if the density and the mass flux in conservation law (14.1) have the form (14.10), then the corresponding expressions for the mutually associated conservation law (14.2) are given by formulas (14.11). In this case equation (14.1) has first order in t and $(n+1)$-st order with respect to x. Such equations often occur in applications (hydrodynamics, plasma physics, the theory of phase transitions, etc.); in these applications $n = 2, 3$, as a rule.

As an example, let us consider the famous Korteweg–de-Vries equation

$$u_t + uu_x + u_{xxx} = 0. \tag{14.12}$$

Besides its value in applications, this equation has extremely rich mathematical content [72, 161, 5, 177]. It has given progress to the theory of solitons, and to the method of the inverse scattering problem as a tool to study nonlinear integrable equations.

As was already pointed out in Section 9, the reciprocity transformation of the conservation law (9.1) is equivalent to passing in it to the mass Lagrangian coordinate ξ, replacing x. An analogous assertion is applicable to a more general conservation law (14.1), (14.10). In order to assign to the variable x' the sense of a Lagrangian coordinate, let us denote $x' = \xi$, $u'(x', t') = \omega(\xi, t)$, and take into account that $t' = t$. According to (14.1), the equation reciprocal to (14.12) with respect to transformation (14.3) has the form

$$\omega_t - \left(\frac{1}{2\omega}\right)_\xi - \left(\frac{1}{2\omega^2}\right)_{\xi\xi\xi} = 0. \tag{14.13}$$

One may say that equation (14.13) is a result of the passage to Lagrangian coordinates in the Korteweg–de-Vries equation (14.12).

In conclusion of this section, let us present a few examples of equations (14.1), (14.10) which can be linearized by the reciprocity transformation (14.3). Observe that Theorem 14.1 is true in the cases when the functions T and F depend on the evolution variable t. Consider the linear third-order equation

$$u_t - u_{xxx} - \mu(t)u_{xx} = 0, \tag{14.14}$$

where μ is an arbitrary function of t. Having passed to Lagrangian coordinates, we have

$$\omega_t + \left(\frac{1}{2\omega^2}\right)_{\xi\xi\xi} + \mu(t)\left(\frac{1}{\omega}\right)_{\xi\xi} = 0. \tag{14.15}$$

Thus, the nonlocal reciprocity transformation linearizes the family of equations (14.15), [137].

The following example is also contained in [137]. The equation

$$v_t + v^{3/2}\left(v^{1/2}v_{xx}\right)_{xx} = 0$$

transforms into the linear one

$$\omega_t + \omega_{\xi\xi\xi\xi} = 0$$

by the composition of the pointwise transformation $v = 1/u^2$ and the reciprocity transformation (14.3) with the notation $x' = \xi$, $1/u \equiv u'(x', t) = \omega(\xi, t)$.

The next construction enables one to extend significantly the family of linearizable equations. Let us consider the linear equation

$$\frac{\partial u}{\partial t} + \sum_{k=0}^{n} a_k(t)\frac{\partial^{k+1}u}{\partial x^{k+1}} = 0,$$

where a_k are arbitrary functions of t. Let us pass in it to Lagrangian coordinates. For the function $\omega(\xi, t) = [u(x, t)]^{-1}$ we get the equation

$$\frac{\partial \omega}{\partial t} - \frac{\partial}{\partial \xi}\left[\omega\sum_{k=0}^{n} a_k(t)\left(\frac{1}{\omega}\frac{\partial}{\partial \xi}\right)^k\left(\frac{1}{\omega}\right)\right] = 0. \tag{14.16}$$

Each equation obtained from (14.16) by means of pointwise or contact transformations is equivalent to a linear one [143, 72, 173].

14.2 Third-order L-invariant equations. In this section we consider the evolution equations of the form

$$u_t = [\Phi\,(u, u_x, u_{xx})]_x\,. \tag{14.17}$$

According to the Kingston–Rogers theorem, the equations reciprocal to (14.17) with respect to L-transformation have the form

$$\omega_t = \left[\Psi\left(\omega, \omega_\xi, \omega_{\xi\xi}\right)\right]_\xi\,, \tag{14.18}$$

where

$$\Psi(u, p, r) = -u\Phi\left(u^{-1}, -u^{-3}p, -u^{-4}r + 3u^{-5}p^2\right).$$

Equation (14.17) we call *L-invariant* if it coincides with the reciprocal equation (14.18). The condition of L-invariance leads to the functional equation for the function Ψ:

$$\Psi(u, p, r) = -u\Psi\left(u^{-1}, -u^{-3}p, -u^{-4}r + 3u^{-5}p^2\right). \tag{14.19}$$

Let us recall that in Section 12 we were studying the second-order L-invariant equations of the form (8.1) and it happened to be that the class of such equations was very broad. In particular, a smooth function $\varphi(u, p)$ defining equation (8.1) can be taken arbitrary in each half-plane $u \geq b > 1$, $p \in \mathbb{R}$, of the plane (u, p).

As for the third-order L-invariant equations, the problem of describing this set (i.e. the set of solutions of functional equation (14.19)) is not yet solved. However, it is quite easy to find solutions of (14.19) with Φ linear in the third argument:

$$\Phi = \lambda(u, p)r + \mu(u, p). \tag{14.20}$$

The substitution of (14.20) into (14.19) gives the equations for the functions λ and μ,

$$\begin{aligned}
\lambda(u, p) &= u^{-3}\lambda(u^{-1}, -u^{-3}p), \\
\mu(u, p) &= -u\mu(u^{-1}, -u^{-3}p) - 3u^{-4}p^2\lambda(u^{-1}, -u^{-3}p),
\end{aligned}$$

which are to be solved sequentially. The scheme of resolving these equations is similar to that described in Section 12 in consideration of functional equation (12.5). Skipping the auxiliary computations, we present the expressions for the functions $\lambda(u, p)$, $\mu(u, p)$ under the assumption $u > 0$:

$$\lambda = u^{-3/2}G(u, u^{-3/2}p), \tag{14.21}$$

$$\mu = -\frac{3}{2}u^{-5/2}p^2G\left(u, u^{-3/2}p\right) + u^{1/2}H(u, u^{-3/2}p).$$

Here $G(u, q)$ and $H(u, q)$ are functions defined in the half-plane $u > 0$ of the plane (u, q) and satisfying the relations

$$G(u, q) = G(u^{-1}, -q), \quad H(u, q) = -H(u^{-1}, q).$$

The simplest L-invariant equation of the type (14.17) can be found by setting in (14.21) $G = -1/2$, $H = 0$. It has the form

$$u_t = \left(u^{-1/2}\right)_{xxx}. \tag{14.22}$$

Equation (14.22) is called the Dym equation. Its L-invariance was established in [136]. This equation possesses some other interesting properties. There exists a differential substitution transforming the solutions of this equation into solutions of the

Korteweg–de-Vries equation (14.12), [72]. Finally, equation (14.22) is closely connected to the equation

$$v_t = (v_{xxx})^{-1/2}, \tag{14.23}$$

which plays the same role in the theory of third-order evolution equations as equation (10.19) among the second-order equations. Namely, given a solution $v(x,t)$ to equation (14.23), the function $u(x,t) = v_{xxx}$ satisfies the Dym equation. Equation (14.23) possesses the widest (8-parameter) group of pointwise transformations among all other equations of the form

$$v_t = F(t, x, v, v_x, v_{xx}, v_{xxx}),$$

which are not equivalent to any linear equation [101].

14.3 Reciprocity transformations for systems of one-dimensional evolution equations. We have been dealing so far with a single one-dimensional evolution equation of divergence form, i.e. with equations expressing some conservation laws. Let us consider now a system of the one-dimensional evolution equations where at least one of the equations is in divergence form.

An example of such a situation is furnished by the system

$$\begin{aligned} u_t &= [\varphi(u, v, u_x)]_x, \\ v_t &= [\alpha(u, v, v_x)]_x + \beta(u, v, u_x, v_x). \end{aligned} \tag{14.24}$$

Introducing Lagrangian coordinates by the relation

$$d\xi = u\,dx + \varphi(u, v, v_x)\,dt,$$

and employing the notation

$$\omega(\xi, t) = [u(x,t)]^{-1}, \quad \chi(\xi, t) = v(x,t),$$

we get the equivalent to (14.24) system of equations for the functions ω, χ:

$$\begin{aligned} \omega_t &= [\psi(\omega, \chi, \omega_\xi)]_\xi, \\ \chi_t &= [\gamma(\omega, \chi, \chi_\xi)]_\xi + \delta(\omega, \chi, \omega_\xi, \chi_\xi), \end{aligned}$$

where

$$\begin{aligned} \psi &= -\omega\varphi\left(\omega^{-1}, \chi, -\omega^{-3}\omega_\xi\right), \quad \gamma = \omega^{-1}\alpha\left(\omega^{-1}, \chi, \omega^{-1}\chi_\xi\right), \\ \delta &= \omega^{-2}\omega_\xi\alpha\left(\omega^{-1}, \chi, \omega^{-1}\chi_\xi\right) - \chi_\xi\varphi\left(\omega^{-1}, \chi, -\omega^{-3}\omega_\xi\right) \\ &\quad + \beta\left(\omega^{-1}, \chi, -\omega^{-3}\omega_\xi, \omega_\xi^{-1}\right). \end{aligned}$$

We see that the Lagrangian coordinate transformation preserves the class of equations (14.24).

Lagrangian coordinates as a tool of description of fluid motions arose practically simultaneously in works by Lagrange and Euler already in the 18-th century. The gas dynamics and the nonlinear elasticity theory are those branches of mechanics where Lagrangian coordinates are used especially frequently and successfully (see, for instance, [148]). In what follows, we derive the equations of the one-dimensional motion of ideal and viscous gases in the Euler and Lagrange variables. For the sake of simplicity we restrict consideration to motions with plane waves. Also, it is always assumed that there are no external mass forces and heat sources.

Let us consider first the equations of motion of the ideal gas in the Euler coordinates. Its state is usually characterized by the density ρ, the velocity u, and the specific entropy s. The pressure p is defined with the help of the caloric state equation

$$p = A(\rho, s), \tag{14.25}$$

where A is a given function of ρ and s. The system of differential equations describing the one-dimensional motion of the ideal gas is constituted by the continuity equation

$$\frac{\partial \rho}{\partial t} + \frac{\partial}{\partial x}(\rho u) = 0, \tag{14.26}$$

the momentum equation

$$\rho\left(\frac{\partial u}{\partial t} + u\frac{\partial u}{\partial x}\right) + \frac{\partial p}{\partial x} = 0, \tag{14.27}$$

and the equation of entropy conservation in the liquid particle

$$\frac{\partial s}{\partial t} + u\frac{\partial s}{\partial x} = 0. \tag{14.28}$$

One should substitute expression (14.25) for p into equation (14.27).

In order to pass in (14.27) to Lagrangian coordinates, let us write the trajectory equation of the liquid particle

$$\frac{dx}{dt} = u(x, t), \tag{14.29}$$

and pose for this equation the Cauchy problem

$$x = \eta \quad \text{as } t = 0. \tag{14.30}$$

η is called the geometrical Lagrangian coordinate. The equation in variations corresponding to (14.29), along with the continuity equation (14.26), lead to the relation

$$\rho\,[x(\eta, t), t]\,\frac{\partial x}{\partial \eta} = \rho_0(\eta), \tag{14.31}$$

where $x(\eta, t)$ is a solution of the Cauchy problem (14.29)–(14.30), and $\rho_0(x)$ is the value of $\rho(x, t)$ at the instant $t = 0$. Passing in equations (14.26)–(14.28) to the variables

η, t with the help of (14.29)–(14.30), and keeping the same notation for the density, velocity, pressure, and entropy, we have:

$$\begin{cases} \partial_t \rho + (\rho^2/\rho_0)\, \partial_\eta u = 0, \\ \rho_0 \partial_t u + \partial_\eta p = 0, \\ \partial_t s = 0. \end{cases}$$

One may simplify the system obtained replacing η by the mass Lagrangian coordinate ξ given by the relation

$$\xi = \int_0^\eta \rho_0(\zeta)\, d\zeta.$$

In the variables ξ, t, the equations of one-dimensional dynamics of the ideal gas take the especially simple form

$$\begin{cases} \partial_t \rho + \rho^2 \partial_\xi u = 0, \\ \partial_t u + \partial_\xi p = 0, \\ \partial_t s = 0 \end{cases} \tag{14.32}$$

(system (14.32) should be completed with the state equation (14.25)). There is a voluminous literature devoted to the theory of initial-boundary value problems for system (14.32) (see the monograph [148] and the references therein).

Let us now turn to the equations of viscous perfect gas dynamics. It is convenient to describe its state in terms of the density ρ, the velocity u, and the absolute temperature θ. The continuity equation here has the same form (14.26). The momentum equation now acquires a right-hand side which characterizes the contribution of the viscous forces and takes the form

$$\rho\left(\frac{\partial u}{\partial t} + u\frac{\partial p}{\partial x}\right) + \frac{\partial p}{\partial x} = \nu\frac{\partial^2 u}{\partial x^2}. \tag{14.33}$$

Instead of (14.28), we now have the energy equation

$$c_v \rho\left(\frac{\partial \theta}{\partial t} + u\frac{\partial \theta}{\partial x}\right) = \kappa\frac{\partial^2 \theta}{\partial x^2} - p\frac{\partial u}{\partial x} + \nu\left(\frac{\partial u}{\partial x}\right)^2, \tag{14.34}$$

where the term $-p\,(\partial u/\partial x)$ expresses the impact of the pressure forces on the change of the heat energy, and $\nu\,(\partial u/\partial x)^2$ is the heat influx caused by the dissipation of the kinetic energy of the gas. System (14.26), (14.33), (14.34) is completed with the thermic state equation, which for perfect gases has the form

$$p = R\rho\theta. \tag{14.35}$$

In equations (14.33)–(14.35) ν, c_v, κ, R are positive constants.

Apart from system (14.26)–(14.28) which is of hyperbolic type (under the physically reasonable assumption $\partial A/\partial \rho > 0$), equations (14.26), (14.33), (14.34) do not have a definite type. This complicates the analysis of the viscous gas equations. It suffices to say that the first results about the existence and uniqueness of solutions to such problems appeared less than 40 years ago. The modern state of the mathematical theory of motions of viscous gases is presented in monograph [9]. The starting point for most of the investigations in the field are the equations of viscous gas dynamics in Lagrangian coordinates. For one-dimensional motions with plane waves these equations have the form

$$\partial_t \rho + \rho^2 \partial_\xi u = 0,$$
$$\partial_t u + \partial_\xi p = \nu \partial_\xi \left(\rho \partial_\xi u\right),$$
$$c_V \partial_t \theta = \kappa \partial_\xi \left(\rho \partial_\xi \theta\right) - p \frac{\partial u}{\partial \xi} + \nu \rho \left(\partial_\xi u\right)^2,$$

where $p(\rho, \theta)$ is a given function.

14.4 Evolution equations in complex domains. An example of such an equation was already given in Section 13.3: it is the nonlinear Schrödinger equation (13.25). In that equation the sought function was complex-valued, while both independent variables were real. As a matter of fact, we were dealing there with system of equations (13.29), (13.30) equivalent to (13.25). This was the system for defining two real-valued functions and, in addition, one of its equations expressed a conservation law. Thus, it was the situation described in the previous section.

In what follows, we consider another class of equations admitting the reciprocity transformation. Let $w(z, t) = u + iv$ be a complex-valued function of the complex variable $z = x + iy$ and the real variable t so that u, v, x, y, are real. Assume w analytic with respect to z, differentiable in t, and satisfying the differential equation

$$\frac{\partial w}{\partial t} = \frac{\partial}{\partial z}\left[S\left(w, \frac{\partial w}{\partial z}, \dots, \frac{\partial^n}{\partial z^n}\right)\right], \tag{14.36}$$

where S is an analytic function of its arguments. Let us introduce the new complex-valued function $\zeta(z, t) = \xi + i\eta$ by the relations

$$d\zeta = w dz + S\left(w, \frac{\partial w}{\partial z}, \dots, \frac{\partial^n}{\partial z^n}\right) dt. \tag{14.37}$$

The last equality makes sense because the right-hand side of (14.37) is a total differential, due to (14.36). Arguing now as in the proof of Theorem 14.1, we arrive at the following conclusion: if $w \neq 0$ is a solution of equation (14.36), then the function

$$\phi(\zeta, t) = [w(\zeta, t)]^{-1}, \tag{14.38}$$

with ζ defined by (14.37) and $\zeta = 0$ for $z = 0$, $t = 0$, satisfies the equation

$$\frac{\partial \phi}{\partial t} = \frac{\partial}{\partial \zeta}\left(\Sigma\left(\phi, \frac{\partial \phi}{\partial \zeta}, \dots, \frac{\partial^n \phi}{\partial \zeta^n}\right)\right). \tag{14.39}$$

Here

$$\Sigma = -\phi\left[\phi^{-1}, \phi^{-1}\frac{\partial}{\partial \zeta}\left(\phi^{-1}, \dots, \left(\phi^{-1}\frac{\partial}{\partial \zeta}\right)^n (\phi)^{-1}\right)\right].$$

Formulas (14.37), (14.38) perform the reciprocity transformation for equations (14.36) and (14.39). It is easy to verify that this transformation is involutive. Let us also remark that in (14.36) an explicit dependence of S on t is admitted. Analogous to the real-valued divergence equations with two independent variables, for equations of the class (14.36) one may pose questions concerning their linearization via the reciprocity transformation, the latent symmetry or L-invariance. We leave these questions aside because the principal difficulties here are analogous to those in the real case and most of the positive results follow in much the same way. Let us just present a simple example of an equation of the class (14.36) equivalent to a linear equation. Namely, the equation

$$w_t = \left(w^{-2} w_z\right)_z \tag{14.40}$$

can be transformed into the equation

$$\phi_t = \phi_{\zeta\zeta}$$

via the transformation (14.37), (14.38). The latter is a nonstationary analog of the remarkable Bitsadze equation $\phi_{\zeta\zeta} = 0$. Passing in (14.40) to the real variables, we obtain an example of a system of quasilinear equations with three independent variables admitting explicit linearization:

$$u_t = -A_{xx} + A_{yy} + 2B_{xy},$$
$$v_t = 2A_{xy} + B_{xx} - B_{yy},$$

where

$$A = u/\left(4(u^2 + v^2)\right), \quad B = -v/\left(4(u^2 + v^2)\right).$$

Equations of the type (14.36) have applications in filtration theory and the Hele–Shaw problem [132, 150, 70]. For example, the one-phase liquid motion in the Hele–Shaw cell under large surface tension is modeled by the equation

$$w_t = \left(w^{-1/2}\right)_{zzz},$$

[70], which is the complex counterpart of the Dym equation (14.22).

In conclusion, let us note that our considerations can be easily extended to the equations

$$\frac{\partial}{\partial z_1}\left[F_1\left(\frac{\partial}{\partial z_1}; \frac{\partial}{\partial z_2}; w\right)\right] + \frac{\partial}{\partial z_2}\left[F_2\left(\frac{\partial}{\partial z_2}; \frac{\partial}{\partial z_1}; w\right)\right] = 0,$$

where z_j, $(j = 1, 2)$, are complex variables, and $F_j(q_1; q_2; w)$ depend on their arguments q_1, q_2 as polynomials, and are analytic in w.

15 A remarkable equation of nonlinear heat conduction

The object of consideration of the present section is the equation

$$u_t = \left(u^{-2}u_x\right)_x . \tag{15.1}$$

This equation was already mentioned in Section 8 (cf. (8.8)). Its multi-dimensional analog was considered in Section 13.2.

There are many reasons to think of equation (15.1) as of a remarkable one. This equation was the first among the other second-order quasilinear parabolic equations which have been linearized via a nonlocal transformation of the type (9.8) [164]. Equation (15.1) accounts for a number of important physical applications. One of them, [146], was already discussed in Section 13. Another application (the one that was found first, historically) is relevant to the process of heat propagation in a number of pure metals [164]. Assuming this process to be one-dimensional and using the notation $v(x, t)$ for the absolute temperature, let us write down the heat conduction equation in the form

$$S(v)v_t = [K(v)v_x]_x . \tag{15.2}$$

Here K is the coefficient of heat conductivity, and S stands for the product of the density of the medium by the specific heat; all these functions are assumed positive for $v > 0$.

Let us perform in equation (15.2) Kirchhoff's transformation: instead of v we introduce the new sought function $u(x, t)$ by the relation

$$u \equiv U(v) \equiv \int_0^v S(w)\,dw.$$

By virtue of (15.2), this function obeys the equation

$$u_t = [C(u)u_x]_x , \tag{15.3}$$

where $C(u) = K[V(u)]\,V'(u)$, and $V(u)$ is the inverse to $U(v)$. Let us claim the fulfilling of the equality

$$C(u) = Au^{-2},$$

where $A > 0$ is a constant. Due to the definitions of $C(u)$ and $U(u)$, this relation can be presented in the form

$$\frac{K(v)}{S(v)} = \frac{A^2}{\left(\int_0^v S(w)\,dw\right)^2}.$$

By differentiation, we transform it to the following:

$$\frac{1}{\sqrt{KS}}\frac{d}{dv}\left(\ln\sqrt{\frac{S}{K}}\right) = A. \tag{15.4}$$

The differential relation (15.4) determines the connection between the functions $K(v)$ and $S(v)$. It is convenient to present this connection in the parametric form:

$$K = K_0 g(v) \exp\left(-A\sqrt{K_0 S_0}\int_0^v g(u)\,du\right),$$
$$S = S_0 g(v) \exp\left(A\sqrt{K_0 S_0}\int_0^v g(u)\,du\right).$$

Here K_0 and S_0 are positive constants, and $g(v)$ is an arbitrary function assuming positive values for $v > 0$.

As was shown by M. Storm [164], for many metals the dependences of thermophysical characteristics on the temperature are such that relation (15.4) with $A > 0$ is fulfilled with an appropriate accuracy. This provides a possibility to model the process of one-dimensional heat transfer in such metals with the help of solutions of the equation

$$u_t = A\left(u^{-2}u_x\right)_x .$$

The latter equation is equivalent to (15.1), which, in turn, can be transformed into the linear equation $\omega_t = \omega_{\xi\xi}$ by transformation (9.8).

Equation (15.1) is distinguished from the other nonlinear heat conduction equations (15.3) from the point of view of group theory. Namely, equation (15.3) admits nontrivial Lie–Bäcklund transformations, i.e. not reducible to pointwise ones, iff $C(u) = (au + b)^{-2}$, where a and b are constants, [28, 44]. Under this condition, equation (15.3) can be transformed by a simple pointwise change of variables. This property of equation (15.1) is the source of the explicit solutions invariant with respect to the Lie–Bäcklund transformation.

Let us also mention two works devoted to the study of initial-boundary value problems for equation (15.1). With the use of the method of differential substitutions the following problem was successfully solved in [82]: to find a solution of equation (15.1) in the quarter-plane $x > 0$, $t > 0$, under the conditions $u = u_0(x)$ as $x > 0, t = 0$, $u = f(t)$ as $x = 0, t > 0$. In [135], for Stefan's problem for equation (15.1) a family of explicit solutions with functional arbitrariness was constructed. This result is presented in Section 16.

In the remainder of this section, we discuss some properties of solutions to the Cauchy problem

$$u(x,0) = u_0(x) \quad \text{as } x \in \mathbb{R} \tag{15.5}$$

for equation (15.1). Along with (15.1) we shall consider the family of equations

$$u_t = \left(u^m u_x\right)_x \tag{15.6}$$

with the parameter $m > -2$. These considerations are chiefly based on the results of paper [66].

In the case $m > 0$ equation (15.1) is often called the *porous medium equation,* and for $m < 0$ it is termed the *fast diffusion equation,* as a rule. If $m = 0$, equation (15.6) becomes linear. We confine ourselves to the study of the nonnegative solutions of the Cauchy problem (15.5)–(15.6) which are of the most interest from the point of view of physical applications. This restriction assumes the nonnegativity of the function u_0 in condition (15.5).

In the theory of the Cauchy problem (15.5)–(15.6), a special role is played by the so-called finite mass solutions:

$$\int_{-\infty}^{\infty} u(x,t)\,dx = Q < \infty. \tag{15.7}$$

It is known (see [66] and references therein), that in the case $m > -2$ such solutions exist "in the large", i.e. for all $t > 0$, under very weak assumptions on the initial data u_0; also, the integral in (15.7) is independent of t. This class of solutions includes, in particular, the self-similar "source-type" solutions

$$w(x,t;\,Q,m) = \left(\frac{Q^2}{t}\right)^{1/(m+2)} f(\zeta), \tag{15.8}$$

where

$$\zeta = x\left(Q^m t\right)^{-1/(m+2)}$$

$$f(\zeta) = \begin{cases} \left(\frac{m}{m+2}\left(\zeta_0^2 - \zeta^2\right)_+\right)^{1/m} & \text{if } m > 0 \\[2ex] \left(\frac{|m|}{m+2}\left(\zeta_0^2 + \zeta^2\right)\right)^{1/m} & \text{if } -2 < m < 0, \end{cases} \tag{15.9}$$

and $s_+ = \max\{s, 0\}$. Here $Q > 0$ is a given constant and the value of ζ_0 is uniquely defined from the relation

$$\int_{-\infty}^{\infty} f(\zeta)\,d\zeta = 1.$$

Letting in (15.8) $m \to \infty$, we get that $w(x,t;\,Q,m) \to G(x,t;\,Q)$, where

$$G(x,t;\,Q) = Q\,(4\pi t)^{-1/2}\exp\left(-x^2/4t\right)$$

is the classical Poisson kernel. Another limit transition shows that

$$w(x,t;\,Q,m) \to Q\delta(0) \quad \text{as } t \to 0,$$

where $\delta(x)$ is the Dirac function.

The structure of solution (15.8) does not allow one to pass in it to the limit as $m \to -2$. Moreover, for equation (15.1), which is the limiting one with respect to (15.6), the source-type solutions do not exist at all. This corresponds to the remarkable result of paper [66] asserting the non-existence of positive classical solutions to equation (15.1) with a finite mass, e.g. with a finite norm $u(\cdot,t)$ in $L^1(\mathbb{R})$ for $t > 0$, even if the mass was finite (and positive) at the instant $t = 0$. (We emphasize that in application to the solutions of equation (15.1) the notion of mass has a conditional character.) Thus, the value of the exponent $m = -2$ is critical for the family of equations (15.6).

In the light of this introduction, it makes sense to study the infinite–mass solutions of the Cauchy problem. The existence of such solutions is established in [66]. To formulate the result, let us introduce the auxiliary functions $M(x)$, $\varphi(x)$ by the relations

$$M(x) = \int_0^x u_0(s)\,ds,$$
$$\varphi(x) = \begin{cases} \inf\{y \ge 0 : \ M(y) = x\} & \text{if } x \ge 0, \\ \sup\{y < 0 : \ M(y) = x\} & \text{if } x < 0. \end{cases} \tag{15.10}$$

Let us observe that $\varphi(x) = M^{-1}(x)$ if $u_0(x) > 0$, but $\varphi(x)$ fails to be continuous if u_0 vanishes within some interval. If $u_0 \in L^1(\mathbb{R})$, then $\varphi(x)$ is defined only on a bounded set but, in any case, φ is monotone non-decreasing and, hence, has one-sided limits at each point of the domain of its definition.

Let us call a function $u(x,t)$ a classical solution of the Cauchy problem (15.1), (15.5) in a strip $S_T = \mathbb{R} \times (0, T)$ if:

1) $u(x,t) \in C^{2,1}(S_T)$ and satisfies (15.1) in S_T;

2) $u(x,t) > 0$ in S_T;

3) $u(x,t) \in C\left([0,T) : L^1_{\text{loc}}(\mathbb{R})\right)$ and (15.5) holds in the sense of $L^1_{\text{loc}}(\mathbb{R})$.

The following result of paper [66] we present without proof.

Theorem 15.1. *Let $u_0 \in L^1_{\text{loc}}(\mathbb{R})$, $u_0 \ge 0$. Assume that the following conditions hold:*

$$\int_{-\infty}^0 u_0(s)\,ds = \int_0^\infty u_0(s)\,ds = \infty; \tag{15.11}$$

there exists $c = \text{const.} > 0$ such that

$$\int_{-\infty}^\infty \exp\left(-cs^2\right) |\varphi(s)|\,ds < \infty, \tag{15.12}$$

where φ is determined by condition (15.10)*; the set of points of discontinuity of the function φ is either empty, or discrete.*

Then problem (15.1), (15.5) *has a classical positive solution in the strip $S_T = \mathbb{R} \times (0, T)$ where*

$$T < \frac{1}{4c}, \tag{15.13}$$

and c is the constant from inequality (15.11). *If* (15.11) *is fulfilled for each $c > 0$, then a solution of problem* (15.1), (15.5) *exists in S_T for all $T > 0$. Moreover,*

$$\int_{-\infty}^0 u(s,t)\,ds = \int_0^\infty u(s,t)\,ds = \infty \tag{15.14}$$

for all $t \in (0, T)$.

Let us make several comments to Theorem 15.1. First of all, observe that (15.10) makes condition (15.11) meaningful. Next, estimate (15.13) of the guaranteed life span of the solution is exact. This is confirmed by the following example [66].

Let us define the function $h(x)$ as the inverse of the function

$$E(x) = \int_0^x \exp\left(y^2\right) dy$$

and set

$$u(x, t; T) = 2\sqrt{T-t}\exp\left(-h^2(x)\right).$$

The function $u(x, t; T)$ is a positive solution of equation (15.1) in the strip S_T, vanishing identically at $t = T$. The corresponding function φ has the form

$$\varphi(x) = E\left(\frac{x}{2\sqrt{T}}.\right)$$

One can directly check that condition (15.11) is fulfilled here for each $c > 1/4T$, but fails if $c = 1/4T$.

In the proof of Theorem 15.1, the key role is played by the Bäcklund transformation (in our terminology "L-transformation") that linearizes equation (15.1). According to (9.8), this transformation has the form

$$\begin{aligned} \omega(\xi, t) &= [u(x, t)]^{-1}, \\ \xi &= \int_0^x u(y, t)\,dy + \int_0^t u^{-2}(0, s)u_x(0, s)\,ds. \end{aligned}$$

The function $\omega(\xi, t)$ satisfies the equation

$$\omega_t = \omega_{\xi\xi}. \tag{15.15}$$

Later on, we will show how one may apply transformation (15.14) to study the question of uniqueness of positive solutions to problem (15.1), (15.5), which in [66] remained open.

Theorem 15.2. *Assume that $u_0 \in C(\mathbb{R})$, $u_0 > 0$ for all x, and that condition* (15.11) *holds. Let $u(x, t)$ be a positive classical solution of problem* (15.1), (15.5) *in the strip S_T satisfying condition* (15.13). *Then this solution is unique.*

Proof. Assume that $u_j(x, t)$, $j = 1, 2$, are two different solutions satisfying the conditions of Theorem 15.2. Then for each of them we may perform the L-transformation (15.14). Denote by $\omega_j(\xi, t)$ the images of u_j under this transformation. By virtue of conditions (15.10), (15.13), each segment $\{x \in \mathbb{R},\ t = \tau \in [0, T)\}$ is transformed thereby into the segment $\{\xi \in \mathbb{R},\ t = \tau\}$, and the mapping of the strip S_T onto the strip $\Sigma_T = \mathbb{R} \times (0, T)$ is a bijection. If we establish the coincidence of the images ω_j of the functions u_j under mapping (15.14), the theorem will be proved.

Each ω_j is a positive classical solution of equation (15.15) satisfying the initial condition

$$\omega(\xi, 0) = \omega_0(\xi) \quad \text{as } \xi \in \mathbb{R}, \tag{15.16}$$

where ω_0 is determined by the relations

$$\omega_0(\xi) = [u_0(x)]^{-1}, \quad \xi = \int_0^x u_0(y)\, dy.$$

Note that $\omega_0 \in C(\mathbb{R})$ and $\omega_0 > 0$ for all ξ.

It follows from the classical theory (see, for example, [52]), that for the positive classical solutions of the Cauchy problem (15.15), (15.16) the uniqueness theorem holds without any restrictions on the character of growth of the function ω as $|\xi| \to \infty$. This implies the equality $\omega_1 = \omega_2$ in Σ_T, whence the desired assertion. □

In conclusion, let us remark that, generally speaking, the conditions of Theorem 15.2 do not guarantee the existence of solutions to problem (15.1), (15.5). Nonetheless, having added to these conditions requirement (15.11), the existence in the strip S_T with $T < 1/4c$ also takes place. In [66], the solvability of the problem discussed was proven under the conditions $u_0 \in C(\mathbb{R})$, $u_0 > 0$ for all x, (15.10), (15.11). Also, the solution satisfied relations (15.13). Thus, the above-listed set of conditions provides the existence and uniqueness of a positive classical solution of problem (15.1), (15.6) in the strip S_T of width $T < 1/4c$.

16 The one-phase Stefan problem: explicit solutions with functional arbitrariness

16.1 Statement of the problem. Let us consider the equation

$$u_t = \left(u^{-2}u_x\right)_x \tag{16.1}$$

in the domain $-l < x < s(t), 0 < t < T$, and subject its solutions to the conditions

$$-u^{-2}u_x = q(t) \qquad \text{as } x = -l, 0 \le t \le T, \tag{16.2}$$

$$u = 1, \quad \frac{ds}{dt} = -u^{-2}u_x \qquad \text{as } x = s(t), 0 \le t \le T, \tag{16.3}$$

$$u = u_0(x) \qquad \text{as } -l \le x \le 0, t = 0; s(0) = 0, \tag{16.4}$$

where $q(t)$ and u_0 are given smooth functions; also, $u_0(0) = 1$ and $u_0(x) > 1$ as $-l \le x < 0$.

Problem (16.1)–(16.4) can be interpreted as the one-phase Stefan problem (cf. Section 4) for the nonlinear heat conduction equation considered in the previous section.

The function u stands for the temperature in the liquid phase; the temperature in the solid phase is assumed constant and equal to the melting temperature $u = 1$. The latent heat of melting is also assumed unit. With the use of rescaling one can reduce to the present case the case of arbitrary constant positive values of the melting temperature, the latent heat of melting, and the product of density by specific heat.

The purpose of this section is to construct an explicit solution to problem (16.1)–(16.4) under a special (two-parameter) dependence of the heat flux q on time on the prescribed part of the boundary $x = -l = \text{const.}$, but with an arbitrary initial distribution of the temperature u_0. The existence of such solutions distinguishes equation (16.1) from the other nonlinear heat conduction equations of the type

$$u_t = [k(u)u_x]_x \,.$$

For this class of equations only the self-similar explicit solutions described in Section 4 are known, and the travelling-wave solutions of the form $u = f(x - ct)$, $c = \text{const.}$ These solutions possess only a parametric arbitrariness.

As is shown in Section 11, the passage to the Lagrangian coordinate ξ in equation (16.1) (coinciding with (11.1)) transforms it into the linear equation for the function $x = X(\xi, t)$:

$$X_t = X_{\xi\xi}. \tag{16.5}$$

Since $X_t = v(X, t)$ and, by virtue of equation (16.1), $v(x, t) = -u_x/u^3$, conditions (16.3) imply the equality

$$\frac{ds}{dt} = v(x, t) \quad \text{as } x = s(t).$$

This means that in Lagrangian coordinates the unknown boundary $x = s(t)$ corresponds to the segment $\{\xi = 0,\ t \in [0, T]\}$.

Relations (16.4) and the first condition of (16.3) along with the equality

$$u(X(\xi, t), t)X_\xi = 1$$

lead to the following conditions for the function X:

$$X = X_0(\xi) \quad \text{as } -a \le \xi \le 0, t = 0, \tag{16.6}$$

$$X_\xi = 1 \quad \text{as } \xi = 0, 0 \le t \le T, \tag{16.7}$$

where X_0 is determined by the formula $u_0(X_0)dX_0 = d\xi$ as $\xi \in [-a, a]$, $X_0(0) = 0$. The constant a is given by

$$a = \int_{-l}^{0} u_0(x)\,dx. \tag{16.8}$$

To complete the statement of the problem for defining X, it remains to transform condition (16.2) to Lagrangian coordinates. To this end, let us differentiate the identity $X(\zeta(t), t) = -l$, where $\xi = \zeta(t)$ is the image of the segment $\{x = -l,\ t \in [0, T]\}$ on the plane (ξ, t). We get thereby

$$X_t + \zeta' X_\xi = 0 \quad \text{as } \xi = \zeta(t).$$

On the other hand, due to (16.2) and the definition of v, we have

$$q(t) = vu = X_t/X_\xi \quad \text{as } \xi = \zeta(t).$$

It follows then that

$$\zeta(t) = -\int_0^t q(\tau)\,d\tau - a \tag{16.9}$$

(here the equality $X_0(-a) = -l$ is taken into account). The required condition has the form

$$X = -l \quad \text{as } \xi = \zeta(t),\ 0 \le t \le T. \tag{16.10}$$

Thus, Stefan's problem (16.1)–(16.4) is reduced to a problem in the fixed domain $\{\zeta(t) < \xi < 0,\ 0 < t < T\}$ for equation (16.5). In this sense, it is analogous to Verigin's problem E_1 formulated in Section 5. Moreover, problem (16.5)–(16.7), (16.10) is linear. The local solvability of this problem is guaranteed by general results of the theory of parabolic equations [97] (under appropriate compatibility conditions on data).

16.2 Solvability conditions. Introduce the notation

$$Q_\tau = \{(\xi, t) :\ \zeta(t) < \xi < 0,\ 0 < t < T\},$$
$$\Omega_\tau = \{(x, t) :\ -l < x < s(t),\ 0 < t < \tau\}.$$

It is evident that Ω_τ is the image of Q_τ under the mapping $(\xi, t) \mapsto (x, t)$. It is assumed in the definition of Q_τ that for $t \in [0, \tau)$ the inequality $\zeta(t) < 0$ is true. According to (16.9), this is clearly true for $\tau > 0$ small enough. If the function q is such that ζ does not vanish within the interval $[0, T]$ (for instance, $q \ge 0$ for $t \in [0, T]$), then the curvilinear part of the boundary of Q_T does not intersect the ray $\{\xi = 0,\ t \ge 0\}$, and one may anticipate that the problem (16.5)–(16.7), (16.10) (and, hence, problem (16.1)–(16.4)) is solvable globally in time. If the equation $\zeta(t) = 0$ has a root $t = t_* > 0$, then the solutions of problem (16.1)–(16.4) cannot be continued to $t > t_*$.

The function $q(t)$ which enters condition (16.2) has the meaning of the heat flux across the known part of the lateral boundary of the domain Q_τ. The case $q < 0$ corresponds to cooling of the liquid phase from the side of this boundary. If the intensity of cooling is large enough, the liquid phase can disappear within a finite time. The instant of disappearance coincides with the minimal positive root t_* of the equation $\zeta(t) = 0$. Formulas (16.8)–(16.9) imply the following simple relation for defining t_* in terms of data of the original Stefan problem:

$$\int_0^{t_*} q(t)dt + \int_{-l}^0 u_0(x)dx = 0.$$

Let us now formulate sufficient conditions for the classical solvability of problem (16.5)–(16.7), (16.10). In what follows, for the functions of a single variable "the prime" denotes their derivatives with respect to the argument.

U.1 The smoothness conditions

$$X_0(\xi) \in C^{3+\alpha}[-a, 0], \quad \zeta(t) \in C^{(3+\alpha)/2}[0, T], \quad 0 < \alpha < 1.$$

U.2 The compatibility conditions:

$$X_0(-a) = -l, \quad X_0''(-a) + \zeta'(0)X_0'(-a) = 0,$$

$$X_0'(0) = 1, \qquad X_0''' = 0.$$

Lemma 16.1. *Let conditions* U.1–U.2 *hold. Unless the function* $\zeta(t)$ *vanishes within the interval* $[0, T]$, *problem* (16.5)–(16.7), (16.10) *has a unique solution*

$$X \in C^{3+\alpha,(3+\alpha)/2}\left(\overline{Q_T}\right).$$

Otherwise this problem has a unique solution $X \in C^{3+\alpha,(3+\alpha)/2}\left(\overline{Q_\tau}\right)$, *where* $\tau > 0$ *is arbitrary and less than* t_*, *and* t_* *is the minimal positive root of the equation* $\zeta(t) = 0$.

The statement of Lemma 16.1 is an explicit implication of the solvability results for linear parabolic equations in Hölder's classes [97]. The seemingly too strong requirements on the smoothness of data in condition U.1 are caused by the desire to construct a classical solution of the original Stefan problem. The function u acquires now the parameter representation

$$u = 1/X_\xi(\xi, t), \quad x = X(\xi, t), \tag{16.11}$$

while the function $s(t)$, which defines the free boundary, is given explicitly:

$$s = X(0, t). \tag{16.12}$$

Having applied the maximum principle for solutions of parabolic equations [52], to the function $\exp\left(-\lambda\xi - \lambda^2 t\right) X_\xi$ with an appropriately chosen constant λ, we establish the estimates

$$e^{-\lambda a} \min\left\{e^{-\lambda^2\tau}, \left[\max_{-l\le x\le 0} u_0\right]^{-1}\right\} \le X_\xi \le 1 \quad \text{in } \overline{Q_\tau}. \tag{16.13}$$

The constant λ can be taken equal to

$$\max\left\{0, \max_{0\le t\le\tau} q(T)\right\}.$$

(If $\zeta(t) < 0$ for all $t \in [0, T]$, then one should replace τ by T in the definition of λ and in the lower bound for X_ξ; then the latter estimate remains true everywhere in $\overline{Q_T}$.)

It follows from (16.11)–(16.13) and Lemma 16.1 that problem (16.1)–(16.4) in any case has a solution $u \in C^{2+\alpha,(2+\alpha)/2}\left(\overline{Q_\tau}\right)$, $s \in C^{(3+\alpha)/2}[0, \tau]$, where $\tau < t_*$. If $\zeta(t) < 0$ for $0 \le t \le T$, then this solution can be continued on the whole domain

$\overline{Q_T}$ and the whole interval $[0, T]$ without any loss of the smoothness of the functions u and s.

In conclusion of this section, let us formulate the compatibility and smoothness conditions that guarantee the classical solvability of problem (16.1)–(16.4) in terms of the input data of this problem:

$$u_0 \in C^{2+\alpha}[-l, 0], \quad q \in C^{(1+\alpha)/2}[0, T],$$
$$-u_0^{-2}(-l)u_0'(-l) = q(0), \quad u_0(0) = 1, \quad u_0''(0) - 3\left[u_0'(0)\right]^2 = 0.$$

16.3 Global in time explicit solutions. Let in condition (16.2) the function $q(t)$ have the form

$$q = \frac{b^2}{\sqrt{b^2 t + a^2}}, \tag{16.14}$$

where the constant a is given by formula (16.8), and the constant b is as yet undefined. Then, according to (16.9), $\zeta = -\left(b^2 t + a^2\right) < 0$ for all $t \geq 0$. Therefore, under the conditions of Lemma 16.1 problem (16.5)–(16.7) is solvable in Q_T for every $T > 0$ and we may suppose in the sequel that $T = \infty$.

The second of conditions U.2 along with (16.4) lead to the equality

$$b = -2au_0^{-2}(-l)u_0'(-l),$$

whence $u_0'(-l) < 0$. This restriction, the inequality $u_0(x) > 1$ for $-l \leq x < 0$, and two other compatibility conditions formulated at the end of item 2, along with the inclusion $u_0 \in C^{2+\alpha}[-l, 0]$, exhaust the bulk of the constraints for the function $u_0(x)$. It is arbitrary in all other respects.

Let us pass in problem (16.5)–(16.7) to the new variables

$$z = b\xi/\sqrt{b^2 t + a^2}, \quad \theta = \ln\left((b^2 t + a^2)/a^2\right),$$
$$X(\xi, t) = Y(z, \theta). \tag{16.15}$$

The function Y satisfies the equation

$$Y_\theta = Y_{zz} + \frac{z}{2}Y_z, \tag{16.16}$$

in the half-strip $\Pi = \{(z, \theta) : -b < z < 0, \ \theta > 0\}$, and the following initial and boundary conditions

$$Y = Y_0 \quad \text{as } -b \leq z \leq 0, \theta = 0, \tag{16.17}$$
$$Y = -l \quad \text{as } z = -b, \theta \geq 0, \tag{16.18}$$
$$Y_z = (a/b)e^{\theta/2} \quad \text{as } z = 0, \theta \geq 0, \tag{16.19}$$

where $Y_0(z) = X_0(az/b)$.

Problem (16.16)–(16.19) admits a separation of variables which simplifies the construction of solutions. Due to (16.15), the corresponding solution of problem (16.5)–(16.7), (16.10) has the form

$$X = F\left(\frac{b\xi}{\sqrt{b^2t+a^2}}\right)\frac{\sqrt{b^2t+a^2}}{b} - l + \sum_{k=1}^{\infty} c_k w_k\left(\frac{b\xi}{\sqrt{b^2t+a^2}}\right)\left(\frac{b^2t+a^2}{a^2}\right)^{-\lambda_k}. \tag{16.20}$$

Here $w_k(z)$ and λ_k are the eigenfunctions and eigenvalues of the Sturm–Liouville problem

$$w'' + \tfrac{z}{2}w' + \lambda w = 0 \quad \text{as } -b < z < 0,$$

$$w(-b) = 0, \quad w'(0) = 0.$$

The system of functions $\{w_k\}$, $k = 1, 2, \ldots$, is complete and orthogonal in $L_2(-b, 0)$ with the weight $\exp\left(z^2/2\right)$. Also, $0 < \lambda_1 < \lambda_2 < \cdots$ and $\lambda_k \to \infty$ as $k \to \infty$. (The functions w_k can be expressed via the parabolic cylindrical functions but now this does not matter.) The function $F(z)$ is

$$F = C\left(e^{-z^2/4} - \frac{z}{2}\int_z^0 e^{-y^2/4}dy\right) + z,$$

where

$$C = \frac{b}{e^{-b^2/4} + \frac{b}{2}\int_0^b e^{-y^2/4}dy}.$$

The constants c_k, $k = 1, 2, \ldots$, are the coefficients in the Fourier expansion of the function $Y_0(z) + l - F(z)$ by the system $\{w_k\}$. The convergence of series (16.20) is ensured by the smoothness of the initial function $Y_0(z)$ and by the compatibility conditions.

Relations (16.11), (16.20) define a solution of the Cauchy problem (16.1)–(16.4), (16.14) with the function $u_0(x)$ arbitrary (up to the above-presented restrictions). Let us recall that $u_0(x)$ and $X_0(\xi)$ are connected by the relation

$$\int_0^{X_0} u_0(x)\,dx = \xi,$$

and that $Y_0(z) = X_0(az/b)$. Making use of equality (16.12), we obtain from (16.20) the expression for the function $s(t)$ which determines the free boundary

$$s = b^{-1}C\left(b^2t + a^2\right)^{1/2} - l + \sum_{k=1}^{\infty} g_k\left(b^2t + a^2\right)^{-\lambda_k}, \tag{16.21}$$

where $g_k = c_k w_k(0)a^{2\lambda_k}$. Let us point out two features. Series (16.21) is convergent (but not analytic) for all $t \geq 0$. The eigenvalues $\lambda_k = \lambda_k(b)$ need not be rational, generally speaking. Therefore, it seems hardly possible to derive the structure of asymptotics of the free boundary as $t \to \infty$ with the use of ordinary considerations.

16.4 Explicit solutions describing the process of phase disappearance. Let us now assume that in condition (16.2) the dependence on $q(t)$ has the form

$$q = -\frac{b^2}{2\sqrt{a^2 - b^2 t}}, \tag{16.22}$$

where $0 \le t < t_* = a^2/b^2$. The heat flux $q(t)$, given at the prescribed boundary $x = -l$, becomes infinite at the instant $t = t_*$, while the function $\zeta(t)$, determined by formula (16.9), vanishes. This means that the lateral boundaries of Ω_{t_*} intersect.

Now we are going to construct a solution to problem (16.1)–(16.4), (16.22) and to study the free boundary behavior as $t \to t_* - 0$.

Analogous to the previous section, the change of variables

$$y = b\xi/\sqrt{a^2 - b^2 t}, \quad \eta = \ln\left(a^2/(a^2 - b^2 t)\right),$$
$$X(\xi, t) = W(y, \eta) \tag{16.23}$$

reduces the problem to the problem in the separated variables:

$$W_\eta = W_{yy} - \frac{y}{2} W_y \quad \text{as } (y, \eta) \in \Pi, \tag{16.24}$$
$$W = W_0(y) \quad \text{as } -b \le y \le 0,\ \eta = 0, \tag{16.25}$$
$$W = -l \quad \text{as } y = -b,\ \eta \ge 0, \tag{16.26}$$
$$W_y = (a/b)e^{-\eta/2} \quad \text{as } y = 0,\ \eta \ge 0, \tag{16.27}$$

where $W_0(y) = X_0(ay/b)$. The restrictions on "an arbitrary" function u_0 in condition (16.4) of the original Stefan problem coincide with those listed in item 3, except for the equality expressing the second of conditions U.2; here it has the form $b = 2au_0^{-2}(-l)u_0'(-l)$. However, problem (16.1)–(16.4), (16.22) is specific: here appears an exceptional value of the parameter b, $b = b_*$, defined as the unique positive root of the equation

$$\frac{b}{2}\int_0^b e^{z^2/4} dz = e^{b^2/4}.$$

(An approximate value b_* equals 1,848.)

A solution $W(y, \eta)$ of problem (16.24)–(16.27) is constructed as a series in eigenfunctions $\{v_k\}$ of the Sturm–Liouville problem

$$v'' - \tfrac{y}{2}v' + \mu v = 0 \quad \text{as } b < y < 0,\ y(b) = 0,\ y'(0) = 0.$$

The corresponding eigenvalues are denoted μ_k. Note that $0 < \mu_1 < \mu_2 < \cdots$, and $\mu_k \to \infty$ as $k \to \infty$. The system $\{v_k\}$ is complete in $L_2(-b, 0)$ with the weight $\exp\left(-y^2/4\right)$.

Assume that $b \ne b_*$. Passing from $W(y, \eta)$ to $X(\xi, t)$ by formulas (16.23), we find a solution to problem (16.5)–(16.7), (16.22)

$$X = l + E\left(\frac{b\xi}{\sqrt{a^2 - b^2 t}}\right)\frac{\sqrt{a^2 - b^2 t}}{b} +$$

$$+\sum_{k=1}^{\infty} d_k v_k\left(\frac{b\xi}{\sqrt{a^2-b^2t}}\right)\left(\frac{a^2-b^2t}{a^2}\right)^{\mu_k} \tag{16.28}$$

in the domain $\left(a^2-b^2t\right)^{1/2} \leq \xi \leq 0,\ 0 \leq t \leq t^*$. Here d_k are the coefficients of the expansion of the function $W_0(y)+l-E(y)$ in the series by the system $\{v_k\}$ and

$$E(y) = A\left(e^{y^2/4} + \frac{y}{2}\int_y^0 e^{z^2/4}dz\right) + y,$$

$$A = \frac{b}{e^{b^2/4} - \frac{b}{2}\int_0^b e^{z^2/4}dz}.$$

A solution of the Stefan problem (16.1)–(16.4), (16.22) is expressed through $X(\xi, t)$ by formulas (16.11), (16.28). This solution describes the process of disappearance of the liquid phase at the moment $t = t_*$. Letting in (16.28) $\xi = 0$, we get a representation for the free boundary trajectory $s(t)$ as the series

$$s + l = b^{-1}A\left(a^2-b^2t\right)^{1/2} + h_1\left(a^2-b^2t\right)^{\mu_1} + \sum_{k=2}^{\infty} h_k\left(a^2-b^2t\right)^{\mu_k}, \tag{16.29}$$

which converges within the interval $[0, t_*]$. (Here we denote $h_k = d_k v_k(0) a^{-2\mu_k}$, $k = 1, 2, \ldots$.) In dependence on the value of the minimal eigenvalue μ_1, the main term of the asymptotics of $s(t)+l$ as $t \to t_* - 0$ will be either the first or the second term of the right-hand side of (16.29). It is easy to show that $\mu_1 < 1/2$ if $b > b_*$ and $\mu_1 > 1/2$ if $b < b_*$. If $b > b_*$ (this situation, according to (16.22), corresponds to "fast" cooling of the liquid phase on the prescribed boundary $x = -l$), then $s(t) + l \sim (t_* - t)^{\mu_1}$ as $t \to t_*$. If $b < b_*$ ("slow" cooling), then $s(t) + l \sim (t_* - t)^{1/2}$ as $t \to t_*$.

It does not cause serious difficulties to consider the special case $b = b_*$. However, here the expression for X, similar to (16.28), becomes more cumbersome: it includes a logarithmic term. Skipping the details, let us write the asymptotics of the function $s(t)$ in a neighborhood of the moment of phase disappearance:

$$\begin{aligned} s(t) + l &= f_1\,(t_* - t)^{1/2}\ln\,(t_* - t)\,t_*^{-1} + f_2\,(t_* - t)^{1/2} \\ &\quad + O\,(t_* - t)^{\mu_2^*} \quad \text{as } t \to t_* - 0. \end{aligned}$$

Here $f_1 > 0$ and f_2 are constants, μ_2^* is the second eigenvalue as $b = b_*$.

16.5 Explicit solutions of the Cauchy–Verigin problem. The Cauchy–Verigin problem was studied in full detail in Chapter I. We show now that if the state equation connecting the pressure and the density in each of the contacting liquids are chosen in a special way, then this problem admits linearization on the plane of Lagrangian coordinates. This feature enables one to construct effectively exact solutions with an arbitrary initial distribution of the density in the liquids.

For the sake of convenience, we begin by repeating the statement of the one-dimensional Cauchy–Verigin problem keeping the notation adopted in Chapter I. It is required to find the functions $h(t)$, $p_1(x,t)$, $p_2(x,t)$ satisfying the relations

$$m\partial_t \rho_1 = \frac{k}{\mu_1}\partial_x(\rho_1 \partial_x p_1) \quad \text{as } x < h(t), t > 0, \tag{16.30}$$

$$m\partial_t \rho_2 = \frac{k}{\mu_2}\partial_x(\rho_2 \partial_x p_2) \quad \text{as } x > h(t), t > 0, \tag{16.31}$$

$$\begin{aligned} \rho_1 &= \varphi_1(p_1) \quad \text{as } x < h(t), t > 0, \\ \rho_2 &= \varphi_2(p_2) \quad \text{as } x > h(t), t > 0, \end{aligned} \tag{16.32}$$

$$p_1 = p_2 \quad \frac{dh}{dt} = \frac{k}{\mu_1}\partial_x p_1 = \frac{k}{\mu_2}\partial_x p_2 \quad \text{as } x = h(t), t > 0, \tag{16.33}$$

$$p_1 = p_{10}(x) \quad \text{as } x \le 0, \quad p_2 = p_{20}(x) \quad \text{as } x \ge 0; \quad h(0) = 0. \tag{16.34}$$

With respect to the functions φ_1, φ_2, p_{10}, p_{20} we assume that conditions U.1, U.2,U.5 of Section 5 hold, providing the unique solvability of problem (16.30)–(16.34) for all $t > 0$.

Assume now that the functions φ_i determining the state equations (16.32) have the form

$$\varphi_i(p_i) = [2q_i(B - p_i)]^{-1/2}. \tag{16.35}$$

Here q_i and B are positive constants and, also,

$$B > \sup_{x \le 0} p_{10}(x), \quad B > \sup_{x \ge 0} p_{20}(x).$$

Following (6.2), let us introduce the Lagrangian coordinate y by the formula

$$y = \int_{h(t)}^{x} \rho(s,t)ds. \tag{16.36}$$

Here ρ equals ρ_1 if $x < h(t)$, or ρ_2 if $x > h(t)$. Let us pass in relations (16.30)–(16.34) to the new independent variables (y,t), replacing (x,t). However, in contrast to Section 6 here we take for the new sought functions $x_1 = X_1(y,t)$ and $x_2 = X_2(y,t)$, where $x = X(y,t)$ is the transformation inverse to (16.36).

A direct checking (which is left to the reader) shows that the functions X_i constitute a solution of the following system of linear equations:

$$m\partial_t X_1 = \frac{k}{q_1\mu_1}\partial^2_{yy}X_1 \quad \text{as } y < 0, t > 0, \tag{16.37}$$

$$m\partial_t X_2 = \frac{k}{q_2\mu_2}\partial^2_{yy}X_2 \quad \text{as } y > 0, t > 0, \tag{16.38}$$

$$X_1 = X_2, \quad \frac{1}{q_1}\partial_y X_1 = \frac{1}{q_2}\partial_y X_2 \quad \text{as } y = 0, t > 0, \tag{16.39}$$

$$X_1 = X_{10}(y) \quad \text{as } y \le 0, t = 0;$$
$$X_2 = X_{20}(y) \quad \text{as } y \ge 0, t = 0. \tag{16.40}$$

The functions X_{i0} are expressed through p_{i0} with the help of equalities (16.35)–(16.36).

Problem (16.37)–(16.40) is the classical problem of heat propagation in an infinite bar with constant but different physical properties in each domain $y < 0$ and $y > 0$. Its explicit solution can be found, for instance, in [169]. Relations (16.32), (16.35), $x = X_1(y,t)$, $\rho_1 X_{1y} = 1$ as $y < 0$, $x = X_2(y,t)$, $\rho_2 X_{2y} = 1$ as $y > 0$, $h(t) = X_1(0,t) = X_2(0,t)$ give a parametric representation of a solution to the Cauchy–Verigin problem (16.30)–(16.34). Let us stress that this solution is constructed under the assumption that the functions $p_{i0}(x)$ in condition (16.34) are arbitrary, although the dependences (16.35) are very rigid.

Chapter III

One-dimensional parabolic equations

This chapter is devoted to the study of smoothness properties and special features of qualitative behavior of solutions of the Cauchy problem for a class of degenerate parabolic equations. We are mainly interested in the study of the free boundaries, i.e., the zero level curves of solutions.

17 Introduction

We study the Cauchy problem for a class of quasilinear parabolic equations with one independent space variable. A substantial interest in these equations arises from the needs of the mathematical modelling of the processes they cover. On the other hand, their nonlinearity leads to some properties of solutions which are in contrast to those intrinsic in the solutions of linear problems, thus making the study very attractive from the purely mathematical point of view.

Throughout the chapter we consider only nonnegative solutions. Because of the physical context in which these equations occur this restriction seems quite natural. Besides, it proves very convenient from the point of view of the method we follow.

17.1 An example. To begin with, let us consider one physical phenomenon whose mathematical description leads to an equation of the type we are interested in. Let us consider a liquid flow through a homogeneous, isotropic, rigid porous medium [20, 129, 113]. Let $\vec{x} = (x, y, z) \in \mathbb{R}^3$, $u(\vec{x}, t)$ be the volumetric soil-moisture contents and $\vec{V}(\vec{x}, t)$ be the velocity of the liquid. If the density of the liquid is constant, the mass conservation law has the form

$$\frac{\partial u}{\partial t} + \operatorname{div} \vec{V} = 0.$$

It is assumed that in the porous medium the Darcy law holds:

$$\vec{V} = -\psi(u)\nabla\Phi,$$

where ψ denotes the hydraulic conductivity of the medium, and Φ is the total potential of the mass forces. For unsaturated flows the total potential can be expressed as the sum of the hydrostatic potential, generated by the capillary suction $S(u)$, and the gravitational potential. Choosing the Cartesian coordinate system so as to make the axis x vertical and pointing upward, we have: $\Phi = S(u) + x$ and

$$\frac{\partial u}{\partial t} = \operatorname{div}\left\{\psi(u)\frac{dS(u)}{du}\nabla u\right\} + \frac{\partial \psi(u)}{\partial x}. \tag{17.1}$$

For flows in the vertical direction equation (17.1) takes the form

$$\frac{\partial u}{\partial t} = \varphi_{xx}(u) + \psi_x(u), \tag{17.2}$$

where

$$\varphi(q) \equiv \int_0^q \psi(\theta)\frac{dS(\theta)}{d\theta}\,d\theta.$$

It follows from (17.1) that for flows in the horizontal plane (y, z)

$$\frac{\partial u}{\partial t} = \Delta_{y,z}\varphi(u), \quad \left(\Delta_{y,z}\varphi(u) \equiv \frac{\partial^2}{\partial y^2} + \frac{\partial^2}{\partial z^2}\right). \tag{17.3}$$

Naturally, from its physical sense $u(\vec{x}, t)$ must be nonnegative.

Equations of the type (17.2)–(17.3) account for a number of other physical applications. They appear in a natural way in the theory of heat conduction through a medium with the heat conductivity depending nonlinearly on the temperature [179, 151, 169], in the theory of the flow of a thin viscous film over an inclined bed [29, 71], in the laminar boundary layer theory [154], in modelling of the processes of heat conduction by radiation in multiply and fully ionized gases [102, 18, 99, 180] and heat conduction by electrons in plasma [180] and in ecological problems concerning population dynamics [114, 115]. Whatever the variety of possible physical applications, equations (17.2), (17.3) are interpreted here as the continuity equation of a flow through a porous medium. This makes it possible to pass to the plane of Lagrangian coordinates and thus to get new equivalent formulations of the problems under discussion.

In descriptions of the processes of water filtration the functions ψ and S are usually defined empirically. As a rule, power dependences are adopted:

$$\psi(u) \equiv K_0 u^\lambda, \quad \psi(u)\frac{dS(u)}{du} \equiv D_0 u^{m-1},$$

where K_0, D_0, m and λ are positive constants. Scaling out the constants, it is easy to transform equation (17.2) to the canonical form

$$u_t = (u^m)_{xx} + (u^\lambda)_x. \tag{17.4}$$

The functions $\varphi(\cdot)$ and $\psi(\cdot)$ are usually subject to the conditions [20, 167]

$$\varphi, \psi \in C^2(0, \infty) \cap C^1[0, \infty),$$

$$\varphi(0) = \varphi'(0) = 0, \quad \varphi'(s) > 0 \text{ as } s > 0,$$
$$\psi(0) = 0, \quad \psi'(s), \psi''(s) > 0 \text{ as } s > 0.$$

For the powers this means

$$m > 1, \quad \lambda > 1. \tag{17.5}$$

In the theory of water filtration equation (17.4) with parameters m, λ satisfying (17.5) serves to describe processes of *soil-moisture infiltration*.

An interesting situation occurs in modelling of *evaporation* from the soil. Let the surface of the soil be so dry that the water losses are restricted only by the rate of the soil-water movement through the porous medium upwards. Under this assumption the hydraulic conductivity should be a smooth concave function and $\psi(u)S'(u)$ should be a smooth nondecreasing function [128, 94, 167]. It is easy to see that for the powers this is true if

$$m \geq 1, \quad 0 < \lambda < 1.$$

Formally calculating $(u^\lambda)_x$, we get that the coefficient of the 1-st order derivative in equation (17.4) becomes singular at the level $u = 0$.

17.2 Finite speed of propagation. Interfaces. There is a curious property separating nonlinear equations of the type (17.2), (17.3) from the linear ones and usually called *finite speed of propagation of disturbances from zero* (the FSP property). For the continuous nonnegative solutions considered below this property means the following. Let $u(x,t)$ satisfy equation (17.4) for $x \in \mathbb{R}$, $t > 0$. If $u(x_0, 0) = 0$, then there exists $t(x_0) > 0$ such that $u(x_0, t) \equiv 0$ for all $t \in [0, t(x_0)]$. Thus, it makes sense to consider the functions

$$\eta(t) = \sup\{x \in \mathbb{R} : \ u(x,t) > 0\}, \quad \zeta(t) = \inf\{x \in \mathbb{R} : \ u(x,t) > 0\}$$

bounding the support of $u(x,t)$ and to pose questions referring to their smoothness and asymptotic behavior. The curves $x = \eta(t)$, $x = \zeta(t)$ we term the *free boundaries* or the *interfaces*.

The FSP property is not displayed by solutions of any linear problem. For instance, given an initial-boundary value problem for the linear heat equation with nontrivial nonnegative input data, the solution is positive everywhere in the problem domain.

Each equation of the type (17.2), (17.3) is formally parabolic wherever its solution $u(x,t)$ is positive but fails to be so when $u(x,t)$ vanishes and that is why is said to be *degenerate*. Generally speaking, such equations need not have classical solutions. Let us illustrate this assertion by the example of a remarkable equation of the type (17.3):

$$u_t = \Delta(u^m), \quad m > 1, \ (x,t) \in \mathbb{R}^N \times (0, \infty), \tag{17.6}$$

usually referred to as the *nonlinear filtration equation* or the *porous medium equation* (in short PME).

These names indicate the physical problem with which this equation associated. Let ρ, $\vec{V}$ and p represent the density, velocity, and pressure of the ideal gas moving through the porous medium. This flow is governed by the mass balance law

$$\sigma\rho_t + \operatorname{div}(\rho\vec{V}) = 0,$$

Darcy's law

$$\mu\vec{V} = -k\nabla p,$$

and the equation of state of the perfect gas

$$p = p_0\rho^{m-1}.$$

The exponent m equals 2 for isothermal processes and is larger than 2 for adiabatic ones. The parameters σ (the porosity of the medium), μ (the viscosity), k (the permeability), and p_0 (the reference pressure) are assumed to be positive constants. Collecting these relations, we get a single equation for the density ρ and arrive at (17.6) after a rescaling.

Equation (17.6) admits a special class of self-similar solutions discovered and studied in [179, 19, 125]. This class of solutions (usually called "the Barenblatt solutions") has the explicit formula

$$U(x,t) = \frac{1}{t^\alpha}\left\{\left(A - B\frac{|x|^2}{t^{2\beta}}\right)_+\right\}^{1/(m-1)} \equiv \frac{1}{t^\alpha}F\left(\frac{|x|}{t^\beta}\right), \tag{17.7}$$

$\sigma_+ \equiv \max\{\sigma; 0\}$. The constants α and β are defined by the relations

$$\beta = \frac{\alpha}{N} = \frac{1}{2 + N(m-1)}.$$

$B = \frac{m-1}{m}\beta$, and A is an arbitrary positive constant distinguishing a concrete solution of this family. This solution takes as initial data the Dirac mass:

$$U(x,t) \to M\delta(x) \quad \text{as } t \to 0,$$

where the constant M depends on A (and N and m). Actually, the constants A and M can be expressed in terms of one another. Transforming the easily verified identity

$$\int_{\mathbb{R}^N} U(s,t)\,ds = M, \quad t \in (0,\infty), \tag{17.8}$$

to the form

$$\omega_N \int_0^\infty F(\zeta)\zeta^{N-1}\,d\zeta = M, \tag{17.9}$$

where ω_N is the volume of the unit ball in $\mathbb{R}^N$, we find the connection between A and M:

$$\omega_N A^{1/2\beta(m-1)} B^{-N/2} \int_0^{\pi/2} (\cos\theta)^{(m+1)/(m-1)}(\sin\theta)^{N-1}\,d\theta = M.$$

It will be convenient to suppose that M is given and A is defined in terms of M. The corresponding solution we will denote by $U(x, t; M)$.

For each t fixed the support of $U(x, t; M)$ is a compact set in $\mathbb{R}^N$, it expands monotonically as t grows and covers $\mathbb{R}^N$ as $t \to \infty$.

The function $U(x, t; M)$ satisfies equation (17.6) in the classical sense at each point of the half-space $\mathbb{R}^N \times (0, \infty)$ except, maybe, the hypersurface

$$|x| = r(t) \equiv (A/B)^{1/2} t^{\beta}, \quad t > 0,$$

bounding the support of $U(x, t; M)$. Since

$$|\nabla U|_{|x|=r(t)} \neq 0,$$

the space derivatives of U fail to be continuous at the hypersurface $|x| = r(t)$ if $m \geq 2$ and, therefore, for such m none of the functions $U(x, t; M)$ is the classical solution of equation (17.6) in the ordinary sense. Meanwhile, this class of self-similar solutions meets all physically reasonable requirements usually presented to the solutions of filtration problems. The function $U(x, t; M)$ defines the density of a gas in the radially symmetric flow through a porous medium arising from a point mass source of intensity M at the instant $t = 0$. Condition (17.8) means that the total mass of the gas remains constant for all $t > 0$. According to Darcy's law, the gas velocity obeys the equality

$$\vec{v}(x, t) = -\frac{m}{m-1} \nabla U^{m-1}(x, t; M),$$

implying that the mass flux across the boundary $|x| = r(t)$ equals zero, and that at this boundary the natural kinematic condition holds:

$$\frac{dr(t)}{dt} = \left(\vec{v}(x, t)_{|_{|x|=r(t)}}, \vec{n} \right),$$

where $\vec{n}$ is the unit outer normal vector to the hypersurface $|x| = r(t)$. The kinematic condition expresses the fact that the boundary of the gas volume advances with a speed equal to the particle velocity of the gas approaching it.

The above-presented example clearly shows that for the study of degenerate equations of the type (17.2), (17.3) the concept of solution should be widened.

17.3 Preliminaries and results. In recent years, the theory of quasilinear parabolic equations degenerate at definite values of solutions or their derivatives has been developing very rapidly. By now it has reached such breadth and depth that, without exaggeration, an attempt at any detailed description of the already known results of this theory would lead us far beyond the scope of the book. That is why we restrict ourselves to indicating a number of monographs and survey works that reflect completely enough the current state of the subject [127, 27, 76, 63]. Works [126, 10] are devoted directly to PME. The classical theory of PME is presented in [51, Ch. 5]. The basic theory of equations of the type (17.3) is constructed in papers [59, 61, 41, 60]. A special theory of asymptotic representations for solutions of problems related to those studied in this chapter is developed in monograph [103].

In what follows, we rely on the very necessary references to the papers substantially connected with the questions under discussion. We cite in this introduction those results of the general theory which are essentially used in our further considerations.

Let us consider the Cauchy problem

$$(\mathrm{E})^* \quad \begin{cases} u_t = \varphi_{xx}(u) + \psi_x(u) & \text{in } S \equiv \mathbb{R} \times (0, T], \\ u(x, 0) = u_0(x) & \text{in } \mathbb{R} \end{cases}$$

with a nonnegative initial function u_0 and $T < \infty$.

Definition 17.1. A function $u(x, t)$ is said to be a weak solution of problem (E)* if it is bounded, nonnegative, and continuous in $\overline{S}$, and for each test function $f(x, t) \in C^\infty(\overline{S})$, vanishing as $t = T$ and for all $|x|$ large enough, the following integral identity holds:

$$\int_S \{f_t u - f_x (\varphi_x(u) + \psi(u))\}\, dx\, dt + \int_{\mathbb{R}} f(x, 0) u_0(x)\, dx = 0. \tag{17.10}$$

In the consideration of problem (E)* we distinguish two cases.

1. Problem $(\mathrm{E}_{m,\lambda})$: the Cauchy problem in the strip S for the equation

$$u_t = (u^m)_{xx} + (u^\lambda)_x,$$

(equation (17.4)), in the range of parameters $m > 1$ and $0 < \lambda < 1$.

With regard to the initial data we assume the following:

$$(\mathrm{A.0}) \quad \begin{cases} u_0(x) > 0 \quad \text{in} \quad (0, a), \\ u_0 \equiv 0 \quad \text{outside} \quad (0, a); \quad (a < \infty), \\ u_0 \quad \text{is strictly monotone as } x \searrow 0, \ x \nearrow a; \end{cases}$$

(A.1) $\left(u_0^{m-\lambda}\right)'(x) \in L^\infty(0, a)$;

(A.2) $\left(u_0^{m-\lambda}\right)'(x) \geq \frac{\lambda - m}{m}$ a.e. in $(0, a)$.

(The notation "a.e." means "almost everywhere".)

2. Problem (E): the Cauchy problem in the strip S for equation (17.2)

$$u_t = \varphi_{xx}(u) + \psi_x(u).$$

The functions $\varphi(\cdot)$, $\psi(\cdot)$ are subject to the following restrictions:

(i) $\varphi(s), \psi(s) \in C[0,\infty) \cap C^{3+\gamma}_{\mathrm{loc}}(0,\infty)$ with some $\gamma \in (0,1]$;

$$\varphi(0) = \psi(0) = 0; \quad \varphi'(s) > 0 \text{ as } s > 0;$$

(ii) $\varphi(s)$ is *almost a power*: there exist positive constants α and β, $\alpha \le \beta$, such that for some $s_0 > 0$

$$\alpha \le \frac{s\varphi''(s)}{\varphi'(s)} \le \beta < 1 + 2\alpha \quad \text{as } s \in (0, s_0); \tag{17.11}$$

(iii) *quasi-homogeneity condition*: there exists a constant $\delta_* \in (0,1)$ such that

$$\overline{\lim_{s\to 0}} \frac{\varphi^{-1}(s)}{\varphi^{-1}(2s)} \le \delta_*;$$

(iv) there exist positive constants M and ν such that

$$\begin{cases} 0 \le s^{-1}\psi(s) \le M & \text{for } s \in (0, s_0); \\ s^{-1}\psi(s) \to 0 & \text{as } s \to 0; \end{cases} \tag{17.12}$$

$$0 \le \varphi^{-1}(s) \left[\frac{\psi\left(\varphi^{-1}(s)\right)}{\varphi^{-1}(s)} \right]'_s \le \nu \quad \text{for } s \in (0, s_0). \tag{17.13}$$

These restrictions become very simple in the power case $\varphi(s) \equiv s^m$, $\psi(s) \equiv s^\lambda$: the exponents of nonlinearity should satisfy the inequalities

$$\lambda \ge m > 1 \tag{17.14}$$

or, if we omit condition (iv), $m > 1$, $\lambda > 1$.

For instance, conditions (i)–(iii) are fulfilled for such functions as

$$\varphi(s) \equiv s^m + s^p, \quad (m \ne p,\ m, p > 1),$$

$$\varphi(s) \equiv s^m \ln \frac{1}{s}, \quad (s \in (0,1),\ m > 1).$$

The initial datum u_0 is assumed to satisfy the conditions:

(B.0) $\begin{cases} u_0(x) > 0 \text{ in } (0,a),\ (a < \infty), \\ u_0(x) \equiv 0 \text{ outside } (0,a); \end{cases}$

(B.1) $\begin{cases} u_0(x) \in C[0,a]; \\ \left(\int_0^{u_0(x)} \frac{\varphi'(s)}{s}\, ds \right)' \in L^\infty(0,a); \end{cases}$

$$(\mathrm{B.2})\quad \begin{cases} \text{there exists some } d_0 > 0 \text{ such that for any } d \in (0, d_0) \\ \text{the set } \{x \in (0, a) : \; u_0(x) > d\} \text{ is an interval.} \end{cases}$$

Let us formulate in advance a few general results referring to equations of the type (17.2).

Proposition 17.1 ([41]). *Let conditions* (A.0)–(A.1) *hold. Then problem* $(\mathrm{E}_{m,\lambda})$ *has a unique weak solution.*

Proposition 17.2 ([61]). *Let*

$$\varphi(s), \psi(s) \in C[0, \infty) \cap C^2(\mathbb{R}^+)$$

and $\varphi''(s)$, $\psi''(s)$ *are locally Hölder continuous in* $\mathbb{R}^+$,

$$\varphi(0) = \psi(0) = 0, \quad \varphi'(s) > 0 \textit{ for } s > 0.$$

For each function $u_0(x)$, *bounded, nonnegative, and continuous in* $\mathbb{R}$, *problem* (E^*) *has a unique weak solution* $u(x, t)$.

Proposition 17.3 ([60]). *Let the conditions of Proposition* 17.2 *be fulfilled and, additionally,*

$$\sup\{x \in \mathbb{R} : \; u_0(x) > 0\} \in (-\infty, \infty).$$

Then for every $t > 0$

$$\sup\{x \in \mathbb{R} : \; u(x, t) > 0\} \in (-\infty, \infty)$$

if and only if there exists a real number σ *such that*

$$\sigma s + \psi(s) > 0 \quad \textit{for all } s \in (0, M)$$

with some M *and*

$$\frac{\varphi'(s)}{\sigma s + \psi(s)} \in L^1(0, M).$$

Propositions 17.1–17.3 are presented without proofs. At this stage they just confirm that under our assumptions on the data weak solutions of problems $(\mathrm{E}_{m,\lambda})$ and (E) do exist and display free boundaries.

Remark 17.1. The above-cited papers [41, 60, 61] use a definition of a solution to problem (E)* different from Definition 17.1. The difference is that instead of the integral identity (17.9) these solutions are defined as satisfying the identity

$$\int_S \{f_t u + f_{xx}\varphi(u) - f_x\psi(u)\}\, dx\, dt + \int_{\mathbb{R}} f(x, 0)u_0(x)\, dx = 0 \tag{17.15}$$

under the same assumptions on the test function f. Such solutions are termed the *generalized solutions.*

However, it is shown in [41] that each *weak* solution of problem (E)* in the sense of Definition 17.1 is also a *generalized* solution, i.e., satisfies identity (17.15). Later on, we shall construct weak solutions of problem (E)* independently. The uniqueness of these solutions will be provided by Propositions 17.1–17.2.

Let us pass to the description of results. We begin with problem $(E_{m,\lambda})$. Our first result addresses the propagation properties of the solutions to this problem.

Theorem 17.1. *Let conditions* (A.0)–(A.2) *be true, and* $u(x,t)$ *be a weak solution of* $(E_{m,\lambda})$.

Then:

1) $\eta(t)$ *is finite for every* $t \in [0,T]$; $\eta(t) \in C[0,T]$, *and for every* $\tau > 0$ $\eta(t)$ *is Lipschitz continuous in* $(\tau, T]$;
2) $\zeta(t) = -\infty$ *for every* $t > 0$;
3) *there exist constants* $B > 0$, $C \in [a, \infty)$, $D > 0$ *such that for all* $t \in [0,T]$

$$-Bt \le \eta(t) \le C - Dt; \tag{17.16}$$

4) $\eta(t)$ *is a monotone decreasing function in* $[0,T]$;
5) *for every* $t \in (0,T]$ *there exists*

$$\lim_{x \nearrow \eta(t)} \left(u^{m-\lambda}\right)_x (x,t) = \frac{\lambda - m}{m}; \tag{17.17}$$

6) *the curve* $x = \eta(t)$ *is constituted by at most two portions: a vertical one for* $t \in [0, t_*]$ *with some finite* t_*, *and a strictly monotone one for* $t \in (t_*, T]$;
7) $\eta(t)$ *satisfies "the differential equation"*

$$\eta'(t) + \lim_{x \nearrow \eta(t)} \left\{ \frac{m}{m-1} \left(u^{m-1}\right)_x + u^{\lambda-1} \right\} = 0 \quad \text{in } \mathcal{D}'(0,T).$$

(*The notation "in* $\mathcal{D}'(0,T)$*" always means "in the sense of distributions"*).

These properties need certain comments. The existence of the right-hand interface and non-existence of the left-hand one in problem $(E_{m,\lambda})$ follow from Proposition 17.3. Yet earlier these properties (for problem $(E_{m,\lambda})$) were established in [40]. In what follows we present a new independent proof.

Let us emphasize the property (17.17). As shown in [73], the solutions of the Cauchy problem for the one-dimensional PME need not be smooth even if they correspond to initial data of infinite smoothness. Given a solution of the Cauchy problem with compactly supported initial data belonging to $C^\infty(\mathbb{R})$, within a finite time the space derivative of the solution becomes discontinuous at the boundary of its support. This property is manifested even more clearly in problem $(E_{m,\lambda})$. It follows from (17.17) that for $m \ge \lambda + 1$ the "jump of discontinuity" of $\left(u^{m-\lambda}\right)_x$ across the line $x = \eta(t)$

appears instantaneously and, moreover, the value of this "jump" remains constant for all $t \in (0, T]$ and does not depend on the nature of the initial function u_0. Earlier this property was predicted in [79] by means of the analysis of formal asymptotic series representing solutions of equation (17.4).

Properties 4) and 6) show that the curve $x = \eta(t)$, having started moving, does not stop. The conditions on u_0 providing that the free boundary has a vertical part are established in [4]. The upper bound t_* of the initial time interval where the free boundary is stable is usually termed the *waiting time*. Due to the upper estimate in (17.16) (this upper estimate was first derived in [40]), all possible waiting times in problem $(\mathrm{E}_{m,\lambda})$ are finite.

The regularity properties of the interfaces are stated in the next assertion.

Theorem 17.2. *Under the assumptions of Theorem* 17.1

1) *for every* $\tau \in (0, T)$
$$\eta(t) \in C^{1,\mu}_{\mathrm{loc}}(\tau, T)$$
with the exponent $\mu = (m - \lambda + 1)/(3m - 3\lambda + 2)$;

2) *for every* $\tau \in (0, T - t_*)$ *it holds* $\eta(t) \in C^{\infty}_{\mathrm{loc}}(t_* + \tau, T)$.

It is curious to compare these properties with those known for the interfaces occurring in solutions of the Cauchy problem for the one-dimensional PME, by now studied with exhaustive completeness. The curves $x = \eta(t)$ and $x = \zeta(t)$ corresponding to a nonnegative, continuous, and bounded function $u_0(x)$, compactly supported in $\mathbb{R}$, are Lipschitz continuous on each interval $(\tau, T]$, $\tau > 0$, [11], can display the waiting time property, [87], and become strictly monotone after their waiting times, [32, 87]. The strictly monotone parts of these curves are infinitely differentiable, [17, 69], and even analytic, [7]. However, as shown in [14] the functions $\eta(t)$ and $\zeta(t)$ are $C^1(0, T)$ only for some narrow enough class of initial data and, generally speaking, the waiting times correspond to the corner points of their graphs. Thus, when adding to PME with one space variable the lower-order term $(u^\lambda)_x$, we eliminate one of two possible free boundaries while the remainder gains some complementary regularity.

Let us pass to problem (E). To simplify the matter, we impose an additional constraint on the initial function $u_0(x)$:

(B.3) a) there exists a constant $C_0 > 0$ such that
$$\left(\int_0^{u_0(x)} \frac{\varphi'(s)}{s} ds\right)' \leq -C_0$$
a.e. in some left neighborhood of the point $x = a$;

b)
$$\left(\int_0^{u_0(x)} \frac{\varphi'(s)}{s} ds\right)' \geq C_0$$
a.e. in some right neighborhood of the point $x = 0$.

These conditions are convenient but not necessary at all. In essence, they mean that we study the interfaces which have already passed their waiting times.

Theorem 17.3. a) *Let conditions* (B.0)–(B.3) a) *and* (17.11)–(17.13) *be fulfilled, and* $u(x,t)$ *be a weak solution of* (E). *Then:*

1) $\eta(t)$ *is strictly monotone increasing in* $(0, T)$*;*
2) *for each* $\tau \in (0, T)$
$$\eta(t) \in C^1_{\mathrm{loc}}[\tau, T];$$
3) *for each* $t, t+\theta \in [\tau, T]$
$$|\eta'(t+\theta) - \eta'(t)| \le L_0 \omega(|\theta|),$$
where L_0 *is a positive constant depending only on* τ, α, β, T, *the constants from conditions* (17.10)–(17.12), *and the* L^1*-norm of* u_0*;*
$$\omega(|\theta|) \equiv \int_0^{\rho(|\theta|)} \frac{\varphi'(s)}{s} ds,$$
and $\rho(|\theta|)$ *is a solution of the equation*
$$\rho^2 \int_0^{\rho} \frac{\varphi'(s)}{s} ds = |\theta|.$$

b) *Let conditions* (B.0)–(B.2), (B.3) b) *and* (17.11)–(17.13) *be fulfilled, and* $u(x,t)$ *be a weak solution of problem* (E). *Then:*

1) $\zeta(t)$ *is strictly monotone decreasing in* $(0, T)$*;*
2) *for* $\zeta(t)$ *the assertions of* a) 2)–3) *hold.*

c) *Under the assumptions of* a)–b) *the following "equations" hold:*
$$\eta'(t) + \lim_{x \nearrow \eta(t)} \left\{ \frac{1}{u} [\varphi_x(u) + \psi(u)] \right\} = 0,$$
$$\zeta'(t) + \lim_{x \searrow \zeta(t)} \left\{ \frac{1}{u} [\varphi_x(u) + \psi(u)] \right\} = 0 \quad \text{in } \mathcal{D}'(0, T).$$

It is easy to calculate that for $\varphi(s) \equiv s^m$, $\psi(s) \equiv s^\lambda$ with the exponents $m > 1$, $\lambda \ge m$ both $\eta(t)$ and $\zeta(t)$ belong to $C^{1,\mu}(0, T)$ with the exponent $\mu = (m-1)/(3m-1)$.

The "equations" of free boundaries and their Lipschitz continuity were stated in [60] for the case when $\varphi(s)$ and $\psi(s)$ satisfied the conditions of Proposition 17.3 and, besides,
$$\frac{\varphi'(s)}{s} \in L^1(0, M) \quad \text{with some } M = \text{const.} > 0.$$

For power nonlinearities with the exponents $m > 1$ and $\lambda \geq m$, the C^1-regularity of the free boundaries and their strict monotonicity after the waiting times were proved in [64].

Under the conditions of Theorem 17.3, the demonstrable regularity of the interfaces is limited by the smoothness of the functions φ and ψ. In the limiting power case the following assertion holds.

Theorem 17.4. *Let* $\varphi \equiv s^m$ *with* $m > 1$, $\psi(s) \equiv s^\lambda$ *with* $\lambda > 1$, *and conditions* (B.0)–(B.2) *be fulfilled.*

a) *If* $\lambda \geq m$ *and condition* (B.3) a) *holds, then*

$$\eta(t) \in C^\infty_{\mathrm{loc}}(\tau, T) \quad \textit{for each } \tau \in (0, T).$$

b) *If* $\lambda > 1$ *and condition* (B.3) b) *holds, then*

$$\zeta(t) \in C^\infty_{\mathrm{loc}}(\tau, T) \quad \textit{for each } \tau \in (0, T).$$

Having proved Theorems 17.1–17.4, we extend this technique to the study of interfaces occurring in solutions of a related equation. We show that the regularity properties of interfaces stated in Theorem 17.4 a) are also displayed by solutions of the Cauchy problem for a class of nonlocal equations arising in the population dynamics. In Subsection 24.8 we indicate how all these considerations can be adapted for the equation

$$u_t = \left(u^m\right)_{xx} + \left[u\Phi\left(\int_{-\infty}^{x} u(y,t)\,dy\right)\right]_x$$

under appropriate assumptions on the function Φ. After the passage to Lagrangian coordinates this seemingly irrelevant equation yields an equation almost coinciding with PME.

Section 25 is devoted to the study of asymptotic properties as $t \to \infty$ of solutions of the Cauchy problem for PME. Earlier this question was intensively treated in the literature. Let us present here a brief description of the strongest of the results obtained so far.

Let $\varphi(s) \equiv s^m$ with $m > 1$ and let $u(x,t)$ be a weak solution of the Cauchy problem for PME in the half-space $\mathbb{R}^N \times \mathbb{R}_+$. (The notion of weak solution will be given later in this introduction.) It is shown in [53] that the following limit relation

$$t^\alpha |u(x,t) - U(x,t;M)| \to 0 \quad \text{as } t \to \infty. \tag{17.18}$$

holds uniformly with respect to x on any set of the form

$$\Sigma = \left\{(x,t) \in \mathbb{R}^N \times \mathbb{R}_+ :\ |x| \leq At^\alpha,\ t > 0\right\},$$

where $A > 0$ is an arbitrary constant and $\alpha = N/(N(m-1)+2)$. In (17.18) $U(x,t;M)$ is the similarity solution of PME selected by the condition

$$M = \left\{\int_{\mathbb{R}^N} u_0(x)\,dx\right\}^{1/(m-1)}.$$

For $N = 1$ relation (17.18) is generalized, [77], to a class of functions $\varphi(s)$ not necessary powers but displaying some of their properties. Suppose that there exists a constant $\lambda > 0$ such that for each $z_0 > 0$ fixed

$$\left\{\frac{\varepsilon\varphi'(\varepsilon z)}{\varphi(\varepsilon)z^{\lambda-1}} - \lambda\right\} z^{\sigma} \to 0 \quad \text{as } \varepsilon \to 0, \ \sigma = \min\{1; \lambda - 1\}$$

uniformly in $z \in [0, z_0]$. Then

$$t^{\alpha}\left|u\left(x, \frac{1}{t^{\lambda/(\lambda+1)}\varphi(t^{-1/(\lambda+1)})}\right) - U(x,t;M)\right| \to 0 \quad \text{as } t \to \infty \tag{17.19}$$

uniformly with respect to $(x, t) \in \Sigma$.

However, relations (17.18)–(17.19) do not allow one to conclude that the similarity solution $U(x, t; M)$ represents the main term of the asymptotics as $t \to \infty$ of the solution to the Cauchy problem for PME. The reason is that the similarity solution itself has the same rate of vanishing as $t \to \infty$: $U(x, t; M) \sim O(t^{-\alpha})$. It is shown in [171] that for $\varphi(s) \equiv s^m$ and $N = 1$ the character of convergence of the solutions for the Cauchy problem to the similarity solution $U(x, t; M)$, with M fixed by the initial condition, is the following:

$$u(x,t) = U(x,t;M) + O(t^{-(1+\alpha)}) \quad \text{as } t \to \infty. \tag{17.20}$$

Paper [22] treats the asymptotic properties of the curves $x = \zeta(t)$, $x = \eta(t)$, bounding the supports of solutions to the Cauchy problem for PME in one space dimension. It is shown that

$$\left|\frac{d^i}{dt^i}(\eta(t) - [r(t) + \eta(0)])\right| \le C_i t^{(i+m\alpha)}, \tag{17.21}$$

(and the same for $\zeta(t)$), where i can take the values 0, 1, 2; $x = r(t)$ is the boundary of support of the corresponding self-similar solution; C_i are positive constants depending only on m and the L^1-norm of u_0.

Paper [124] deals with the asymptotic properties of expansion of the supports of weak solutions to the problem

$$(\mathrm{P}) \quad \begin{cases} u_t = \Delta\varphi(u) & \text{in } \mathbb{R}^N \times \mathbb{R}_+, \\ u(x,0) = u_0(x) & \text{in } \mathbb{R}^N \end{cases}$$

in the class of compactly supported and nonnegative initial data. Let us adopt the notation

$$\Omega(t) = \{x \in \mathbb{R}^N : u(x,t) > 0\},$$
$$R_M(t) = \sup\{|x| : x \in \Omega(t)\},$$
$$R_m(t) = \inf\{|x| : x \in \Gamma(t) \equiv \partial\Omega(t)\}$$

and assume the fulfillment of the following conditions:

$$\varphi \in C^1[0,\infty), \quad \varphi(0) = 0;$$

there exist constants a, b such that

$$1 < a \le \frac{s\varphi'(s)}{\varphi(s)} \le b < \infty \quad \text{for } s > 0.$$

Then there exist absolute positive constants C_1 , C_2 such that as

$$C_1 t^{\lambda_a/N} \le R_m(t) \le R_M(t) \le C_2 t^{\lambda_b/N} \quad \text{as } t \to \infty, \tag{17.22}$$

$$\lambda_p = \frac{N}{N(p-1)+2}.$$

In what follows, we also consider the question of the asymptotic behavior of the radially symmetric solutions to problem (P). We distinguish two cases.

a) $\varphi(s) \equiv s^m$ with $m > 1$. In this case we derive estimates analogous to (17.21) with $i = 0$ but valid for an arbitrary natural number N; for $N = 1, 2$ the analogs of relations (17.20) and (17.21) with $i = 1$ are proved.

b) $\varphi(s)$ is "almost a power":

$$\varphi(s) \in C^1[0,\infty) \cap C^{2+\gamma}(0,\infty), \quad \varphi(0) = \varphi'(0) = 0,$$

$$\alpha \le \frac{s\varphi''(s)}{\varphi'(s)} \le \beta \quad \text{as } s > 0 \tag{17.23}$$

with some constants $0 < \alpha \le \beta < \infty$ and $\gamma \in (0, 1)$; we show that under these assumptions both the upper and lower bounds for the supports of the axially symmetrical solutions to (P) behave as $t \to \infty$ like solutions of the equation

$$y'(t) = N y^{(N-2)/N} \varphi'\left(\frac{N}{\delta y(t)}\right), \quad \delta = \text{const.} > 0.$$

The estimates obtained in case b) are then applied for certain specification of estimates (17.22).

The main object of investigation in Section 25 is the following one-dimensional problem:

$$u_t = \frac{1}{r^{N-1}} \frac{\partial}{\partial r}\left(r^{N-1} \frac{\partial \varphi(u)}{\partial r}\right) \tag{17.24}$$

$$\text{in } H = (0,\infty) \times (0, T], \quad T < \infty,$$

$$\left.\left\{r^{N-1} \frac{\partial \varphi(u)}{\partial r}\right\}\right|_{r=0} = 0 \quad \text{as } t \in [0, T], \tag{17.25}$$

$$u(r, 0) = u_0(r) \quad \text{for } r \ge 0 \tag{17.26}$$

which will be termed *problem* (P_1).

The initial data are assumed to satisfy the conditions

$$\text{(C.1)}\quad \begin{cases} u_0(r) > 0 & \text{for } r \in [0,a),\ (a < \infty), \\ u_0(r) \equiv 0 & \text{for } r > a; \\ u_0(r) \in C[0,a]; \end{cases}$$

(C.2) u_0 is monotone decreasing in $[0, a)$;

$$r^{N-1}\frac{\varphi_r'(u_0)}{u_0} \in C[0,a];$$

(C.3) there exist constants $M_0 > 0$, $r_0 \in [0, a)$ such that

$$\frac{1}{u_0}\varphi_r'(u_0) \le -M_0 \quad \text{in } [r_0, a].$$

The solution to this problem is understood in the following sense.

Definition 17.2. A function $u(x,t)$ is said to be a weak solution of problem (P_1) in H if it is bounded, nonnegative, and continuous in $\overline{H}$,

$$r^{N-1}\frac{\partial\varphi(u)}{\partial r} \in L^2(H),$$

and for each test function $f(r,t) \in C^\infty\left(\overline{H}\right)$, vanishing for $t = T$ and for all r large enough, the following integral identity holds:

$$\int_H \left\{ f_t u - \frac{\partial f}{\partial r}\frac{\partial\varphi(u)}{\partial r} \right\} r^{N-1}\,dr\,dt + \int_0^\infty r^{N-1} u_0 f(r,0)\,dr = 0. \tag{17.27}$$

Definition 17.3. A function $u(x,t)$ is said to be a weak solution of problem (P) in the strip $S = \mathbb{R}^N \times \mathbb{R}_+$ if it is bounded, nonnegative, and continuous in $\overline{S}$,

$$\frac{\partial}{\partial x_i}\varphi(u) \in L^2(S), \quad i = 1, \dots, N,$$

and for an arbitrary test function $f(x,t) \in C^\infty\left(\overline{S}\right)$, vanishing for $t = T$ and for all $|x|$ large enough, the following integral identity holds:

$$\int_S \{ f_t u - \nabla f \nabla\varphi(u)\}\,dx\,dt + \int_{\mathbb{R}^N} u_0(x) f(x,0)\,dx = 0. \tag{17.28}$$

Evidently, each weak solution of problem (P_1) is also a weak solution of problem (P). The existence of a weak solution to problem (P_1) will be proven independently; the uniqueness of this solution will then be provided by the following general assertion.

Proposition 17.4 ([51, Ch. 5]). *Let $\varphi(s)$ be continuous in $[0,\infty)$ and strictly monotone increasing as $s > 0$, $\varphi(0) = 0$. For each nonnegative function*

$$u_0 \in L^1(\mathbb{R}^N) \cap L^\infty(S)$$

there exists a unique solution of the equation

$$u_t - \Delta\varphi(u) = 0 \quad \text{in } \mathcal{D}'(S) \tag{17.29}$$

belonging to

$$C\left([0,T); L^1(\mathbb{R}^N)\right) \cap L^\infty(S)$$

such that $u(x,0) = u_0(x)$ *in* $\mathbb{R}^N$.

We establish that this weak solution of problem (P_1) possesses the following asymptotic properties.

Theorem 17.5. *Let* $\varphi(s) \equiv s^m$, $m > 1$, *and let* $u_0(x)$ *satisfy conditions* (C.1)–(C.3). *Let* $u(R,t)$ *be the corresponding weak solution of problem* (P_1) *and*

$$\eta(t) \equiv \sup\{r \in \mathbb{R} : \ u(r,t) > 0\}.$$

There exist finite positive constants λ, K, *and* A_i, $i = 1, 2$, *depending only on* u_0 *and* m, *such that for all* $t > 0$

1) *for each natural number N*

$$\left|\eta^N(t) - A_1(\lambda+t)^\alpha\right| \le A_1 K (\lambda+t)^{\alpha-1}; \tag{17.30}$$

2) *for* $N = 1, 2$

$$\left|(\eta^N)'(t) - A_2(\lambda+t)^{\alpha-1}\right| \le A_2 K (\lambda+t)^{\alpha-2}. \tag{17.31}$$

Remark 17.2. 1) In Section 25, we establish also the limit relation similar to (17.20) for the axially symmetrical solutions of problem (P_1) with $N = 1, 2$. Its rigorous formulation requires certain preliminary work and therefore is omitted here.

2) Condition (C.3) guarantees that the corresponding solutions of (P_1) have zero waiting times. One can omit this condition and show that each solution to (P_1) acquires this property within a finite time.

Now, let us relax restrictions on the function $\varphi(\cdot)$ and assume that it need not be a power but only "almost a power".

Theorem 17.6. *Let conditions* (17.23) *and* (C.1)–(C.3) *be fulfilled. Then there exist constants* δ, λ, A, B, *depending only on* α, β, M_0, *such that for all* t, *beginning with some* $t_0 > 0$,

$$A \le \frac{\eta^N(t)}{y(\lambda+t)} \le B, \tag{17.32}$$

where $y(t)$ *is the solution of the Cauchy problem*

$$y'(t) = N y^{(N-2)/N}(t)\varphi'\left(\frac{N}{\delta y(t)}\right) \quad \text{as } t > 0, \tag{17.33}$$

$$y(0) = 1.$$

To extend some of these assertions to the case when solutions to problem (P) need not be axially symmetric we make recourse to the following result of the general theory.

Proposition 17.5 ([51, Ch. 5], Comparison principle). *Let the conditions of Proposition* 17.4 *hold. Let* $u_0, v_0 \in L^1(\mathbb{R}^N)$ *and* $u_0 \geq v_0$. *Then the corresponding solutions of problem* (P) *satisfy the inequality* $u(x,t) \geq v(x,t)$ *in* $\mathbb{R}^N \times \mathbb{R}_+$.

The proof of this proposition relies on the constructive method of proof of the existence theorem for problem (P). Having the possibility to bound from below and from above the initial function of the general problem (P) by axially symmetric initial functions, one can then refer to Proposition 17.5 and Theorem 17.6 to get the following result.

Theorem 17.7. *Let* $u_0(x) \in C(\mathbb{R}^N)$. *Assume it is nonnegative, compactly supported in* $\mathbb{R}^N$, *and* $u_0(0) > 0$. *Then there hold the inequalities*

$$A' \leq \frac{R_m(t)}{y(\lambda+t)} \leq \frac{R_M(t)}{y(\lambda+t)} \leq B' \quad as\ t \to \infty,$$

where A', B', λ *are positive constants, depending only on* α, β, N, *and properties of* u_0; $y(t)$ *is the solution of problem* (17.33).

All assertions formulated here and later on as theorems are proven by means of passing to Lagrangian coordinates. As the result of this coordinate transformation we obtain initial-boundary value problems equivalent to the original ones but with the problem domains known beforehand. The lateral boundaries of these domains, merely by the definition of the change of variables, are the images of the free boundaries occurring in the original problems. In particular, the question of the asymptotic behavior as $t \to \infty$ of solutions to problem (P) is reduced thereby to the question of the asymptotic behavior of solutions to the first boundary value problem for a nonlinear parabolic equation with nonlocal coefficients posed in a vertical half-strip. For PME such a problem was studied in [15, 16], and in what follows we benefit from some ideas of these works.

In the concluding sections of this chapter, we study the conditions for uniqueness of the unbounded solution for one special, non-classical, initial-boundary value problem for the equation

$$\omega_t = \left[\nu(\omega)\omega_\xi\right]_\xi,$$

(Section 26), and present an example of exact solution to a special Stefan-type problem where the free boundary advancement is controlled by a nonlinear flux condition (Section 27).

18 Lagrangian coordinates in one-dimensional evolution equations

For the convenience of the reader, we start by recalling some information from continuum mechanics. The mathematical concept corresponding to the physical intuitions about the motion of continua is the notion of continuous transformation of Euclidean space into itself. The parameter t describing this transformation is identified with time. For the initial instant of motion we take $t = 0$. Here we are interested only in one-dimensional motions. Let us introduce a rectangular coordinate system (x, t) and identify each point $x \in \mathbb{R}$ with a particle moving along with the continuum. Assume that at the initial instant of the motion this particle is located at the point ξ, and occupies the position x at the instant t. This defines the transformation

$$x = X(\xi, t)$$

characterizing the continuum motion and defining the trajectory of the particle starting from the point ξ. On the other hand, this relation defines the transformation of the initial domain occupied by the continuum into the domain it occupies at the instant t, and the inverse one.

Let the state of the medium be characterized by the density $\rho(x, t)$ and the velocity $V(x, t)$, and let $h(x, t)$ stand for the intensity of the mass sources or sinks. Denote by $I(t)$ an arbitrary volume of the medium containing the same particles at every instant of the motion. As we assumed, there exists a bijective mapping $I(0) \mapsto I(t)$, and the position (the trajectory) of each particle can be determined from the relation (the trajectory equation)

$$X_t = V[X(\xi, t), t].$$

The mass balance law of the motion is easy to derive. In the most general form it is given by the relation

$$\frac{d}{dt}\int_{I(t)} \rho(y, t)\, dy = \int_{I(t)} h(y, t)\, dy.$$

Formally performing the differentiation in t in the left-hand side of this equation, we get

$$\begin{aligned}
\frac{d}{dt}\int_{I(t)} \rho(y, t)\, dy &= \frac{d}{dt}\int_{I(0)} \rho(X(\xi, t), t) X_\xi(\xi, t)\, d\xi \\
&= \int_{I(0)} \left[(\rho_t + \rho_X V) X_\xi + \rho V_\xi\right] d\xi \\
&= \left(\text{since } dx = X_\xi d\xi + X_t dt, \text{ we have } \tfrac{\partial}{\partial \xi} = X_\xi \tfrac{\partial}{\partial x}\right) \\
&= \int_{I(0)} \left[(\rho_t + \rho_X V) X_\xi + \rho V_X X_\xi\right] d\xi \\
&= \int_{I(t)} \left[\rho_t + (\rho V)_x\right] dx. \qquad (18.1)
\end{aligned}$$

Comparing the last and the first lines of (18.1), and recalling that the volume $I(t)$ was taken to be arbitrary, we arrive at the following equation

$$\rho_t + (\rho V)_x = h(x,t), \tag{18.2}$$

usually called the *mass balance law* in the differential form.

18.1 Equations in divergence form. Abstracting from the physical content of problem (E)*, let us consider it as a problem of the one-dimensional motion of a liquid with density $\rho = u(x,t)$ and velocity

$$V(x,t) = -\frac{\partial}{\partial x}\left\{\int_0^{u(x,t)} \frac{\varphi'(s)}{s}\,ds\right\} - \frac{\psi(u(x,t))}{u(x,t)}.$$

This formal replacement transforms equation (17.2) to the mass balance law (18.2) with $h \equiv 0$. The function u_0 serves to define the initial characteristics of the motion. Let us describe the imaginary flow by means of Lagrangian coordinates. It is convenient to use the *mass Lagrangian coordinate*

$$\xi = \int_{-\infty}^{x} u_0(s)\,ds : [0,a] \mapsto [0,b] \equiv I(0).$$

This relation is always invertible since $u_0 > 0$ inside $(0,a)$. The mechanical meaning of the variable ξ is transparent: since $u(x,t)$ is decreed to be the density of the continuum, ξ represents the mass of the substance within the interval $(0,x)$ at the initial instant of motion. Note that under the foregoing assumptions on initial data of problems (E)* and ($E_{m,\lambda}$) the value of b (the initial mass) is always finite. For the new unknowns we choose

$$x = X(\xi,t),$$

the trajectory of the liquid particle labeled by its initial position, and the density in the particle

$$v(\xi,t) \equiv u[X(\xi,t),t].$$

The flow is now governed by the following relations:

a) the equations of the trajectories

$$X_t(\xi,t) = V(X(\xi,t),t) \quad \text{as } \xi \in (0,b),\ t > 0,$$

$$X(\xi,0) \in [0,a];$$

b) the mass conservation law, which is now given by the first line of (18.1)

$$\frac{d}{dt}\big(v(\xi,t)X_\xi(\xi,t)\big) = 0 \quad \text{as } \xi \in (0,b),\ t > 0$$

and implies, after integration in t,

$$v(\xi,t)X_\xi(\xi,t) = 1 \quad \text{as } \xi \in (0,b),\ t > 0.$$

Differentiating the trajectory equation in ξ, the mass balance law in t, and comparing the results, we arrive at the equation for the density $v(\xi,t)$:

$$\frac{\partial}{\partial t}\left(\frac{1}{v}\right) + \frac{\partial}{\partial \xi}\left(\frac{\partial\varphi(v)}{\partial\xi} + \frac{\psi(v)}{v}\right) = 0 \tag{18.3}$$

$$\text{in} \quad Q = \{(\xi,t):\ \xi \in (0,b),\ t \in (0,T]\}.$$

The lateral boundaries of the rectangle Q correspond to the free boundaries of the liquid volume, which is why a natural constraint is to pose the boundary conditions

$$v(0,t) = v(b,t) = 0 \quad \text{as } t \in (0,T]. \tag{18.4}$$

And finally, the initial conditions for $v(\xi,t)$ are generated by function $u_0(x)$ and we get

$$v(\xi,0) = v_0(\xi) \quad (\equiv u_0(X(\xi,0))) \quad \text{as } \xi \in (0,b). \tag{18.5}$$

By definition, $X(\xi,0)$ is to be defined from the equation

$$\xi = \int_0^{X(\xi,0)} u_0(s)\,ds$$

and the conditions $X(b,0) = a$, $X(0,0) = 0$.

Problem (18.3)–(18.5) is later termed *problem* (L), respectively, *problem* ($L_{m,\lambda}$) if $\varphi(s) \equiv s^m$, $\psi(s) \equiv s^\lambda$.

Let us formulate the conditions on the initial function v_0. Due to assumptions (A.0)–(A.2), in the case of problem ($L_{m,\lambda}$) we have:

$$\text{(D.0)} \quad \begin{cases} v_0(\xi) > 0 \quad \text{as } \xi \in (0,b); \\ v_0(0,t) = v_0(b,t) = 0; \\ v_0 \text{ is strictly monotone as } \xi \nearrow b,\ \xi \searrow 0; \end{cases}$$

$$\text{(D.1)} \quad \left(v_0^{m-\lambda+1}\right)'(\xi) \in L^\infty(0,b).$$

Condition (D.1) is a byproduct of the definition of the coordinate transformation: at the instant $t = 0$

$$\frac{d}{dx} = \frac{1}{X_\xi}\frac{d}{d\xi} \equiv v_0(\xi)\frac{d}{d\xi},$$

whence

$$\left(u_0^{m-\lambda}\right)'(x) = \frac{m-\lambda}{m-\lambda+1}\left(v_0^{m-\lambda+1}\right)'(\xi).$$

Likewise, condition (A.2) gives

(D.2) $\left(v_0^{m-\lambda+1}\right)'(\xi) \geq -(m-\lambda+1)/m$ a.e. in $(0, b)$.

Let us also point out one additional useful property implied directly by the definition of the coordinate transformation:

$$\int_0^b \frac{ds}{v_0(s)} \equiv \int_0^a \frac{u_0(x)}{u_0(x)} dx = a. \tag{18.6}$$

Under assumptions (B.0)–(B.3) the initial data of problem (E) generate a function v_0 with the following properties:

(F.0) $v_0(\xi) > 0$ as $\xi \in (0, b)$; $v_0(0, t) = v_0(b, t) = 0$;

(F.1) $v_0 \in C[0, b]$; $\varphi_\xi(v_0) \in L^\infty(0, b)$;

(F.2) there exists $d_1 > 0$ such that for each $d \in (0, d_1)$ the set

$$\{\xi \in (0, b) : v_0(\xi) > d\}$$

is an interval;

(F.3) there exists a positive constant σ such that

$$\varphi(v_0(\xi)) \geq \sigma\xi \quad \text{in a neighborhood of the point } \xi = 0,$$

$$\varphi(v_0(\xi)) \geq \sigma(b - \xi) \quad \text{in a neighborhood of the point } \xi = b.$$

Also, combining (F.1) with (17.12) and recalling that v_0 is bounded, we have

(F.4) $\varphi_\xi(v_0) + \frac{\psi(v_0)}{v_0} \in L^\infty(0, b)$.

The original problem (E)* is replaced now by problem (L) for equation (18.3) which is also degenerate parabolic but is posed in an already known domain. However, in order to formulate conditions (18.3)–(18.5) we have used only a *formal resemblance* between equation (17.2) and the mass conservation law of some liquid motion. Of course, this is not a justification of the legitimacy of the passage to Lagrangian coordinates. As we have seen, the very possibility of Lagrangian description of liquid motions is based on some mechanical axioms which may appear to be meaningless in a concrete situation. To circumvent this difficulty we proceed as follows. Whatever the origin of problems (L) and ($L_{m,\lambda}$), we treat them as independent mathematical objects and prove that under the assumptions posed on the data each of them has at least one classical solution. Thereafter we present the explicit formulas restoring weak solutions to problems (E) and $(E_{m,\lambda})$ from the obtained solutions of (L) and $(L_{m,\lambda})$. Taking into account the uniqueness of weak solutions of (E) and $(E_{m,\lambda})$, one then has to conclude that the solutions of (E) and $(E_{m,\lambda})$ constructed via this procedure in fact coincide with those whose existence and uniqueness are guaranteed by Propositions 17.1–17.2.

In other words, the Lagrangian reformulation results in a new parameter representation to the solutions of free boundary problems (E) and $(E_{m,\lambda})$.

Let us pass to the rigorous formulation of the results. For problems (L) and $(L_{m,\lambda})$ we prove the following existence theorems.

Theorem 18.1. *Let* $m > 1$, $0 < \lambda < 1$. *Under assumptions* (D.0)–(D.2) *problem* $(L_{m,\lambda})$ *has at least one classical solution*

$$v(\xi, t) \in C\left(\overline{Q}\right) \cap C^{\infty}(Q)$$

strictly positive in $(0, b) \times [0, T]$ *and such that*

$$v_{\xi}(\xi, t) \leq 0 \quad \text{as } (\xi, t) \in Q \cap \{\xi \geq \xi_1\}, \tag{18.7}$$

$$v_{\xi}(\xi, t) \geq 0 \quad \text{as } (\xi, t) \in Q \cap \{\xi \leq \xi_0\} \tag{18.8}$$

with some ξ_0, ξ_1 *depending only on the data.*

Theorem 18.2. *Let conditions* (17.11)–(17.13) *and* (F.0)–(F.4) *be fulfilled. Then problem* (L) *has at least one classical solution*

$$v(\xi, t) \in C\left(\overline{Q}\right) \cap C^{2,1}_{\xi,t}(Q)$$

strictly positive in $(0, b) \times [0, T]$. *Also,*

$$\varphi(v(\xi, t)) \in C\left(\overline{Q}\right) \cap C^{3+\delta,(3+\delta)/2}_{\xi,t}(Q) \quad \text{with some } \delta \in (0, 1].$$

Having proved Theorems 18.1, 18.2, we now consider the function

$$X(\xi, t) = X(\xi, 0) - \int_0^t \left[\frac{\partial \varphi(v)}{\partial \xi} + \frac{\psi(v)}{v}\right](\xi, \theta)\, d\theta \tag{18.9}$$

which, as we shall show, is well-defined up to the lateral boundaries of Q. Next, we define the set

$$\Omega(t) = \{x \in \mathbb{R} : \ x = X(\xi, t), \xi \in (0, b)\}$$

and the function

$$\overline{u}(x, t) = \begin{cases} v(\xi, t) & \text{for } x \in \Omega(t), \\ 0 & \text{for } x \in \mathbb{R} \setminus \Omega(t), \ t \in [0, T]. \end{cases} \tag{18.10}$$

The justification of the legitimacy of the Lagrangian coordinate transformation is completed by proving the following assertion:

Theorem 18.3. *Under the conditions of Theorem* 18.1 *(Theorem* 18.2*) the function* $\bar{u}(x,t)$ *is a weak solution of problem* ($E_{m,\lambda}$) ((E)).

In the light of our future purposes, the most substantial byproduct of this assertion is that now the functions $\eta(t)$ and $\zeta(t)$ acquire the representations

$$\eta(t) = a - \lim_{\xi \nearrow b} \int_0^t \left[\frac{\partial \varphi(v)}{\partial \xi} + \frac{\psi(v)}{v} \right] d\xi, \tag{18.11}$$

$$\zeta(t) = - \lim_{\xi \searrow 0} \int_0^t \left[\frac{\partial \varphi(v)}{\partial \xi} + \frac{\psi(v)}{v} \right] d\xi, \tag{18.12}$$

which prove very convenient. In fact, the question of the free boundary behavior is no longer independent. It is reduced to the study of the vanishing rates of $v(\xi, t)$ near the lateral boundaries of Q.

This procedure offers a way to construct solutions to problems (E) and ($E_{m,\lambda}$). As we have seen, we need not take care of the uniqueness of solutions to problems (L) and ($L_{m,\lambda}$) because the uniqueness of solutions to the original problems ($E_{m,\lambda}$), (E) is already guaranteed by the results of the general theory. It is worth noting that the problem of uniqueness of solutions to these problems is far from trivial. In the special case of the equation $U_t = U^2(U^m)_{\xi\xi}$, $m > 1$ (the Lagrangian counterpart of the one-dimensional PME), the uniqueness of a strictly positive solution of the Cauchy problem was stated in [24].

The above-described scheme of Lagrangian coordinate transformation derives from ideas of papers [23, 109, 65]. Paper [23] deals with the problem of stability of the similarity solutions to PME in one space dimension. In [65] a Lagrangian coordinate was introduced in order to make stationary the free boundary occurring in a solution of the one-dimensional PME and to render it a vertical segment in the plane of Lagrangian coordinates. This provided the possibility of constructing an effective algorithm for numerical treatment. The justification of the method relied on preliminary information about the solutions of the original Cauchy problem. These ideas were given further development in [162, 163]. In particular, it was established in [162] that the original Cauchy problem for PME with one space variable and its Lagrangian counterpart are equivalent: each solution of the Cauchy problem for PME generates a solution of the corresponding problem (L) and conversely. In [64, 25] Lagrangian coordinates were introduced to study regularity properties of the free boundaries occurring in solutions of the Cauchy problem for equation (17.4) in the range of parameters $m > 1$, $\lambda \geq m$, and a special equation generalizing the one-dimensional PME. The legitimacy of the coordinate transformation was also provided there by certain a priori information about the solutions under investigation. The results presented in Theorems 17.1 and 17.2 were obtained in [157, 42, 156, 155].

18.2 Some equations not in divergence form. All previous constructions referred to the case when the original equation (posed in Euler coordinates) had divergence form, that is, was already presented in the form of a conservation law.

Let us consider the Cauchy problem for the nonlinear diffusion-reaction equation

$$u_t = \left(u^m\right)_{xx} + f(u), \quad m > 1, \text{ in } S \equiv \mathbb{R} \times \mathbb{R}^+. \tag{18.13}$$

Following [159], let us construct the Lagrangian counterpart of equation (18.13). Assume that the initial function u_0 is continuous, $\operatorname{supp} u_0 = (0, a)$, $a < \infty$, and that the absorption term $f(u)$ is subject to appropriate growth and smoothness conditions.

Observe that, if we identify a nonnegative solution of equation (18.13) $u(x, t)$ with the density of some continuum and declare

$$-\frac{m}{m-1}\left(u^{m-1}\right)_x$$

the velocity, the reaction term must be interpreted as the density of the mass forces, and equation (18.13) provides the mass balance law.

Let us take the quantity

$$\xi := \int_0^x u_0(s)\, ds \; : (0, a) \mapsto (0, b) \equiv I(0) \tag{18.14}$$

for the new independent space variable.

The mass balance law now takes the form (see (18.1)–(18.2))

$$v(\xi, t) := u(X(\xi, t), t),$$

$$\left(vX_\xi\right)_t = vX_\xi \frac{f(v)}{v} \quad \text{in } Q := I(0) \times (0, T].$$

This equation can be easily integrated, whence

$$X_\xi = \frac{1}{v} \exp\left(\int_0^t \frac{f(v)}{v}(\xi, t)\, dt\right) \quad \text{in } Q. \tag{18.15}$$

Formula (18.15) gives the rule to recalculate the derivatives in x

$$\frac{\partial}{\partial x} = \frac{1}{X_\xi} \cdot \frac{\partial}{\partial \xi} = vA(\xi, t)\frac{\partial}{\partial \xi},$$

$$A(\xi, t) := \exp\left(-\int_0^t \frac{f(v)}{v}(\xi, \theta)\, d\theta\right),$$

which, in its turn, shows that the velocity of the motion is now given by the expression

$$V\left(X(\xi, t), t\right) = w_\xi(\xi, t)A(\xi, t), \quad w := v^m.$$

Now, differentiating the trajectory equation in ξ, (18.15) in t, and comparing the results, we get the following equation to define the density as a function of the mass Lagrangian coordinate:

$$w_t = mw^{(m+1)/m} A(\xi, t)\frac{\partial}{\partial \xi}\left(A(\xi, t)\frac{\partial w}{\partial \xi}\right) + mw^{(m-1)/m} f\left(w^{1/m}\right) \tag{18.16}$$

in Q. As above, to this equation we add the boundary conditions

$$w(0,t) = w(b,t) = 0 \quad \text{as } t \in [0,T] \tag{18.17}$$

and the initial conditions

$$w(\xi,0) = w_0(\xi) := u_0(X(\xi,0)) \quad \text{in } [0,b] \tag{18.18}$$

where $X(\xi,0)$ is to be recalculated by (18.14).

Not going deeply into detail, let us present here the result of paper [159] by S. I. Shmarev and J. L. Vazquez discussing the regularity properties of interfaces in solutions of the Cauchy problem for the *reaction-diffusion equation* (18.13). We assume that $m > 1$, and that the reaction term $f(u)$ is subject to the following conditions:

(H.1) $f(s) : \mathbb{R}^+ \mapsto \mathbb{R}$;

$$f(0) = 0; \quad f(s) \in C^1([0,\infty)) \cap C^{4+\sigma}_{\text{loc}}(\mathbb{R}^+), \quad \sigma \in (0,1];$$

(H.2) the function

$$\frac{f'(s)}{s^{m-1}} - \frac{f(s)}{s^m}$$

is locally bounded from above in $\mathbb{R}^+$;

(H.3) if $f(s)$ is nonpositive for small s, then there exists s_0 such that $sf'(s) \le (1 + \varphi(s))f(s)$ in $[0,s_0]$, where $\varphi(s) > 0$ in $(0,s_0)$ and $\underline{\lim}_{s\to 0} s^{1-m}\varphi(s) = \infty$;

These conditions are trivially satisfied in the cases when $f(u)$ is a power

$$f(u) = \begin{cases} -u^\lambda & m > 1,\ \lambda > 1, \\ \text{or} & \\ u^\lambda & m > 1,\ \lambda \ge m. \end{cases}$$

Note that the latter case is of special interest because the corresponding solutions *blow up*, that is, become infinite within a finite time.

We assume that the initial data satisfy the conditions

$$u_0(x) > 0 \quad \text{in (0,a)}, \quad u_0 \equiv 0 \quad \text{elsewhere}, \tag{18.19}$$

$$\left\| \left(u_0^{m-1}\right)_x \right\|_{L^\infty(\mathbb{R})} < K, \quad K = \text{const.}, \tag{18.20}$$

$$\left(u_0^{m-1}\right)_x \le -\delta, \quad \delta = \text{const.} > 0, \tag{18.21}$$

a.e. in a neighborhood of the point $x = a$.

Theorem 18.4 ([159]). *Let conditions* (18.19)–(18.21) *be fulfilled and* $f(s)$ *satisfy conditions* (H.1)–(H.3). *Then there exists a classical solution* $u(x,t)$ *of equation* (18.13) *such that*

$$\eta(t) \in C^{1,(m-1)/(3m-1)}_{\text{loc}}(0,T).$$

18.3 A commentary. This is an appropriate moment to explain: why do we need Lagrangian coordinates? Indeed, the questions we are concerned with in this chapter were already given answers for some related equations, so it would be natural to undertake an attempt to employ the familiar techniques instead of introducing coordinates of a new type.

The answer may seem rather technical, but in essence it refers to the different meaning of the time derivative in the coordinates of Euler and Lagrange.

Let us consider the Cauchy problem for PME

$$u_t = (u^m)_{xx}, \quad m > 1. \tag{18.22}$$

This equation describes the behavior of the density u of some continuous medium. If we view this equation as Darcy's law, the pressure in the medium is given by

$$p = \frac{m}{m-1} u^{m-1},$$

and $V = -p_x$ represents the velocity. It is proven that under appropriate conditions on initial data the velocity V is bounded in the problem domain. An equivalent form of equation (18.22) is

$$p_t = (m-1)\, pp_{xx} + p_x^2.$$

The main tool of investigation of the regularity properties of solutions and interfaces in equation (18.22) (see [12, 17]) was the following a priori estimate usually called the Aronson–Bénilan estimate:

$$p_{xx} \geq -\frac{C}{t}, \quad C = \frac{1}{m+1}, \tag{18.23}$$

first derived in [12]. The same estimate happens to be true for a number of related equations for which similar studies of the regularity properties of solutions were undertaken, say, for the equations

$$\begin{aligned} u_t &= (u^m)_{xx} + (u^\lambda)_x, \quad m > 1,\ \lambda \geq m, \\ u_t &= (u^m)_{xx} + f(u), \quad f(u) \equiv -u^\lambda, \quad m > 1,\ \lambda \geq m, \end{aligned} \tag{18.24}$$

see [64] and [67]. However, the arguments that had worked perfectly in application to PME failed in the case of equation (18.24) in the range of the exponent $1 < \lambda < m$: the familiar procedure to derive estimate (18.23) for the "pressure" p proved impossible. This was all the more surprising since none of the known qualitative properties of the solutions to (18.24) with $\lambda > 1$ indicated that the differential properties in the cases $\lambda < m$ and $\lambda \geq m$ might be different.

The Lagrangian coordinates seem to be the most natural to use in deriving estimates of this sort. It is shown in [159] that the Lagrangian counterpart of (18.24) can be given the form

$$\frac{v_t}{v} - \frac{f(v)}{v} = -Av\frac{\partial V}{\partial \xi} = -\frac{\partial V}{\partial x} = \frac{\partial^2 p}{\partial x^2}. \tag{18.25}$$

The basic estimate in the study of the regularity of solutions in [159] is similar to the Aronson–Bénilan estimate (18.23). The point is that instead of estimating the right-hand side of (18.25) one can obtain an estimate of the form

$$\frac{(v^m)_t}{v^m} \geq -\frac{L_0}{t} - L_1 \quad \text{in } Q.$$

It is worth noting that for the equations under study a similar estimate in Eulerian coordinates,

$$\frac{\partial u^m}{\partial t} \geq -(C_2 + \frac{K_2}{t})u^m,$$

was obtained earlier by F. Simondon[1]. However, this formula does not allow for an estimate of p_{xx} as in (18.25) due to the different meaning of the time derivative. Indeed, in Eulerian coordinates

$$p_{xx} = \frac{(u^m)_t}{mu^m} - \frac{f(u)}{u} - \frac{p_x^2}{(m-1)p},$$

and the last term is not controlled from below, hence the need for the Lagrangian approach which dispenses with this term by incorporating it into the total time derivative.

19 Analysis of the problem in Lagrangian terminology

This section contains the proof of solvability of problem ($L_{m,\lambda}$).

19.1 Regularization. A classical solution to problem ($L_{m,\lambda}$) can be constructed as the pointwise limit of a sequence of functions $\{w_n(\xi, t)\}$ defined by the following conditions:

$$Lw_n \equiv w_{nt} - mw_n^{\alpha(m+1)} w_{n\xi\xi} + (1-\lambda)w_n^{\lambda\alpha} w_{n\xi}(1 + m\alpha w_{n\xi}) = 0 \quad \text{in } Q; \qquad (19.1)$$

$$\alpha = \frac{1}{m - \lambda + 1},$$

$$w_n = \frac{1}{n} \quad \text{as } \xi = 0,\ \xi = b; \qquad (19.2)$$

$$w_n(\xi, 0) = w_{0n}(\xi) \quad \text{as } \xi \in [0, b]. \qquad (19.3)$$

We will term this problem (L_n). The operator L in (19.1) is obtained from (18.3) by the formal replacement of $v(\xi, t)$ by $w_n^\alpha(\xi, t)$. We choose the sequence $\{w_{0n}\}$ approximating $w_0 \equiv v_0^{1/\alpha}$ by the conditions

[1] F. Simondon "Strong solutions and stabilization for a degenerate parabolic problem", to appear.

$$\text{(G.1)} \quad \begin{cases} w_{0n} \in C^\infty[0,b], & w_{0n}^\alpha \searrow v_0 \text{ as } n \to \infty, \\ \frac{1}{n} \le w_{0n} \le 1 + M_0, & M_0 \equiv \max_{[0,b]} v_0^{1/\alpha}, \\ |w_{0n}'| \le M_1, & M_1 = \text{const.} > 0; \end{cases}$$

$$\text{(G.2)} \quad \begin{cases} w_{0n} = \frac{1}{n} + \frac{1}{m\alpha}(b-\xi) \quad \text{in a neighborhood of the point } \xi = b; \\ w_{0n} \equiv \frac{1}{n} \quad \text{in a neighborhood } S_0^{(n)} \text{ of the point } \xi = 0; \\ \text{there exist neighborhoods } V_0 \text{ and } V_1 \text{ of the points } \xi = 0 \text{ and } \xi = b \\ \text{such that } w_{0n} \text{ are strictly monotone in } V_1 \text{ and } V_0 \setminus S_0^{(n)}; \end{cases}$$

(G.3) $w_{0n}'(\xi) \ge -1/m\alpha$ as $\xi \in [0, b]$.

Let us show a way to construct such a sequence. Define the function

$$U_{0n}(\xi) = \max\left\{\frac{1}{n};\ w_0(\xi)\right\}.$$

By (G.1), the graph of U_{0n} has horizontal parts adjoining the points $\xi = 0$, $\xi = b$ and the length of each of these parts is not less than q/n, where q is the Lipschitz constant of w_0. Pick an integer n_0 so large that $1 + \frac{1}{m\alpha} < n_0 M_0$ and fix $n \ge n_0$. Let ξ_n be the maximum of the points where

$$U_{0n}(\xi) = \frac{1}{n}\left(1 + \frac{1}{m\alpha}\right).$$

Now set

$$\overline{U}_{0n}(\xi) = \begin{cases} \frac{1}{n} & \text{for } \xi \in (-\infty, 0), \\ U_{0n}(\xi) & \text{for } \xi \in [0, \xi_n), \\ \frac{1}{n}(1 + \frac{1}{m\alpha}) & \text{for } \xi \in [\xi_n, b - \frac{1}{n}), \\ \frac{1}{n} + \frac{1}{m\alpha}(b - \xi) & \text{for } \xi \in [b - \frac{1}{n}, b + \frac{m\alpha}{2n}), \\ \frac{1}{2n} & \text{for } \xi \in [b + \frac{m\alpha}{2n}, \infty) \end{cases}$$

and define for $\xi \in [0, b]$

$$w_{0n}(\xi) = \int_{\mathbb{R}} \omega_{m\alpha/4n}(\xi - \zeta)\overline{U}_{0n}(\zeta)\,d\zeta,$$

where ω_ε is the standard even averaging kernel of the form

$$\omega_\varepsilon(s) = C_\varepsilon \begin{cases} \exp\left(-\frac{1}{1-(s/\varepsilon)^2}\right) & \text{as } |s| < \varepsilon, \\ 0 \quad \text{as } |s| \ge \varepsilon \end{cases}$$

with the constant C_ε chosen by the condition $\|\omega_\varepsilon\|_{L^1(R)} = 1$. It easy to verify that such a sequence satisfies all the properties listed in (G.1)–(G.3). Note also that the initial and boundary data of problems (L_n) satisfy the infinite-order compatibility conditions at the corner points of Q.

By the maximum principle, each of the possible classical solutions of (L_n) can take values only in the interval $[\frac{1}{n}, M_0 + \frac{1}{n}]$. But in this range of values equation (19.1) can be viewed as a *nondegenerate uniformly parabolic* equation, which indeed has a unique classical solution $w_n(\xi, t)$ [97, Ch. 5, 6], [52, Ch. 3]. Referring to [97, Ch. 5] we get the enclosure

$$w_n \in C^\infty\left(\overline{Q}\right).$$

These functions constitute the sequence $\{w_n\}$ which, also by the maximum principle, is monotone decreasing at each point of $\overline{Q}$.

Our immediate aim is to show that the sequence $\{w_n\}$ converges to a function $w(\xi, t)$ which, indeed, is a classical solution of problem ($L_{m,\lambda}$).

Notation. Throughout the chapter we denote

$$B_\tau = Q \cap \{t = \tau\}, \quad \tau \in (0, T], \quad B_0 = (0, b) \times \{0\},$$

$$\partial Q \text{ is the lateral boundary of } Q,$$

$$\Gamma = \partial Q \cup B_0 \text{ is the parabolic boundary of } Q.$$

19.2 Level curves. Positivity properties of solutions. We shall rely on the following fact from the theory of functions of several real variables. Let $f(\xi, t) \in C^\infty\left(\overline{Q}\right)$ be an arbitrary function, positive in $\overline{Q}$. Set

$$\inf_Q f = \mu > 0, \quad \sup_Q f = M < \infty$$

and denote

$$\Lambda_\varepsilon(f) = \{(\xi, t) \in Q : \; f(\xi, t) = \varepsilon\}.$$

Lemma 19.1. *For a.e. $\varepsilon \in (\mu, M)$ the set $\Lambda_\varepsilon(f)$ is constituted by smooth curves; each of these curves is either a closed curve in $Q \setminus B_T$, or begins and ends at $\Gamma \cup B_T$.*

The proof can be found in [152]. In application to the solutions of uniformly parabolic equation (19.1) the assertion of Lemma 19.1 can be specified as follows.

Proposition 19.1. *Let $l_\varepsilon(w_n) \in \Lambda_\varepsilon(w_n)$. If $l_\varepsilon(w_n)$ begins in B_t, it can cross each of B_τ with $\tau \in (0, t)$ at most once.*

Proof. Contrary to Proposition 19.1, let us assume the existence of a curve $l_\varepsilon(w_n) \in \Lambda_\varepsilon(w_n)$ contained in $Q \cap \{t < \tau\}$ and connecting two points of B_τ. Then, by the "weak" maximum principle, [52, Ch. 2], $w_n \equiv \varepsilon$ everywhere in the domain bounded by the curve $l_\varepsilon(w_n)$ from below and having the top face at the line B_τ, which is impossible due to the "strong" maximum principle [52, Ch. 2]. □

Having had recourse to Proposition 19.1, one can show now that the monotonicity of w_{0n} in ξ (condition (G.2)) is preserved for all $t \in (0, T]$.

Lemma 19.2. *There exist an integer n_0 and positive constants θ_0, θ_1 such that for all $n \geq n_0$*

$$w_{n\xi} \leq 0 \quad in\ \overline{Q^{(1)}}, \quad Q^{(1)} = Q \cap \{\xi \geq \theta_1\}, \tag{19.4}$$

$$w_{n\xi} \geq 0 \quad in\ \overline{Q^{(0)}}, \quad Q^{(0)} = Q \cap \{\xi \leq \theta_0\}. \tag{19.5}$$

The proof is based on the following assertions.

Lemma 19.3.

$$\forall n \in \mathbb{N} \quad \max_{\overline{Q}} |w_{n\xi}| \leq 1 + \max\left\{\frac{1}{m\alpha}; M_1\right\}. \tag{19.6}$$

Proof. Introduce the functions

$$\omega_1 = \frac{1}{n} + \frac{1}{m\alpha}(b - \xi), \quad \omega_0 = \frac{1}{n} + M_2\xi.$$

It is easy to verify that

$$L\omega_i \geq 0 \quad \text{in } Q, \quad \omega_i \geq w_n \text{ on } \Gamma, \quad i = 1, 2.$$

Thus, by the maximum principle,

$$\omega_i \geq w_n \quad \text{in } \overline{Q}. \tag{19.7}$$

Inequalities (19.7) are equivalent to the following: for each $t_0 \in (0, T]$

$$w_n(\xi, t_0) - \frac{1}{n} \geq M_2\xi, \quad \frac{1}{n} - w_n(\xi, t_0) \geq -\frac{1}{m\alpha}(b - \xi).$$

Letting here $\xi \to 0$, $\xi \to b$, we get

$$w_{n\xi}(b, t_0) \geq -\frac{1}{m\alpha}, \quad w_{n\xi}(0, t_0) \leq M_2. \tag{19.8}$$

Furthermore,

$$w_{n\xi}(b, t_0) \leq 0, \quad w_{n\xi}(0, t_0) \geq 0$$

since each point of ∂Q realizes for w_n its minimal value in $\overline{Q}$. Combining these inequalities with (19.8), we get

$$|w_{n\xi}| \leq M_2 \quad \text{at } \partial Q. \tag{19.9}$$

Now that the estimate for $t = 0$ is already contained in condition (G.1), (19.9) holds everywhere on Γ.

It follows from (19.1) that the function $V_n(\xi, t) = 1 + m\alpha w_{n\xi}$ satisfies the uniformly parabolic equation

$$V_{nt} - m w_n^{\alpha(m+1)} V_{n\xi\xi} - w_n^{\lambda\alpha}[(m - 2\lambda + 1)V_n - (m + \lambda)]V_{n\xi} + \frac{\lambda(1-\lambda)}{m} w_n^{\alpha\lambda - 1}(V_n - 1)^2 V_n = 0 \quad \text{in } Q.$$

By the maximum principle, all the points of local positive maximum and negative minimum of V_n can be located only on Γ, and the desired estimate (19.6) follows from (19.9). □

Corollary 19.1.

$$w_{n\xi}(\xi,t) \geq -\frac{1}{m\alpha} \quad in\ \overline{Q}. \tag{19.10}$$

Corollary 19.2. *Estimates* (19.6) *mean the Lipschitz equicontinuity of the sequence* $\{w_n\}$ *in* ξ. *The results about the character of continuity of solutions to parabolic equations* [93, 62] *imply that* $\{w_n\}$ *is Hölder equicontinuous in* t. *It then follows from the Ascoli–Arzelà theorem that the sequence* $\{w_n\}$ *contains a subsequence converging to some function* $w(\xi,t) \in C\left(\overline{Q}\right)$. *(By agreement we will assume that this subsequence merely coincides with the origin* $\{w_n\}$, *otherwise we may confine our considerations to this subsequence.) Thus, we obtain the function*

$$w(\xi,t) \equiv \lim_{n\to\infty} w_n(\xi,t) \quad in\ \overline{Q}.$$

Lemma 19.4. *For each* $\tau \in (0,T]$, $\delta > \max\{1; (1-\lambda)/\lambda\}$

$$I \equiv \int_0^b \frac{\xi^\delta(b-\xi)^\delta}{w_n^\alpha(\xi,\tau)}\,d\xi + \int_{Q\cap\{t<\tau\}} \frac{\xi^\delta(b-\xi)^\delta}{w_n^\alpha(\xi,\tau)}\,d\xi\,dt \leq C \tag{19.11}$$

uniformly in n.

Proof. Let us rewrite equation (19.1) in the form

$$\left(\frac{1}{w_n^\alpha}\right)_t + \left(\left(w_n^{\alpha m}\right)_\xi + w_n^{\alpha(\lambda-1)}\right)_\xi = 0, \tag{19.12}$$

multiply it by the function $\psi(\xi,t) = (\tau+1-t)\xi^\delta(b-\xi)^\delta$, and integrate by parts in $Q\cap\{t<\tau\}$:

$$I = (\tau+1)\int_0^b \frac{\xi^\delta(b-\xi)^\delta}{w_{0n}^\alpha(\xi)}\,d\xi - \int_{Q\cap\{t<\tau\}} \psi_\xi w_n^{\alpha(\lambda-1)}\left(1+m\alpha w_{n\xi}\right)d\xi\,dt. \tag{19.13}$$

By virtue of (18.6) and due to the choice of w_{0n}, the first term of the right-hand side of (19.13) is bounded from above uniformly in n. Taking into account (19.7) and applying Young's inequality, we estimate the second term as follows:

$$\left|\int_{Q\cap\{t<\tau\}} \psi_\xi w_n^{\alpha(\lambda-1)}\left(1+m\alpha w_{n\xi}\right)d\xi\,dt\right| \leq \varepsilon(1+M_2)\int_{Q\cap\{t<\tau\}} \frac{-\psi_t}{w_{0n}^\alpha}\,d\xi\,dt$$
$$+\,C(\varepsilon,\lambda)(1+M_2)\int_{Q\cap\{t<\tau\}} \frac{|\psi_\xi|^{1/\lambda}}{(-\psi_t)^{(1-\lambda)/\lambda}}\,d\xi\,dt. \tag{19.14}$$

A straightforward computation shows that

$$\int_{Q\cap\{t<\tau\}} \frac{|\psi_\xi|^{1/\lambda}}{(-\psi_t)^{(1-\lambda)/\lambda}}\, d\xi\, dt$$
$$\leq \tau(1+T)^{1/\lambda}\delta^{1/\lambda}b^{(2\delta+(2\delta-1)(1-\lambda))/\lambda} B(\delta - \frac{1}{\lambda}+1, \delta - \frac{1}{\lambda}+1) \equiv C_0$$

where $B(\cdot,\cdot)$ is Euler's function. Then it follows from (19.13)–(19.14) that

$$I \leq \varepsilon(1+M_2)I + C_0,$$

and we arrive at (19.11), letting $\varepsilon = (1+M_2)/2$. □

Corollary 19.3. *There exists $\varepsilon_0(T) > 0$ such that each B_τ, $\tau \in [0,T]$, contains at least one point $\xi(n)$ where $w_n(\xi(n),\tau) > \varepsilon_0$. (Had we assumed the contrary, we would get a contradiction to* (19.11).*)*

We are now in the position to prove Lemma 19.2.

Proof. Recall that $w_{0n} \to w_0$ uniformly in $C[0,b]$, w_0 is strictly monotone as $\xi \to 0$, $\xi \to b$, and is strictly positive in $(0,b)$. These properties allow one to choose some $\delta_0 \in (0,\varepsilon_0)$ so that for all integers n, beginning with some n_*, w_{0n} would take values in the interval $I \equiv (\delta_0/4, \delta_0/2)$ only if $\xi \in V_0 \setminus S_n^{(0)}$ or $\xi \in V_1$ (cf. (G.2)). Let $l_\varepsilon(w_n)$, $\varepsilon \in I$, be an arbitrary level curve of w_n beginning in V_1. This curve can terminate only in B_T or in $V_0 \setminus S_n^{(0)}$. Observe that in our choice of ε the second of these two possibilities is not realized. Indeed: consider the domain bounded by the vertical lines $\xi = b$, $\xi = 0$, the curve $l_\varepsilon(w_n)$, and the parts of B_0 connecting the points $\xi = 0$, $\xi = b$ to the points where $l_\varepsilon(w_n)$ intersects B_0; by the maximum principle, everywhere in this domain $\frac{1}{n} \leq w_n \leq \varepsilon < \varepsilon_0$, which contradicts Corollary 19.3. Thus, each curve $l_\varepsilon(w_n) \subset \Lambda_\varepsilon(w_n)$ with $\varepsilon \in I$ connects $V_1 \cup \left(V_0 \setminus S_n^{(0)}\right)$ to B_T. Let us check that the level curves of w_n cannot accumulate to ∂Q as $n \to \infty$. By Lemma 19.3, everywhere along each curve $l_\varepsilon(w_n)$, $\varepsilon \in I$, exiting from V_1,

$$w_n - \frac{1}{n} \equiv \varepsilon - \frac{1}{n} \leq M_2(b-\xi).$$

By the choice of the interval I, $\varepsilon - \frac{1}{n} \geq \frac{\delta_0}{4} - \frac{1}{n}$. Hence, all level curves of w_n are bounded away from the vertical line $\xi = b$ uniformly in $\varepsilon \in I$ and $n > n_*$. Now set

$$\theta_1 = \sup\{\xi \in (0,b) : w_n(\xi,t) \in I,\ n > n_*,\ t \in [0,T]\}.$$

Likewise, all level curves of w_n with $\varepsilon \in I$ and $n > n_*$, exiting from $V_0 \setminus S_n^{(0)}$, continue up to B_T and are, uniformly in $\varepsilon \in I$ and n, bounded away from the lateral boundary of Q $\xi = 0$. Thus, we can set

$$\theta_0 = \inf\{\xi \in (0,b) : w_n(\xi,t) \in I,\ n > n_*,\ t \in [0,T]\}.$$

The sets $Q^{(0)}$ and $Q^{(1)}$ are defined. Let us verify that each w_n with n large enough is monotone with respect to ξ in $Q^{(0)}$ and $Q^{(1)}$. For example, let us consider $Q^{(1)}$. Given some $\theta \in (0, T]$, let us assume the existence of points $\theta_1 < \xi_1 < \xi_2 < \xi_3$ such that $w_n(\xi_1, \theta) < w_n(\xi_2, \theta)$, $w_n(\xi_3, \theta) < w_n(\xi_2, \theta)$. By continuity, there exist level curves of w_n $l_\alpha(w_n)$, $l_\beta(w_n)$, $l_\gamma(w_n)$ such that $\alpha < \beta, \gamma < \beta$, but

$$\xi_\alpha \equiv B_\theta \cap l_\alpha(w_n) < \xi_\beta \equiv B_\theta \cap l_\beta(w_n) < \gamma \equiv \xi_\gamma \equiv B_\theta \cap l_\gamma(w_n).$$

By virtue of Proposition 19.1, these three curves connect B_θ to V_1 and, also,

$$l_\alpha(w_n)|_{t=0} < l_\beta(w_n)|_{t=0} < l_\gamma(w_n)|_{t=0},$$

which contradicts (G.2). □

Corollary 19.4. *The limit function $w(\xi, t)$ is strictly positive in Q.*

Assuming the contrary, there exists a point (ξ_0, t_0) where w vanishes. Then there exists a sequence of level curves $\{l_{\varepsilon_n}(w_n)\}$ such that

$$\varepsilon_n \to 0, \quad \eta_n \equiv B_0 \cap l_{\varepsilon_n}(w_n) \to \xi_0 \quad \text{as } n \to \infty.$$

Without loss of generality one can assume that each $l_{\varepsilon_n}(w_n)$ starts, say, in V_1. Then it follows from the maximum principle that $w_n \in \left(\frac{1}{n}, \varepsilon_n\right)$ everywhere in the domain bounded by the vertical line $\xi = b$, the curve $l_{\varepsilon_n}(w_n)$ (monotone in t), and the part of B_0 that connects the point $\xi = b$ to the point of intersection of $l_{\varepsilon_n}(w_n)$ and B_0. However, this is impossible since

$$\infty > C \geq \int_{\eta_n}^{b} \frac{\xi^\delta (b-\xi)^\delta}{w_n^\alpha(\xi, t_0)}\, d\xi \geq \frac{1}{\varepsilon_n^\alpha} \int_{\eta_n}^{b} \xi^\delta (b-\xi)^\delta\, d\xi \to \infty \quad \text{as } n \to \infty.$$

19.3 Solvability of the problem in Lagrangian formulation. We are now in a position to prove Theorem 18.1. Let us take an arbitrary function $f(\xi, t) \in C^\infty\left(\overline{Q}\right)$ with support bounded away from Γ, multiply equation (19.12) by f, and integrate by parts in Q:

$$\int_Q \left\{ \frac{f_t}{w_n^\alpha} + f_\xi \left(\left(w_n^{\alpha m}\right)_\xi + w_n^{\alpha(\lambda-1)} \right) \right\} d\xi\, dt = 0. \tag{19.15}$$

It follows from the uniform convergence of w_n to w, and the strict positivity of w in Q, that in the first and third terms of (19.15) one can pass to the limit as $n \to \infty$ since they already have integrable majorants. Given n fixed, let us multiply (19.12) by $w_n^{\alpha m}$ and integrate by parts in Q. We have:

$$\begin{aligned}
\left\| \left(w_n^{\alpha m}\right)_\xi \right\|_{L^2(Q)}^2 &= \int_0^T \left[\left(\left(w_n^{\alpha m}\right)_\xi + w_n^{\alpha(\lambda-1)} \right) w_n^{\alpha m} \right]_0^b dt \\
&\quad - \frac{1}{\alpha(\lambda-1)} \int_0^b w_n^{\alpha(m-1)}(\xi, t)\, d\xi + \frac{1}{\alpha(m-1)} \int_0^b w_{0n}^{\alpha(m-1)}(\xi)\, d\xi \\
&\leq \frac{(1+M_0)^{\alpha m}}{\alpha(m-1)} a.
\end{aligned} \tag{19.16}$$

Uniform estimate (19.16) means that the sequence $\{(w_n^{\alpha m})_\xi\}$ contains a subsequence converging weakly in $L^2(Q)$ to some function U. Next, by virtue of the uniform convergence $w_n \to w$,

$$U \equiv (w^{\alpha m})_\xi \quad \text{in } \mathcal{D}'(Q).$$

Passing in (19.15) to the limit by the chosen subsequence we get the integral identity for the limit function w and may conclude that in each subdomain Q' of Q, bounded away from Γ by a positive distance, w can be treated as a weak solution of the linear uniformly parabolic equation

$$w_t = mw^{\alpha(m+1)}w_{\xi\xi} - (1-\lambda)w^{\lambda\alpha}w_\xi(1+m\alpha w_\xi)$$

with continuous and bounded coefficients. It follows from results about the interior smoothness of solutions to linear parabolic equations [97, Ch. 5] that $w(\xi, t) \in C^\infty(Q'')$ in each subdomain Q'' of Q', bounded away from the parabolic boundary of Q'.

It remains to observe that the initial and boundary conditions for w are fulfilled by the choice of the approximating sequence of data. The positivity and monotonicity properties of w are stated in Lemma 19.2 and Corollary 19.4.

The proof of Theorem 18.1 is completed. □

20 Uniform estimates. The inverse transformation

To verify invertibility of the coordinate transformation, and to prove Theorem 18.3, we need special estimates for the constructed solution w of problem $(\mathrm{L}_{m,\lambda})$. We deduce them by stating the corresponding uniform in n estimates for the solutions of problems (L_n).

Let us introduce the functions

$$\omega_n(\xi, t) \equiv w_n^\alpha\left((w_n^{\alpha m})_\xi + w_n^{\alpha(\lambda-1)}\right)_\xi, \quad n = 1, 2, \ldots.$$

By virtue of (19.1), $\omega_n = \alpha w_{nt}/w_n$, whence

$$\omega_n = 0 \quad \text{on } \Gamma. \tag{20.1}$$

Fix some

$$\xi_1 \in \left[\theta_1, \max\left\{0; b - T^{1/2}\right\}\right)$$

and denote

$$Q^{(3)} = Q \cap \{\xi > \xi_1\}.$$

Lemma 20.1. *For all integers n, beginning with some n_1,*

$$\omega_n \geq -\frac{CT}{(\xi-\xi_1)^2 t} \quad \text{in } \overline{Q^{(3)}}, \tag{20.2}$$

with a finite constant $C > 0$ not depending on n.

Proof. Recall that $w_{n\xi} \le 0$ in $\overline{Q^{(3)}} \subset \overline{Q^{(1)}}$. Let us consider the function

$$\phi(\xi, t) = -\frac{CT}{t(\xi - \xi_1)^2}.$$

Obviously, $\phi < \omega_n$ on the parabolic boundary of $Q^{(3)}$. It satisfies in $Q^{(3)}$ the equation

$$\begin{aligned} M_n\omega_n \equiv \omega_{nt} - m w_n^{\alpha(m+1)}\omega n\xi\xi - w_n{}^{\lambda\alpha}\left(\alpha m^2 w_{n\xi} + \lambda - 1\right)\omega_{n\xi} \\ - (1-\lambda)w_n^{\lambda\alpha-1}w_{n\xi}\omega_n - (1+m)\omega_n^2 = 0. \end{aligned}$$

Let us calculate $M_n\phi$:

$$\begin{aligned} M_n\phi = \frac{CT}{t^2(\xi-\xi_1)^2} - \frac{6mCT}{(\xi-\xi_1)^4}w_n^{\alpha(m+1)} - (1+m)\frac{C^2T^2}{t^2(\xi-\xi_1)^4} \\ + (1-\lambda)w_n^{\lambda\alpha-1}\frac{CTw_{n\xi}}{t(\xi-\xi_1)^2} - \frac{2CT}{t(\xi-\xi_1)^3}w_n^{\alpha\lambda}\left(\alpha m^2 w_{n\xi} + \lambda - 1\right). \end{aligned}$$

Taking into account the choice of ξ_1 and the preceding a priori estimates, it is easy to derive from this that

$$\begin{aligned} M_n\phi &\le \frac{CT}{t^2(\xi-\xi_1)^2}\Bigg(1 - \frac{(1+m)C}{2} \\ &\quad + \left(2\frac{w_n^{\alpha\lambda}}{\xi-\xi_1}\left(1-\lambda-\alpha m^2 w_{n\xi}\right) - (1+m)\frac{CT}{2(\xi-\xi_1)^2}\right)\Bigg) \\ &< \frac{CT}{t^2(\xi-\xi_1)^4}\left[2\left(1-\lambda+\alpha m^2\right)b^{1+\lambda\alpha} - \frac{(1+m)Cb^2}{2}\right] \\ &\quad + \frac{CT}{t^2(\xi-\xi_1)^2}\left(1 - \frac{(1+m)C}{2}\right). \end{aligned}$$

Thus, $M_n\phi < 0$ in $Q^{(3)}$ provided

$$C > \max\left\{\frac{2}{1+m}; \frac{4(1-\lambda+\alpha m^2)}{1+m}b^{\lambda\alpha-1}\right\},$$

and (20.2) follows by the maximum principle. □

Corollary 20.1. *The concrete values of C and ξ_1 are unimportant, which is why we can take $\xi_2 = (b+\xi_1)/2$, set*

$$Q_1 = Q^{(3)} \cap \{\xi > \xi_2\},$$

and instead of (20.2) *use the estimate*

$$\omega_n \ge -\frac{L}{t} \quad \text{in } \overline{Q_1}, \quad L = \frac{4CT}{(b-\xi_1)^2}. \tag{20.3}$$

We proceed to study some properties of the functions

$$W_n(\xi, t) = w_n^{\alpha(\lambda-1)}(1 + m\alpha w_{n\xi}).$$

By (19.1) and (19.2)

$$\frac{\partial W_n}{\partial \xi} = 0 \quad \text{on } \partial Q, \tag{20.4}$$

and everywhere in $\overline{Q}$

$$\frac{\partial W_n}{\partial \xi} = \alpha \frac{w_{nt}}{w_n^{1+\alpha}} \equiv \frac{\omega_n}{w_n^\alpha}. \tag{20.5}$$

Lemma 20.2. *There exists a bounded constant $\mu > 0$ such that in $\overline{Q^{(3)}}$*

$$\int_0^t W_n(\xi, \theta)\, d\theta \le a + \mu t, \tag{20.6}$$

$$\int_0^t W_n(\xi, \theta)\, d\theta \ge 0 \tag{20.7}$$

uniformly in n.

Proof. Fix some $\xi_* \in (\xi_2, b]$. It follows from (20.5) that for each $t \in [0, T]$ and $n \ge n_1$

$$W_n(\xi_*, t) - W_n(\xi_2, t) = \int_{\xi_2}^{\xi_*} \frac{\omega_n(s, t)}{w_n^\alpha(s, t)}\, ds. \tag{20.8}$$

The strict positivity of $w(\xi, t)$, along with (19.6), imply that $|W_n(\xi_2, t)|$ are bounded uniformly in $n \in \mathbb{N}$, $t \in [0, T]$. Let μ be a constant bounding $|W_n(\xi_2, t)|$. Integrating (20.8) in t, we get:

$$\begin{aligned}
\int_0^t W_n(\xi_*, \theta)\, d\theta &= \int_0^t W_n(\xi_2, \theta)\, d\theta + \int_{\xi_2}^{\xi_*} \left\{ \alpha \int_0^t \frac{w_{nt}}{w_n^{1+\alpha}}\, d\theta \right\} ds \\
&\le \mu t - \int_{\xi_2}^{\xi_*} \left\{ \int_0^t \frac{\partial}{\partial t}\left(\frac{1}{w_n^\alpha}\right) d\theta \right\} ds \\
&= \mu t + \int_{\xi_2}^{\xi_*} \left(\frac{1}{w_{0n}^\alpha(s)} - \frac{1}{w_n^\alpha(s, t)} \right) ds \\
&\le a + \mu t.
\end{aligned} \tag{20.9}$$

Estimate (20.7) follows directly from (19.10). □

Lemma 20.3. *Given $t_0 \in (0, T]$, there exists a constant $d > 0$ not depending on n and such that for each $n > n_1$ the interval $I \equiv (t_0/4, t_0/2)$ contains at least one point t_n where*

$$W_n(b, t_n) \le d.$$

Proof. Let us argue by contradiction. Assuming the contrary, we get: given an arbitrary integer p, there exists an integer $n(p)$ such that $W_{n(p)}(b,t) > p$ for all $t \in I$. Take a function $\varphi(t) \in C_0^\infty(I)$ such that $0 \le \varphi \le 1$, $|\varphi'| \le q/t_0$ with some $q = \text{const.}$ and

$$\varphi(t) = \begin{cases} 1 & \text{if } t \in (5t_0/16, 7t_0/16), \\ 0 & \text{if } t \notin (9t_0/32, 15t_0/32), \end{cases}$$

multiply (20.8) by $\varphi(t)$, and integrate by parts in I. Then

$$\begin{aligned} &\int_0^T W_{n(p)}(b,\theta)\varphi(\theta)\,d\theta \\ &= \int_0^T W_{n(p)}(\xi_2,\theta)\varphi(\theta)\,d\theta + \alpha \int_{\xi_2}^b \int_0^T \varphi(\theta)\frac{w_{nt}(s,\theta)}{w_n^{1+\alpha}(s,\theta)}\,d\theta \\ &= \int_0^T W_{n(p)}(\xi_2,\theta)\varphi(\theta)\,d\theta + \alpha \int_{\xi_2}^b \int_0^T \varphi'(\theta)\frac{d\theta\,ds}{w_n^\alpha(s,\theta)} \\ &\le \mu T + \frac{q}{t_0}\int_{\xi_2}^b\int_0^T \frac{d\theta\,ds}{w_n^\alpha(s,\theta)}. \end{aligned} \tag{20.10}$$

The second term on the right-hand side of (20.10) can be estimated as follows. Integration of equation (19.12) over the rectangle $(\xi_2, b) \times (0,\theta)$,

$$\int_{\xi_2}^b \frac{ds}{w_n^\alpha(s,\theta)} = \int_{\xi_2}^b \frac{ds}{w_{0n}^\alpha(s)} - \int_0^\theta [W_n(s,\tau)]\Big|_{\xi_2}^b\,d\tau,$$

and application of (20.6)–(20.7) lead to the inequality

$$\int_{\xi_2}^b \frac{ds}{w_n^\alpha(s,\theta)} \le a + (a + \mu T) \equiv \gamma. \tag{20.11}$$

Substituting (20.11) into (20.10), we get

$$h(p) \equiv \int_0^T W_{n(p)}(b,\theta)\varphi(\theta)\,d\theta \le \mu T + \alpha\frac{q\gamma T}{t_0} < \infty,$$

which is impossible since

$$h(p) \ge \frac{pt_0}{8} \to \infty \quad \text{as } p \to \infty.$$

□

Lemma 20.4. *For each $t_0 \in (0,T]$ the functions W_n are equibounded from above in the closure of the domain $D(t_0) \equiv Q_1 \cap \{t > t_0/4\}$.*

Proof. Fix an integer n. Let I be the interval introduced in the conditions of Lemma 20.3 and $t_n \in I$ be that value of t where $W_n(b,t_n) \le d$. Let us consider W_n as a solution of the linear uniformly parabolic equation following from (19.1)

$$\mathcal{L}(w_n)W_n \equiv W_{nt} - mw_n^{\alpha(m+1)}W_{n\xi\xi} - \left[m\left(w_n^{\alpha(m+1)}\right)_\xi + (\lambda - 1)w_n^{\lambda\alpha}\right]W_{n\xi} = 0. \tag{20.12}$$

Recalling (20.4) and referring to the parabolic version of Hopf's principle (lemma about the sign of the nontangential derivative) [52, Ch. 2, Th. 14], one has: in $\overline{D(t_0)}$ the function W_n cannot assume its local positive maximum on the line $\xi = b$. Next, $W_n \le \mu$ as $\xi = \xi_2$, and finally, it follows from (20.3), (20.8), (20.11) that

$$W_n(\xi, t_n) = W_n(b, t_n) - \alpha \int_\xi^b \frac{\omega_n(s, t_n)}{w_n^\alpha(s, t_n)}\,ds \le d + \frac{L\gamma}{t_n} \le d + \frac{4L\gamma}{t_0}.$$

By the maximum principle, the solutions of (20.12) attain their maximal values only on the parabolic boundary of $D(t_n)$, whence

$$W_n(\xi, t) \le d + \frac{4L\gamma}{t_0} + \mu \quad \text{in } \overline{D(t_0)}.$$

□

One can justify now the parameter representation for weak solutions of problem ($\mathrm{E}_{m,\lambda}$) asserted in Theorem 18.1. To begin with, let us observe that the inclusion $w \in C^\infty(Q)$ implies as $n \to \infty$

$$W_n(\xi, t) \to W(\xi, t) \equiv w^{\alpha(\lambda-1)}\left(1 + m\alpha w_\xi\right),$$
$$\omega_n(\xi, t) \to \omega(\xi, t) \equiv \alpha w^\alpha W_\xi$$

uniformly on compact subsets of Q; the limit functions W and ω obey estimates (20.3), (20.6)–(20.7) and satisfy Lemma 20.4.

1) Let us check that the function $X(\xi, t)$ defined by (18.9) for $(\xi, t) \in Q$ can be continued to $\overline{Q} \setminus \{(0, 0)\}$ by continuity. It follows from the basic relation $X_\xi = 1/w^\alpha$ and (20.5) that

$$X_\xi = \frac{1}{v_0} - \int_0^t W_\xi d\theta = \frac{1}{w^\alpha} > 0 \quad \text{in } Q.$$

The definition of X and (20.6)–(20.7) imply the uniform boundedness of X in $\overline{Q_1}$. Thus, for each $t \in [0, T]$ there exists

$$\lim_{\xi \to b} X(\xi, t) = X(b, t). \tag{20.13}$$

Next, by (19.5) for each $t > 0$ fixed

$$X(\xi, t) = X(\xi, 0) - \int_0^t w^{\alpha(\lambda-1)}\left(1 + m\alpha w_\xi\right) d\theta$$
$$\le a - \int_0^t w^{\alpha(\lambda-1)}\, d\theta \le a - t\,(M_2\xi)^{\alpha(\lambda-1)} \to -\infty \quad \text{as } \xi \to 0.$$

For each $\xi \in (0, b)$ the function $W(\xi, t)$ is bounded uniformly in t. Therefore, for each $\xi \in (0, b)$ there exists a finite limit

$$X(\xi, 0) = \lim_{t \to 0} X(\xi, t).$$

Fix some $t \in (0, T]$ and consider the function

$$\Phi(\xi) = \delta X(\xi, t) + W(\xi, t), \quad \delta = \text{const.} > 0.$$

Set $\delta = 1 + \frac{L}{t}\alpha$ with the constant L from (20.3). Then $\Phi(\xi)$ becomes monotone, since

$$\Phi'(\xi) = \frac{1}{w^\alpha}[\delta + \alpha\omega(\xi, t)] \geq \frac{1}{w^\alpha} > 0.$$

Due to the boundedness of Φ, for each $t > 0$ the functions $\Phi(\xi)$ and, respectively, $W(\xi, t)$ have finite limits as $\xi \to b$.

Let us adopt the notation

$$W(b, t) = \lim_{\xi \to b} W(\xi, t), \quad t \in (0, T].$$

Differentiating (18.9) we get the equality

$$X_t(\xi, t) = -W(\xi, t) \quad \text{in } Q.$$

As was already shown, for each $t \in (0, T]$ X and W have finite limits as $\xi \to b$. Then, by (20.13), $X(b, t)$ has the weak derivative $X_t(b, t) \in L^2(0, T)$ and

$$X_t(b, t) + W(b, t) = 0 \quad \text{in } \mathcal{D}'(0, T). \tag{20.14}$$

The function $X(b, t)$ is recovered now by integration

$$X(b, t) = a - \int_0^t W(b, \theta)\, d\theta.$$

Also, $X(b, t) \to a$ as $t \to 0$. Finally, $X(\xi, 0) \to a$ as $\xi \to b$ just by definition, and we conclude that $X(\xi, t)$ is continuous in $\overline{Q} \setminus \{(0, 0)\}$.

2) We have shown that $X(\xi, t)$ is continuous in the closure of each domain of the form

$$Q_\varepsilon = Q \cap \{t > \varepsilon\}, \quad \varepsilon \in (0, T).$$

Let $f(x, t)$ be an arbitrary function satisfying the conditions of Definition 17.1. Define the function $\Phi(X(\xi, t), t) \equiv f(x, t)$. Since $X(0, t) = -\infty$ for each $t > 0$, and due to the restrictions on the test functions $f(x, t)$, $f(x, t) \equiv 0$ for all x less than some x_0. Hence, there exists $\xi_* > 0$ such that $\Phi(X(\xi, t), t) \equiv 0$ as $\xi \in (0, \xi_*)$, $t \geq \varepsilon$, and the functions Φ, Φ_X, Φ_t are continuous in $\overline{Q} \setminus \{(0, 0)\}$. Taking the total derivative in t, and making use of the trajectory equation just derived, we get the relation

$$\frac{d\Phi}{dt} = \Phi_t + \Phi_X X_t = \Phi_t - \Phi_X W,$$

whence

$$
\begin{aligned}
-\int_R f(x,\varepsilon)u_0(x)\,dx &= -\int_0^b \Phi(X(\xi,\varepsilon),\varepsilon)\,d\xi \\
&= \int_0^b \int_\varepsilon^T \frac{d}{dt}\left(\Phi(X(\xi,t),t)\right) d\xi\,dt \\
&= \int_{Q_\varepsilon} \{\Phi_t - \Phi_X W\}\,dt\,d\xi. \qquad (20.15)
\end{aligned}
$$

Let us rewrite (20.15) in the form

$$
\int_R f(x,\varepsilon)u_0(x)\,dx + \int_{Q_\varepsilon} \Phi_t\,dt\,d\xi = \int_{Q_\varepsilon} \Phi_X W\,d\xi\,dt \qquad (20.16)
$$

and remark that the definition of $X(\xi,t)$ (18.9), along with equation (18.3) taken in the form (20.5), imply the equality

$$
v(\xi,t)X_\xi(\xi,t) = 1 \quad \text{in } Q.
$$

The left-hand side of (20.16) was already proven to have a finite limit as $\varepsilon \to 0$. Letting $\varepsilon \to 0$ and passing in (20.16) to the variables (x,t) by formulas (18.10), (20.17), we arrive at identity (17.9) for $\overline{u}(x,t)$.

The continuity of $\overline{u}(x,t)$ now follows by the continuity of $X(\xi,t)$ and $v(\xi,t)$. The inclusion $\left(\overline{u}^m\right)_x \in L^2(S)$ is a byproduct of the uniform boundedness of w_ξ in Q, and the initial conditions for $\overline{u}(x,t)$ are fulfilled by the method of construction.

Theorem 18.3 is proved completely in the case of problem $(\mathrm{E}_{m,\lambda})$ with $\lambda \in (0,1)$.

Remark 20.1. In the proof of Theorem 18.3 given in Section 24 we present an alternative way of proving the same assertion.

21 Some starting properties of the interface

We can profit now by representation (18.11) and uniform estimates of Sections 19–20 to prove Theorem 17.1.

1) *Two-sided estimates for the interface location.* As was already established, for each $t \in [0,T]$ the function $X(\xi,t)$ is monotone in ξ. Therefore,

$$
X(\xi,t) \to \eta(t) \quad \text{as } \xi \to b, \qquad X(\xi,t) \to \zeta(t) \quad \text{as } \xi \to 0.
$$

From (18.11) and estimates (20.6)–(20.7) we immediately get the estimate

$$
a \ge \eta(t) \ge -\mu t
$$

with the constant $\mu > 0$ introduced and estimated in the proof of Lemma 20.2.

Remark 21.1. The upper estimate in (17.16) was earlier obtained in [40] by means of comparison of solutions of problem ($E_{m,\lambda}$) with a travelling-wave solution of equation (17.4).

2) *Non-existence of the left-hand interface* is established, in fact, in the proof of Theorem 18.3: for each $t > 0$, $\zeta(t) < X(\xi, t) \to -\infty$ as $\xi \to 0$.

3) *Monotonicity of the interface.* This property follows immediately from representation (18.11) and inequality (19.10). Given arbitrary $t, t + \Delta t \in (0, T]$,

$$\eta(t + \Delta t) - \eta(t) = -\int_t^{t+\Delta t} W(b, \theta)\, d\theta \le 0.$$

4) *Instantaneous loss of initial regularity.* For each $t \in (0, T]$ $W(\xi, t)$ has a finite limit as $\xi \to b$, while $w \to 0$ as $\xi \to b$. Assume that $W(b, t) = D \ge 0$ for some t. Given an arbitrary $\varepsilon > 0$, one can choose $\gamma > 0$ such that

$$|W(\xi, t) - D| < \varepsilon \quad \forall \xi \in (b - \gamma, b)$$

and, correspondingly,

$$\left|1 + m\alpha w_\xi\right| \le (D + \varepsilon) w^{\alpha(1-\lambda)} \le (D + \varepsilon) M_2^{\alpha(1-\lambda)} (b - \xi)^{\alpha(1-\lambda)}.$$

Hence, for each $t \in (0, T]$

$$\lim_{\xi \to b} w_\xi(\xi, t) = -\frac{m - \lambda + 1}{m}, \tag{21.1}$$

whence (17.17) after returning to the original coordinates (x, t).

5) *Lipschitz continuity of the interface* is a combination of "the interface equation" (20.14) and the uniform boundedness of $W(\xi, t)$ on intervals of the form $[\tau, T]$, $\tau > 0$: for all $t, t + \Delta t \in (\tau, T]$

$$|\eta(t + \Delta t) - \eta(t)| = \left|\int_t^{t+\Delta t} W(b, \theta)\, d\theta\right| \le |\Delta t| \max_{\overline{D(t)}} |W(\xi, t)|.$$

6) *Strict monotonicity of the interface.* First of all, let us observe that the upper estimate in (17.16) and estimate (20.7) yield: the curve $x = \eta(t)$ cannot remain vertical for all $t \in [0, T]$. At the same time, the lower estimate in (17.16) means that this curve cannot contain a horizontal part of infinite length. Let us show that the curve $x = \eta(t)$ is constituted by at most two portions: the vertical one, corresponding to the interval $[0, t_*)$ where

$$t_* = \sup\{\tau > 0 : \ W(b, \theta) = 0 \ \forall \theta < \tau\},$$

and the strictly monotone part $[t_*, T]$ where $W(b, t) > 0$.

Assume that $W(b, t_0) \ge D = \text{const.} > 0$ at $t_0 \in (t_*, T)$. Let us check first that W can vanish in $\overline{Q}$ only on Γ. Assuming the contrary, there is a point $(\xi', t') \in Q$ where $W = 0$. Define the domain

$$Q(\varepsilon) = (\varepsilon, b - \varepsilon) \times (\varepsilon, T]$$

with $\varepsilon > 0$ chosen so that $(\xi', t') \in Q(\varepsilon)$. W satisfies in $Q(\varepsilon)$ the equation

$$\mathcal{L}(w)W = 0$$

with the parabolic operator $\mathcal{L}(\cdot)$ defined in (20.12). Due to the strict positivity of w in $Q(\varepsilon)$, this equation can be treated there as a uniformly parabolic one. Since W is nonnegative in $Q(\varepsilon)$, (ξ', t') is a point of local minimum of W in $Q(\varepsilon)$. But then, by the strong maximum principle, $W \equiv 0$ everywhere in $\overline{Q(\varepsilon)} \cap \{t \le t'\}$ which contradicts the limit relation just derived

$$\lim_{\substack{(\xi,t)\in Q_0 \\ (\xi,t)\to(0,t)}} W(\xi, t) = \infty.$$

Thus, $W > 0$ in Q, and $W(b, t_0) \ge D > 0$ by assumption. Now we are going to show that $W(b, t) > 0$ for all $t \ge t_0$. By the continuity of $W(\xi, t_0)$ at the point $\xi = b$, there exists $\rho \in (0, b)$ such that $W(\xi, t_0) \ge D/2$ for all $\xi \in (\rho, b]$. Let us consider the sequence $\{W_n(\xi, t)\}$. From (20.3) we get: for each $\xi_1, \xi_2 \in (\rho, b)$

$$\begin{aligned} W_n(\xi_2, t_0) &\ge W_n(\xi_1, t_0) + \alpha \int_{\xi_1}^{\xi_2} \left.\frac{w_{nt}}{w_n^{1+\alpha}}\right|_{t=t_0} ds \\ &\ge W_n(\xi_1, t_0) - \frac{K_0}{t_0}(b - \xi_1)^{1-\alpha} \end{aligned} \tag{21.2}$$

with an absolute constant $K_0 > 0$ not depending on n. Now fix an integer $n' \ge n_1$ and a point ξ_1 such that

$$\frac{K_0}{t_0}(b - \xi_1)^{1-\alpha} \le \frac{D}{8}, \quad W_n(\xi_1, t_0) \ge \frac{D}{4} \quad \forall n \ge n'.$$

The last choice is always possible since $W_n \to W$ uniformly on each compact subset of Q separated from Γ. From (21.2) we now get:

$$\forall n \ge n', \ \xi \in (\xi_1, b) \quad W_n(\xi, t_0) \ge \frac{D}{8} > 0. \tag{21.3}$$

By the positivity of W in Q, there exists a constant $\theta > 0$ such that $W(\xi_1, t) \ge \theta$ for all $t \in [t_0, T]$ and, respectively, a number $n'' \ge n'$ such that for all $n \ge n''$

$$W_n(\xi_1, t) \ge \frac{\theta}{2} > 0 \quad \text{uniformly in } t \in [t_0, T].$$

We have now: for $n \ge n''$ the functions W_n satisfy equation (20.12), the boundary condition (20.4), and conditions (21.2)–(21.3). Hence, by the maximum principle,

$$W_n(\xi, t) \ge \min\left(\frac{\theta}{2}, \frac{D}{8}\right) > 0 \quad \text{in } [t_0, T] \times [\xi_1, b]. \tag{21.4}$$

Letting $n \to \infty$ in (21.4), we get the desired inequality for the limit function $W(\xi, t)$.

7) *Equation of the interface* merely coincides with (20.14) after the inverse passage to the variables (x, t).

22 Estimates for the time derivative and the higher-order derivatives

22.1 Local variables. As follows from (21.1), $w(\xi, t)$ is strictly monotone near the line $\xi = b$ and hence can be taken for the new independent space variable. Let us introduce the new variable η and the new sought function $y(\eta, t)$ by the relations

$$\begin{cases} b - \xi = y(\eta, t), \\ \eta = w(b - y(\eta, t), t). \end{cases} \tag{22.1}$$

We perform this change in the domain

$$D(t_0) = \left\{(\xi, t) \in Q_1 : t \in (t_0, T],\ w_\xi \in \left(-\frac{2}{m\alpha}, -\frac{1}{m\alpha}\right)\right\},$$

where $t_0 \in (0, T)$ is arbitrary and fixed. Let us derive the differential equation for $y(\eta, t)$. It follows from (22.1) that

$$0 = \frac{d\eta}{dt} \equiv w_t - w_\xi y_t,$$

whence

$$w_t = w_\xi y_t. \tag{22.2}$$

Further,

$$\frac{\partial w}{\partial \xi} \frac{\partial (b - y(\eta, t))}{\partial \eta} = 1, \quad \text{that is,} \quad w_\xi = -\frac{1}{y_\eta}. \tag{22.3}$$

Observe that due to (22.3)

$$y_\eta \to m\alpha \quad \text{as } \eta \to 0. \tag{22.4}$$

Substituting (22.2)–(22.3) into equation (19.1), we find

$$y_t = m\eta^{\alpha(m+1)} \frac{\partial}{\partial \eta}\left(-\frac{1}{y_\eta}\right) - (1 - \lambda)\eta^{\lambda\alpha}\left(1 - \frac{m\alpha}{y_\eta}\right). \tag{22.5}$$

This equation will be considered in the domain

$$H_0 \equiv (0, \eta_0) \times (t_0, T]$$

with η_0 chosen arbitrarily but so as to satisfy the condition $\frac{1}{2}m\alpha \le y_\eta \le 2m\alpha$.

Conforming with the mechanical terminology, one may term this change of variables the passage to the von Mises variable often used in boundary layer theory.

22.2 Preliminary estimation of the time derivative.

Lemma 22.1.

$$\forall t \in (t_0, T] \quad \lim_{\eta \to 0} \eta^{-\lambda\alpha} y_t(\eta, t) = 0. \tag{22.6}$$

For the proof, we need special estimates strengthening inequalities (20.3). By Lemma 20.4, for each $t_0 > 0$ the functions W_n are uniformly bounded in n in the closure of the rectangle $D(t_0)$. Hence, in $\overline{D(t_0)}$ the functions $w_{n\xi}$ are uniformly bounded in n away from zero and infinity (at least, for large n), and

$$-\frac{1}{m\alpha} \leq w_{n\xi}(\xi, t) \leq -\frac{1}{m\alpha} + dn^{-\alpha(1-\lambda)}, \quad d = \text{const.} > 0.$$

Fix an integer n and choose w_n for the new independent space variable. Set

$$\begin{cases} b - \xi = y_n(\eta, t), \\ \eta + \frac{1}{n} = w_n(b - y_n(\eta, t), t). \end{cases}$$

Similarly to (22.2)–(22.3)

$$w_{nt} = w_{n\xi} y_{nt}, \quad w_{n\xi} = -\frac{1}{y_{n\eta}}$$

and from equation (19.1) we derive:

$$y_{nt} - m\left(\frac{1}{n} + \eta\right)^{\alpha(m+1)} \frac{\partial}{\partial \eta}\left(-\frac{1}{y_{n\eta}}\right) + (1-\lambda)\left(\frac{1}{n} + \eta\right)^{\lambda\alpha}\left(1 - \frac{m\alpha}{y_{n\eta}}\right) = 0. \quad (22.7)$$

This equation will also be considered in the domain H_0. Without loss of generality, we assume that η_0 is subject to the following condition:

$$\frac{1}{2}m\alpha \leq y_{n\eta} \leq 2m\alpha \quad \text{as } \eta \in [0, \eta_0]$$

uniformly in n.

Lemma 22.2. *There exists a constant $C_* > 0$ such that*

$$y_t \leq C_* \frac{\eta}{t - t_0} \quad \text{in } H_0, \quad (22.8)$$

$$\frac{\partial}{\partial \eta}(y_t) \leq C_* \frac{1}{t - t_0} \quad \text{in } H_0. \quad (22.9)$$

Proof. 1) Consider the function

$$S(\eta, t) \equiv y_{nt}(\eta, t).$$

It satisfies in H_0 the equation

$$\begin{aligned} \mathcal{P}S \equiv S_t &- m\frac{m}{y_{n\eta}^2}\left(\frac{1}{n} + \eta\right)^{1+\lambda\alpha} \frac{\partial^2 S}{\partial \eta^2} \\ &+ \frac{2}{y_{n\eta}}\left[S + (1-\lambda)\left(\frac{1}{n} + \eta\right)^{\lambda\alpha}\left(1 - \frac{m\alpha}{y_{n\eta}}\right)\right]\frac{\partial S}{\partial \eta} \\ &+ (1-\lambda)\left(\frac{1}{n} + \eta\right)^{\lambda\alpha} \frac{m\alpha}{y_{n\eta}^2}\frac{\partial S}{\partial \eta} = 0 \end{aligned}$$

and the condition $S(0,t)=0$. For $t\in[t_0,T]$ the function $S(\eta_0,t)$ is bounded uniformly in t and n, and $S(t_0,\eta)$ is bounded by a finite constant depending, perhaps, on n. Let us consider the function

$$\theta(\eta,t)=C_1\frac{\eta}{t-t_0}$$

with a constant $C_1>m\alpha$. Taking C_1 as large as we please, we derive that

$$\theta(\eta,t)\geq S(\eta,t)\quad\text{at the parabolic boundary of } H_0 \text{ uniformly in } n\,.$$

Next,

$$\begin{aligned}\mathcal{P}\theta &\equiv C_1\frac{\eta}{(t-t_0)^2}\left[\frac{2C_1}{y_{n\eta}}-1\right]\\ &\quad+C_1\frac{1-\lambda}{t-t_0}\left(\frac{1}{n}+\eta\right)^{\lambda\alpha}\left[\frac{2}{y_{n\eta}}\left(1-\frac{m\alpha}{y_{n\eta}}\right)+\frac{m\alpha}{y_{n\eta}^2}\right]\\ &> C_1\frac{\eta}{(t-t_0)^2}\left[\frac{2C_1}{m\alpha}-1\right]>0\quad\text{in } H_0.\end{aligned}$$

Hence,

$$y_{nt}(\eta,t)\leq C_1\frac{\eta}{t-t_0}\quad\text{in } \overline{H_0},\tag{22.10}$$

and since $y_{nt}\in C^\infty\left(\overline{H_0}\right)$,

$$(y_{nt})_\eta(0,t)=\lim_{\eta\to 0}\frac{1}{\eta}\left[y_{nt}(\eta,t)-y_{nt}(0,t)\right]\leq\frac{C_1}{t-t_0}.\tag{22.11}$$

2) Let us construct now a local upper barrier for the function

$$\sigma(\eta,t)\equiv\frac{(y_{nt})_\eta}{y_{n\eta}^2}.$$

Successively differentiating equation (22.7) in t and η, we get the equation

$$\begin{aligned}\left[(y_{nt})_\eta\right]_t &= m\frac{\partial}{\partial\eta}\left[\left(\frac{1}{n}+\eta\right)^{1+\lambda\alpha}\frac{\partial}{\partial\eta}\left(\frac{1}{y_{n\eta}^2}(y_{nt})_\eta\right)\right]\\ &\quad-(1-\lambda)\left(\frac{1}{n}+\eta\right)^{\lambda\alpha}m\alpha\frac{\partial}{\partial\eta}\left(\frac{1}{y_{n\eta}^2}(y_{nt})_\eta\right)\\ &\quad-(1-\lambda)\lambda\alpha\left(\frac{1}{n}+\eta\right)^{\lambda\alpha-1}\frac{m\alpha}{y_{n\eta}^2}(y_{nt})_\eta\,.\end{aligned}\tag{22.12}$$

With the help of the identity

$$\frac{\left[(y_{nt})_\eta\right]_t}{y_{n\eta}^2}=\left[\frac{(y_{nt})_\eta}{y_{n\eta}^2}\right]_t+\frac{2}{y_{n\eta}^3}(y_{nt})_\eta^2$$

we derive from (22.12) the following equation for the function $\sigma(\eta, t)$:

$$\mathcal{R}\sigma \equiv \sigma_t - \frac{m}{y_{n\eta}^2}\frac{\partial}{\partial\eta}\left[\left(\frac{1}{n}+\eta\right)^{1+\lambda\alpha}\frac{\partial\sigma}{\partial\eta}\right] + (1-\lambda)\frac{m\alpha}{y_{n\eta}^2}\left(\frac{1}{n}+\eta\right)^{\lambda\alpha}\frac{\partial\sigma}{\partial\eta}$$
$$+ (1-\lambda)\lambda\alpha\left(\frac{1}{n}+\eta\right)^{\lambda\alpha-1}\frac{m\alpha}{y_{n\eta}^2}\sigma + 2y_{n\eta}\sigma^2 = 0 \quad \text{in } H_0\,.$$

By virtue of (22.11)

$$\sigma(0,t) \le \frac{4C_1}{m\alpha(t-t_0)} \qquad \text{as } t \in [t_0, T].$$

Next, $\sigma(\eta_0, t)$ is bounded uniformly in $t \in [t_0, T]$ and n, and $\sigma(\eta, t_0)$ is finite. Introduce the function

$$q(t) \equiv \frac{C_2}{t-t_0}$$

with a constant $C_2 > 1/m\alpha$ chosen so that

$$q(t) \ge \sigma(\eta, t) \qquad \text{at the parabolic boundary of } H_0.$$

In H_0,

$$\begin{aligned}\mathcal{R}q(t) &\equiv -\frac{C_2}{(t-t_0^2)} + 2y_{n\eta}\frac{C_2^2}{(t-t_0^2)} \\ &\quad + (1-\lambda)\lambda\alpha\left(\frac{1}{n}+\eta\right)^{\lambda\alpha-1}\frac{m\alpha}{y_{n\eta}^2}\frac{C_2}{t-t_0} \\ &> \frac{C_2}{(t-t_0^2)}[m\alpha C_2 - 1] > 0.\end{aligned} \tag{22.13}$$

Thus, $\sigma(\eta, t) \le q(t)$ in $\overline{H_0}$ and

$$(y_{nt})_\eta(\eta, t) \le \frac{4(m\alpha)^2 C_2}{t-t_0} \quad \text{in } \overline{H_0}. \tag{22.14}$$

Letting $C_* = \max\left(C_1; 4C_2(m\alpha)^2\right)$ and passing in (22.14), (22.10) to the limit as $n \to \infty$, we get (22.8), (22.9). □

Proof of Lemma 22.1. Let us fix arbitrary $t \in (t_0, T]$ and $\sigma \in (0, \eta_0/2)$. The following chain of relations holds:

$$-\left[\frac{1}{y_\eta(2\sigma,t)} - \frac{1}{y_\eta(\sigma,t)}\right] \equiv \int_\sigma^{2\sigma}\left[-\frac{1}{y_\eta(\sigma,t)}\right]d\eta \equiv$$

$$
\begin{aligned}
&\equiv \int_\sigma^{2\sigma} \frac{1}{m\eta^{1+\alpha\lambda}}\left[\eta^{1+\alpha\lambda}\frac{\partial}{\partial\eta}\left[-\frac{1}{y_\eta(\sigma,t)}\right]\right.\\
&\quad \left. - (1-\lambda)\eta^{\lambda\alpha}\left(1-\frac{m\alpha}{y_\eta}\right) - \frac{C_*\eta}{t-t_0}\right]d\eta\\
&\quad + \frac{C_*}{m(t-t_0)}\int_\sigma^{2\sigma}\frac{d\eta}{\eta^{\lambda\alpha}} + \frac{1-\lambda}{m}\int_\sigma^{2\sigma}\frac{1}{\eta}\left(1-\frac{m\alpha}{y_\eta}\right)d\eta\\
&= \int_\sigma^{2\sigma}\frac{1}{m\eta^{1+\alpha\lambda}}\left[y_t - \frac{C_*\eta}{t-t_0}\right]d\eta\\
&\quad + \frac{C_*}{m(t-t_0)}\int_\sigma^{2\sigma}\frac{d\eta}{\eta^{\lambda\alpha}} + \frac{1-\lambda}{m}\int_\sigma^{2\sigma}\frac{1}{\eta}\left(1-\frac{m\alpha}{y_\eta}\right)d\eta\\
&\equiv I_1 + I_2 + I_3.
\end{aligned}
$$

Due to the boundedness of $W(\xi, t)$ in $\overline{D(t_0)}$,

$$
\frac{1}{\eta}\left(1-\frac{m\alpha}{y_\eta}\right) \le C\eta^{\alpha(1-\lambda)-1}
$$

with some finite constant $C > 0$, which yields the inequality

$$
I_3 \le \frac{C}{m\alpha}2^{\alpha(1-\lambda)}\sigma^{\alpha(1-\lambda)}.
$$

Evidently,

$$
I_2 \le \frac{C_*}{m(1-\lambda\alpha)}\frac{1}{t-t_0}2^{1-\lambda\alpha}\sigma^{1-\lambda\alpha}.
$$

Let us estimate I_1. By Lemma 22.2, for each $t \in (t_0, T]$ the function $y_t - C_*\eta/(t-t_0)$ is nonpositive and monotone decreasing as a function of η. Hence ,

$$
\begin{aligned}
I_1 &\le \frac{1}{m}\left[y_t - \frac{C_*\eta}{t-t_0}\right]_{|\eta=\sigma}\int_\sigma^{2\sigma}\frac{d\eta}{\eta^{1+\lambda\alpha}}\\
&= \frac{1}{m}\left[y_t - \frac{C_*\eta}{t-t_0}\right]_{|\eta=\sigma}\frac{1-2^{-\lambda\alpha}}{\lambda\alpha}\sigma^{-\lambda\alpha}\\
&\le 4\frac{1-2^{-\lambda\alpha}}{m\lambda\alpha}\sigma^{-\lambda\alpha}y_t(\sigma,t).
\end{aligned}
$$

Combining the estimates for $I_1 - I_3$, we finally get:

$$
\begin{aligned}
\sigma^{-\lambda\alpha}y_t(\sigma,t) \ge &-\frac{C_*}{t-t_0}\frac{\lambda\alpha}{1-\lambda\alpha}\frac{2^{1-\lambda\alpha}}{1-2^{-\lambda\alpha}}\sigma^{1-\lambda\alpha}\\
&- \frac{1-\lambda}{m}\frac{2^{(1-\lambda)\alpha}}{1-2^{-\lambda\alpha}}C\sigma^{\alpha(1-\lambda)}\\
&- \frac{m\alpha\lambda}{1-2^{-\lambda\alpha}}\left[\frac{1}{y_\eta(2\sigma,t)} - \frac{1}{y_\eta(\sigma,t)}\right] \to 0 \quad \text{as } \sigma \to 0.
\end{aligned}
$$

On the other hand, it directly follows from (22.8) that

$$\sigma^{-\lambda\alpha} y_t(\sigma, t) \le \frac{C_*}{t - t_0}\sigma^{1-\lambda\alpha} \to 0 \quad \text{as } \sigma \to 0.$$

These limit relations imply (22.6). □

22.3 An improved estimate. Relying on the results of the previous subsection, we are now in position to prove that, actually,

$$y_t(\eta, t) = \mathcal{O}(\eta) \quad \text{as } \eta \to 0, \tag{22.15}$$

and even

$$y_t(\eta, t) = \mathcal{O}(\eta^{1+\alpha(1-\lambda)}) \quad \text{as } \eta \to 0 \tag{22.16}$$

if $t > t_*$. For the sake of brevity we always suppose that the value of η_0 is chosen satisfying the following conditions: everywhere in $\overline{H_0}$

1) $\left|\frac{2}{y_\eta}\left[y_t \eta^{-\lambda\alpha} + (1-\lambda)\left(1 - \frac{m\alpha}{y_\eta}\right)\right]\right| \le \frac{1-\lambda}{8m\alpha}$;

2) $|y_t \eta^{-\lambda\alpha}| \le \frac{1-\lambda}{32}$;

These conditions on η_0 are easy to satisfy with the help of (22.4) and (22.6).

Define the parabolic operator

$$Ph(\eta, t) \equiv h_t - m\frac{\eta^{1+\lambda\alpha}}{y_\eta^2} h_{\eta\eta} \tag{22.17}$$

$$+ \frac{2}{y_\eta}\left[y_t + (1-\lambda)\eta^{\lambda\alpha}\left(1 - \frac{m\alpha}{y_\eta}\right)\right] h_\eta + (1-\lambda)\eta^{\lambda\alpha}\frac{m\alpha}{y_\eta^2} h_\eta.$$

Evidently, y_t satisfies in H_0 the equation

$$P\,(y_t) = 0$$

following from (22.5) by differentiation in t.

Lemma 22.3. *Let $q(\eta, t)$ be a function satisfying the conditions*

a)

$$Pq(\eta, t) = f(\eta, t) \quad \textit{in } H_0$$

where

$$|f(\eta, t)| \le C_0 \eta^{\alpha(2-\lambda)} \quad \textit{in } \overline{H_0},\ C_0 = \text{const.} > 0, \tag{22.18}$$

b) *$q(\eta, t)$ is bounded in $\overline{H_0}$ and*

$$|q|\eta^{-\alpha\lambda} \to 0 \quad \textit{as } \eta \to 0. \tag{22.19}$$

Then

$$|q| \le C_1 \frac{\eta}{t - t_0} \quad \textit{in } \overline{H_0}$$

with an absolute constant $C_0 > 0$.

Proof. Let us construct upper and lower barriers for the function

$$h(\eta, t) = (t - t_0)q(\eta, t).$$

Introduce the functions

$$z_\pm(\eta) = \pm\theta\eta, \quad \theta = \text{const.} > 0.$$

By (22.17)

$$Ph(\eta, t) = q(\eta, t) + (t - t_0) f(\eta, t),$$

and a direct computation gives:

$$Pz_\pm = \pm \left\{ m\alpha \frac{1-\lambda}{y_\eta^2} + \frac{2}{y_\eta} \left[y_t \eta^{-\lambda\alpha} + (1-\lambda) \left[1 - \frac{m\alpha}{y_\eta} \right] \right] \right\} \eta^{\alpha\lambda}.$$

By the choice of η_0

$$|Pz_\pm| \ge \theta \frac{1-\lambda}{8m\alpha} \eta^{\lambda\alpha}.$$

Observe that always $\alpha(2-\lambda) > \alpha\lambda$, furnishing the inequality $|f| \le C_0 \eta^{\alpha\lambda}$ for all η small enough. Due to (22.18)–(22.19), there exists a constant $\gamma > 0$ such that

$$|Q| \le \gamma \eta^{\lambda\alpha},$$

whence $|Ph| \le (\gamma + C_0 T)\, \eta^{\lambda\alpha}$. Choosing now

$$\theta \equiv \theta' = 1 + (\gamma + C_0 T) \frac{8m\alpha}{1-\lambda},$$

we obtain

$$Pz_+ > Ph, \quad Pz_- < Ph \quad \forall (\eta, t) \in H_0.$$

Evidently, $z_+ \ge h$, $z_- \le h$ as $t = t_0$ and $\eta = 0$. Let us take

$$\theta \equiv \theta'' = \frac{T}{\eta_0} \max_{[t_0, T]} |q|.$$

Such a choice implies

$$z_+ \ge h, \quad z_- \le h \quad \text{as } \eta = \eta_0$$

and the proof is completed by setting $\theta = \theta' + \theta''$. □

In order to prove (22.16), we assume that $t_0 > t_*$, where the point t_0 defines the lower face of the rectangle $D(t_0)$, and t_* is "the waiting time". Then, as shown in Section 21, there exists a constant $D > 0$ such that

$$W(b, t) \ge 2D \quad \forall t \in [t_0, T].$$

Since $W(\xi, t)$ is continuous as a function of ξ, one can always assume η_0 so small that

$$W(\xi, t) \equiv \eta^{\alpha(\lambda-1)} \left(1 - \frac{m\alpha}{y_\eta} \right) \ge D \quad \text{in } \overline{H_0}. \tag{22.20}$$

Lemma 22.4. *Assume the conditions of Lemma* 22.3 *and* (22.20). *Then there exists a constant* $C_2 > 0$ *such that*

$$|q| \leq C_2 \frac{\eta^{1+\alpha(1-\lambda)}}{t-(t_0+\delta)} \quad in\ \overline{H_1} \tag{22.21}$$

where $H_1 = H_0 \bigcap \{t > t_0 + \delta\} \bigcap \{\eta < \eta_1\}$,

$$\eta_1 = \min\left\{\eta_0; \left(\frac{D\delta(1-\lambda)}{4C_1}\right)^{1/(1-\alpha)}\right\}, \qquad 0 < \delta < T - t_0.$$

Proof. We follow the scheme of proof of Lemma 22.3. Consider the function

$$g(\eta, t) = (t - (t_0 + \delta))\, q(\eta, t)$$

satisfying the equation

$$Pg(\eta, t) = q + (t - (t_0 + \delta))\, f.$$

Recall that by Lemma 22.3

$$|g| \leq \frac{T}{\delta} C_1 \eta.$$

Introduce the functions

$$s_\pm = \pm C_2 \eta^{1+\alpha(1-\lambda)}$$

with a constant C_2 to be defined. In H_1

$$\begin{aligned} Ps_\pm &= \pm C_2 \left\{ y_t \eta^{-\alpha} + (1-\lambda)\eta^{\alpha(\lambda-1)} \left(1 - \frac{m\alpha}{y_\eta}\right)\right\} \\ &\quad \times (1 + \alpha(1-\lambda)) \frac{2\eta^{\alpha(2-\lambda)}}{y_\eta}. \end{aligned}$$

Observe that y_t satisfies the conditions of Lemma 22.3 and, hence,

$$\left|y_t \eta^{-\alpha}\right| \leq C_1 \frac{\eta^{1-\alpha}}{\delta} \quad \text{in } \overline{H_1}.$$

Next, due to the choice of η_0 and η_1 we have

$$\begin{aligned} &\left| y_t \eta^{-\alpha} + (1-\lambda)\eta^{\alpha(\lambda-1)}\left(1 - \frac{m\alpha}{y_\eta}\right)\right| \eta^{\alpha(2-\lambda)} \\ &\quad \geq (1-\lambda)\left[D - \frac{C_1}{\delta(1-\lambda)}\eta_1^{1-\alpha}\right]\eta^{\alpha(2-\lambda)} \\ &\quad \geq \frac{(1-\lambda)D}{4}\eta^{\alpha(2-\lambda)}. \end{aligned} \tag{22.22}$$

Thus,

$$|Ps_{\pm}| \geq \frac{(1-\lambda)D}{4}\eta^{\alpha(2-\lambda)}C_2(1+\alpha(1-\lambda)) \quad \text{in } H_1.$$

Making use of the inequality $1 > \alpha(2-\lambda)$, we get that

$$|Pg| \leq \left(\frac{C_1}{\delta} + C_0 T\right)\eta^{\alpha(2-\lambda)} \quad \text{in } H_1.$$

Let us choose now

$$C_2 = \left(\frac{C_1}{\delta} + C_0 T\right)\left[\frac{4m\alpha}{D(1-\lambda)(1+\alpha(1-\lambda))} + \eta_1^{\alpha(\lambda-1)}\right].$$

Then

$$Ps_+ \geq Pg, \quad Ps_- \leq Pg \quad \text{in } H_1,$$

$$s_+ \geq g, \quad s_- \leq g \quad \text{on the parabolic boundary of } H_1$$

and inequality (22.21) follows by the maximum principle. □

Corollary 22.1. *In the variables* (ξ, t), *the estimate* (22.15) *takes the form*

$$|w_t| \leq K_0 \frac{w}{t-t_0} \quad \text{in } \overline{G_R(t_0)} \tag{22.23}$$

where $G_R(t_0) = \{(\xi, t) \in Q_1 : t \in (t_0, T],\ \xi \in (R, b)\}$.

22.4 Estimating the higher-order derivatives. To estimate the function y_t and its derivatives near the line $\eta = 0$, we compile the equations for the functions $D_t^r(y_t)$, apply appropriate rescaling transformations rendering these equations uniform parabolic (locally, in a neighborhood of an arbitrary fixed point), and then make recourse to the well-known results about the interior smoothness of solutions to uniformly parabolic equations.

Lemma 22.5. *Let under the conditions of Lemma* 22.4

$$|f_\eta| \leq C_3 \eta^{\alpha(2-\lambda)-1} \quad \textit{in } \overline{H_0}. \tag{22.24}$$

Then:

$$|q_\eta| \leq C_4 \eta^{\alpha(1-\lambda)}, \tag{22.25}$$

$$|q_{\eta\eta}| \leq C_5 \eta^{\alpha(1-\lambda)-1}, \tag{22.26}$$

$$|q_t| \leq C_6 \eta^{\alpha} \quad \textit{in } \overline{H_2}, \tag{22.27}$$

where

$$H_2 = H_1 \cap \left\{t > t_0 + \delta + \delta^2\right\} \cap \{\eta < \min(2t_0, \eta_1/2)\}.$$

Proof. Fix an arbitrary point $(\eta', t') \in H_2$ and introduce the new variables (ζ, τ) and the rectangle $G \subset H_2$ by the relations

$$\begin{cases} G = \left\{(\eta, t) : |\eta - \eta'| \le \left(\frac{\eta'}{2}\right)^s, t' - \left(\frac{\eta'}{2}\right)^p \le t \le t'\right\}, \\ \eta = \eta' + \left(\frac{\eta'}{2}\right)^s \zeta, \quad t = t' + \left(\frac{\eta'}{2}\right)^p \tau, \\ s = 1, \quad p = 1 - \lambda\alpha. \end{cases}$$

Set $Q(\zeta, \tau) = q(\eta, t)$. On the plane of the variables (ζ, τ), G becomes the rectangle $J = (-1, 1) \times (-1, 0)$, and equation (22.17) becomes

$$\begin{aligned} Q_\tau &- m\frac{\eta^{\alpha(m+1)}}{y_\eta^2}\frac{1}{(\eta'/2)^{2s-p}}Q_{\zeta\zeta} \\ &+ \frac{2}{y_\eta}\left[y_t + (1-\lambda)\eta^{\lambda\alpha}\left(1 - \frac{m\alpha}{y_\eta}\right)\right]\frac{1}{(\eta'/2)^{s-p}}Q_\zeta \\ &+ (1-\lambda)\eta^{\lambda\alpha}\frac{m\alpha}{y_\eta^2}\frac{1}{(\eta'/2)^{s-p}} = (\eta'/2)^p f. \end{aligned} \tag{22.28}$$

For the sake of brevity we keep the previous notation for the coefficients and the right-hand side of this equation but consider them now as functions of the variables (ζ, τ). In our choice of the parameters s, p equation (22.28) is uniformly parabolic in J and the function Q is uniformly bounded there due to (22.21). Referring to [97, Th. 5.3.1], we get

$$\left|Q_\zeta(0,0)\right| \le C_7 \max_{\overline{J}} |Q|,$$

that is,

$$|q_\eta| \le C_8\left(\frac{\eta'}{2}\right)^{1+\alpha(1-\lambda)}\left(\frac{\eta'}{2}\right)^{-s} = C_9(\eta')^{\alpha(1-\lambda)}.$$

Since the point (η', t') was taken arbitrarily, (22.25) follows.

Denote $S(\eta, t) = q_\eta$. Differentiating equation (22.17) in η, we deduce the equation for S: in G

$$\begin{aligned} S_t &- m\frac{\eta^{\alpha(m+1)}}{y_\eta^2}S_{\eta\eta} + \frac{4}{y_\eta}\left\{y_t + (1-\lambda)\eta^{\lambda\alpha}\left(1 - \frac{m\alpha}{y_\eta}\right)\right\}S_\eta \\ &+ (1-\lambda)\eta^{\alpha\lambda}\frac{m\alpha}{y_\eta^2}S_\eta - m\alpha(m+1)\frac{1}{y_\eta^2}\eta^{\alpha\lambda}S_\eta = F(\eta, t), \end{aligned}$$

where

$$\begin{aligned} F = &-\frac{2}{y_\eta}\left\{q_\eta + (1-\lambda)\eta^{\alpha(1-\lambda)}\left(1 - \frac{m\alpha}{y_\eta}\right)\eta^{\alpha-1}\alpha\lambda\right. \\ &\left. - \alpha(1-\lambda)\eta^{\lambda\alpha}\left(1 - \frac{m\alpha}{y_\eta}\right)\right\}q_\eta + \end{aligned}$$

$$+\frac{2(1-\lambda)\alpha}{\eta y_\eta}\left(y_t+(1-\lambda)\eta^{\alpha\lambda}\left(1-\frac{m\alpha}{y_\eta}\right)\right)q_\eta$$
$$-(1-\lambda)\lambda\alpha\frac{m\alpha}{y_\eta^2}\eta^{\alpha\lambda-1}q_\eta+f_\eta. \tag{22.29}$$

Applying to F estimates (22.24), (22.25), and (22.21), (the last estimate is used to evaluate y_t), we have:

$$|F|\le C_{10}\eta^{\alpha-1}\quad \text{in } \overline{G}. \tag{22.30}$$

Set $\sigma(\zeta,\tau)\equiv S(\eta,t)$. σ satisfies in G the uniformly parabolic equation following from (22.29)

$$\sigma_\tau-\frac{m}{y_\eta^2}\left(\frac{2\eta}{\eta'}\right)^{\alpha(m+1)}\sigma_{\zeta\zeta}+A(\zeta,\tau)\sigma_\zeta=\varphi(\zeta,\tau),$$

where the coefficients and the right-hand side are simply taken from (22.29)–(22.30) and rewritten in the variables (ζ,τ). Besides,

$$|A|\le C_{11}\eta'^{\lambda\alpha},\quad |\varphi|\le C_{12}\eta'^{(1-\lambda)\alpha}\quad \text{in } \overline{J}.$$

As before, we get now

$$\left|\sigma_\zeta(0,0)\right|\le C_{13}\max_{\overline{J}}|\sigma|,$$

whence (22.26). To obtain (22.27) it suffices to express q_t from equation (22.17) and then to apply (22.25)–(22.26) and (22.21). □

Adopt the following notation:

$$q^{(k)}\equiv D_t^k y(\eta,t),\quad k=1,2,\dots,$$
$$H_{k+1}=H_k\cap\left\{t>t_0+\sum_1^{k+1}\delta^i\right\}\cap\left\{\eta<\eta_1 2^{-(k+1)}\right\}. \tag{22.31}$$

The value of δ is taken such that

$$t_0+\sum_1^\infty\delta^i<T.$$

Lemma 22.6. *For each $p=0,1,2,\dots$ the following estimates hold*

$$\left|q^{(p)}\right|\le L_p\frac{\eta^{\alpha(1-\lambda)+1}}{t_0+\sum_1^{3p}\delta^i}\quad \text{in } \overline{H_{3p}}$$
$$\left\{\left|\frac{\partial q^{(p)}}{\partial\eta}\right|;\eta\left|\frac{\partial^2 q^{(p)}}{\partial\eta^2}\right|\right\}\le L_p\eta^{\alpha(1-\lambda)}, \tag{22.32}$$
$$\left|q^{(p+1)}\right|\le L_p\eta^\alpha\quad \text{in } \overline{H_{3(p+1)}},$$

where L_p are absolute positive constants.

Proof. For $p = 0$ these estimates coincide with those already derived in Lemmas 22.4–22.5. Suppose that they are true for some $k > 1$. For $q^{(k+1)}$ we obtain, differentiating the equation $Py_t = 0$ k times in t:

$$Pq^{(k+1)} = \Phi \quad \text{in } \overline{H_{3(k+1)}}, \tag{22.33}$$

where

$$\begin{aligned}\Phi \equiv & -m\eta^{\alpha(m+1)} \sum_1^k D_t^i \left(\frac{1}{y_\eta^2}\right) D_t^{k-i} \left(\frac{\partial^2 q^{(1)}}{\partial \eta^2}\right) \\ & - 2m\eta^{\alpha(m+1)} \sum_1^k D_t^i \left(\frac{1}{y_\eta^3}\frac{\partial^2 q^{(1)}}{\partial \eta^2}\right) D_t^{k-i} \left(\frac{\partial q^{(1)}}{\partial \eta}\right) \\ & - (1-\lambda)\eta^{\lambda\alpha} m\alpha \sum_1^k D_t^i \left(\frac{1}{y_\eta^2}\right) D_t^{k-i} \left(\frac{\partial q^{(1)}}{\partial \eta}\right) \\ \equiv & \; I_1 + I_2 + I_3. \end{aligned} \tag{22.34}$$

By the induction hypothesis

$$\begin{aligned}|I_1| &\le M\eta^{\alpha(m+1)} \sum_1^k \left|\frac{\partial^2}{\partial \eta^2}\left(q^{(k-i+1)}\right)\right| \times \left(\sum_{j=1}^i \frac{1}{y_\eta^{2+j}} \left|q_\eta^{(1)}\right|^{j-i} \left|q_\eta^{(i-j+1)}\right|\right) \\ &\le M\eta^{\alpha(2-\lambda)}.\end{aligned}$$

Likewise, we estimate

$$\{|I_1|\,;\,|I_2|\} \le M\eta^{\alpha(2-\lambda)} \quad \text{in } \overline{H_{3(k+1)}}.$$

The constant M here depends only on $L_1, \ldots, L_k$, D, and the constants appearing in the proofs of Lemmas 22.4 and 22.5.

Observe now that I_2 can be rewritten in the form

$$I_2 = -\sum_1^k D_t^i \left\{\frac{2}{y_\eta}\left[y_t + (1-\lambda)\eta^{\lambda\alpha}\left(1 - \frac{m\alpha}{y_\eta}\right)\right]\right\} D_t^{k-i}\left(\frac{\partial q^1}{\partial \eta}\right),$$

and that I_1, I_2 do not contain the terms $\partial^2 q^{(k)}/\partial\eta^2$. Thus, Φ_η is a linear combination of the functions $q^{(p)}$, $p = 1, \ldots, k$, and their derivatives up to second order. Computing Φ_η and using the induction hypothesis (22.32), we get the inequality

$$\left|\Phi_\eta\right| \le M\eta^{\alpha(2-\lambda)-1}.$$

To complete the induction step it remains to apply Lemmas 22.4, 22.5 to the solution of equation (22.33). □

23 The interface regularity

The estimates of Section 22 now allow one to establish the infinite smoothness of the right-hand side of representation (18.11).

23.1 Global $C^{1+\alpha}$-smoothness. We begin with the proof of $C^{1+\alpha}$-regularity of the interface. Due to the representation (18.11), this amounts to establishing the C^α-regularity of $W(b,t)$. To this end, we adapt for our situation the barrier construction originated in [93, 62].

Let us take an arbitrary $t_* \in (0,T]$ and choose $R \in (0,b)$ so as to satisfy the inequality

$$\left| w_\xi + \frac{1}{m\alpha} \right| < \varepsilon, \qquad \varepsilon = \frac{1}{\frac{1}{m\alpha} + m(1+\alpha\lambda)} \tag{23.1}$$

in the rectangle

$$G_R(t_*) = \{(\xi,t) \in Q_1 : \xi \in (R,b),\ t \in (t_*,T]\}.$$

By virtue of (22.2), such a choice is always possible.

It follows from (22.23) and (20.5) that for each fixed $t_0 \in (t_*,T]$ the function $W(\xi,t_0)$ admits the estimate

$$\begin{aligned} |W(\xi,t_0) - W(b,t_0)| &\le \alpha \int_\xi^b \left.\frac{|w_t|}{w^{1+\alpha}}\right|_{t=t_0} ds \le \alpha \int_\xi^b \frac{ds}{w^\alpha(s,t_0)} \\ &\le \frac{K_1}{t_0 - t_*}(b-\xi)^{1-\alpha} \end{aligned} \tag{23.2}$$

with an absolute positive constant K_1. To deduce (23.2), we made use of (23.1) taken in the form

$$\frac{1}{m\alpha}(b-\xi) \ge w(\xi,t_0) \ge \left(\frac{1}{m\alpha} - \varepsilon\right)(b-\xi) \quad \forall (\xi,t) \in \overline{G_R(t_*)}.$$

Let us consider the functions

$$\begin{aligned} U^\pm(\xi,t) &= \pm[W(b,t_0) - W(\xi,t)] + \frac{K_2}{t_0 - t_*}\left(3\rho^{1-\alpha} - (b-\xi)^{1-\alpha}\right) \\ &+ \frac{2K_3}{\rho^2}\left[K_4(t-t_0) + (b-\xi)^2\right] \end{aligned}$$

with the constants

$$K_2 = \frac{L}{1-\alpha} + K_0 + K_1, \quad K_3 = \sup_{G_R(t_0)} |W|,$$

where L is taken from (20.3) and K_0 from (22.23). As for the constants K_4 and $\rho \in (0, b-R)$, we will choose them later on.

Let us check that the functions $U^{\pm}$ are both nonnegative in $\overline{G_R(t_0)}$. Let $\mathcal{L}(\cdot)$ be the parabolic operator from (20.12). Calculating $\mathcal{L}(w)U^{\pm}$, we get

$$\begin{aligned}\mathcal{L}(w)U^{\pm} &= \frac{2K_3}{\rho^2}\Big[K_4 - 2mw^{1+\lambda\alpha} \\ &\quad + 2w^{\lambda\alpha}\big(m(1+\alpha\lambda)w_\xi + \lambda - 1\big)(b-\xi)\Big] \\ &\quad + K_2(1-\alpha)\frac{w^{\alpha\lambda}}{(b-\xi)^\alpha}\Big[(1-\lambda) - m\alpha\frac{w}{b-\xi} - m(1+\alpha\lambda)w_\xi\Big].\end{aligned}$$

Making use of (21.1), it is easy to conclude that the second term of this expression is positive for ξ close to b. As for the first term, since the choice of the constant K_4 is at our disposal, we may take it so large that

$$\mathcal{L}(w)U^{\pm} > 0 \quad \text{in } G_{b-\rho}(t_0).$$

Next, taking into account (23.2) and recalling the definition of K_2 we find that

$$U^{\pm}(\xi, t_0) \geq -\frac{K_1}{t_0 - t_*}(b-\xi)^{1-\alpha} + \frac{K_2}{t_0 - t_*}\left(3\rho^{1-\alpha} - (b-\xi)^{1-\alpha}\right) > 0.$$

Further, with the use of (22.23) we get

$$\begin{aligned}\frac{\partial U^{\pm}}{\partial \xi} &= \pm\frac{\partial W}{\partial \xi} + (1-\alpha)\frac{K_2}{t_0 - t_*}\frac{1}{(b-\xi)^\alpha} \\ &\quad - \frac{4K_3(b-\xi)}{\rho^2} \to \infty \quad \text{as } \xi \to b.\end{aligned}$$

Finally, it follows by the choice of K_3 that

$$U^{\pm}(b-\rho, t) \geq 2K_3 \pm [W(b, t_0) - W(b-\rho, t_0)] > 0.$$

By the maximum principle, we conclude that

$$U^{\pm} \geq 0 \quad \text{in } \overline{G_{b-\rho}(t_0)}.$$

This is equivalent to the following inequalities: everywhere in $\overline{G_{b-\rho}(t_0)}$

$$|W(b, t_0) - W(\xi, t)| \leq \frac{3K_2}{t_0 - t_*}\rho^{1-\alpha} + \frac{2K_3}{\rho^2}\Big[(t - t_0)K_4 + (b-\xi)^2\Big].$$

In particular, letting $\xi \to b$ and fixing an arbitrary $t_1 > t_0$, we get

$$|W(b, t_0) - W(b, t_1)| \leq \frac{3K_2}{t_0 - t_*}\rho^{1-\alpha} + \frac{2K_3K_4}{\rho^2}(t_1 - t_0). \tag{23.3}$$

Recall now that (23.3) is true *for every* $\rho \in (0, b-R)$. Let us then choose ρ in a special way. Namely, let us set

$$\rho = (t_1 - t_0)^{1/(3-\alpha)}.$$

As a result we obtain: for all $t_0 < t_1 < t_0 + (b - R)^{3-\alpha}$

$$|W(b, t_0) - W(b, t_1)| \leq \sigma (t_1 - t_0)^{1/(3-\alpha)}$$

with the constant $\sigma > 0$ depending on K_2, K_3 and on the difference $t_0 - t_*$.

The last inequality is derived under the assumption that t_0 is already fixed and $t_1 > t_0$ is to be defined in the sequel. The same arguments are true in the case when t_1 is given and fixed, but t_0 can vary. Indeed, all constants that appeared in the preceding were chosen once and for all for the whole domain $G_{b-R}(t_*)$. Keeping them, we merely confine the needed arguments to a smaller rectangle $G_{b-R}(t_* + \varepsilon)$, $\varepsilon > 0$.

23.2 Infinite differentiability of the moving interface. Now we assume $t > t_*$, providing the existence of some $D > 0$ such that

$$W(b, t) \geq 2D > 0 \quad \forall t \in (t_0, T].$$

Then we can apply to the functions $q^{(k)}(\eta, t)$ estimates (22.32) with arbitrary $k = 1, 2, \dots$.

Let us introduce the notation

$$V^{(k)}(\eta, t) = D_t^k W(\xi, t), \quad k = 1, 2, \dots,$$

and agree to keep wherever possible the notation of Section 22.

We consider the functions $V^{(l)}(\eta, t)$, $l = 1, 2, \dots, s$, with some fixed integer s in the domains $H_{3(s+1)}$. To prove Theorem 17.3 amounts to checking that $V^{(l)}(\eta, t)$, $l = 1, 2, \dots, s$, have finite limits as $H_{3(s+1)} \ni \eta \to 0$. Let us argue by induction.

Step 1. Everywhere in Q

$$D_t W(\xi, t) = -\alpha(1 - \lambda)\frac{w_t}{w} W + m\alpha w^{\alpha(\lambda-1)} \left(w_\xi\right)_t$$

and in $H_{3(s+1)}$

$$V^{(1)}(\eta, t) = -\alpha \frac{\lambda - 1}{y_\eta}\frac{y_t}{y} V^{(0)}(\eta, t) + \frac{m\alpha}{y_\eta^2}\eta^{\alpha(\lambda-1)}\frac{\partial q^{(1)}}{\partial \eta}. \tag{23.4}$$

The function $V^{(0)} \equiv W$ is bounded in $H_{3(s+1)}$ for each t and takes on a finite limiting value as $\eta \to 0$. Then (22.32) with $p = 0$ implies that this is also true for $V^{(1)}(\eta, t)$ if

$$t > t_0 + \sum_{1}^{3(s+1)} \delta^i \equiv t_s.$$

Step k. Differentiating (23.4) in t, we find

$$\begin{aligned} V^{(k+1)} &\equiv -\alpha(\lambda - 1)\sum_{0}^{k} V^{(i)} \left\{\frac{1}{\eta} D_t^{k-i}\left(\frac{y_t}{y_\eta}\right)\right\} \\ &\quad + m\alpha\eta^{\alpha(\lambda-1)}\sum_{0}^{k} D_t^i\left(\frac{1}{y_\eta^2}\right)\frac{\partial}{\partial \eta}\left(q^{(k+1-i)}\right). \end{aligned} \tag{23.5}$$

Each function $V^{(i)}$, $i = 1, 2, \dots, k$, is bounded in $\overline{H_{3(s+1)}}$ and has finite limits as $\eta \to 0$. The derivatives $D_t^{k-i}\left(y_t/y_\eta\right)$ are represented by linear combinations of the functions

$$D_t^p y, \quad \frac{\partial}{\partial \eta}\left(D_t^l y\right), \quad p, l \in \{1, 2, \dots, k-i\},$$

divided by y_η^n, $n \in \{1, 2, \dots, k-i+1\}$. Also, it is important that each of these combinations contains $D_t^p y$ with $p \geq 1$. Referring to (22.32), we find that the first term in (23.5) is bounded in $\overline{H_{3(s+1)}}$ and has finite limits as $\eta \to 0$. Treating the second term, we come to the same conclusion. Returning to the variables (ξ, t), we get the desired inclusion

$$W(b, t) \in C^{s-1}(t_1, T) \cap \mathcal{O}^s(t_1, T)$$

where $t_1 > t_0 > t_*$ are arbitrary and s is an arbitrary integer.

This completes the proof of Theorem 17.3.

24 Regularity of interfaces for a generalized porous medium equation

In this section, we study the smoothness of interfaces occurring in solutions of certain equations generalizing the porous medium equation. Unlike problem $(\mathrm{E}_{m,\lambda})$, the problems we are concerned with here are united by the FSP property displayed by their solutions. Roughly speaking, we prove that in such solutions the interfaces have basically the same smoothness properties as in the case of the PME.

We begin with the study of problem (E) for the one-dimensional diffusion-advection equation. The corresponding Lagrangian statement (L) of this problem was already deduced in Section 18.1.

24.1 Solvability of Lagrangian statement. Remaining in the confines of the scheme already realized in Sections 19–20 we utilize certain corrections in its concrete realization, though. First of all, this is due to the fact that the functions φ and ψ are no longer infinitely differentiable but possess only a finite degree of smoothness. On the other hand, many pieces of our further considerations simplify because "the velocity of the material point" introduced in Section 18 proves to be bounded through initial data everywhere in the problem domain S.

Let us proceed to construct a solution to problem (L). It is convenient to introduce the new sought function $w(\xi, t) \equiv \varphi(v(\xi, t))$. Let us formulate the problem for w. In what follows, we repeatedly use the notation

$$\sigma(s) = \left(\varphi^{-1}(s)\right)^2 \varphi'\left(\varphi^{-1}(s)\right),$$

$$\theta(s) = \sigma(s)\left(\frac{\psi(\varphi^{-1})}{\varphi^{-1}(s)}\right)'_s \equiv \left[q^2\left(\frac{\psi(q)}{q}\right)'_q\right]\Bigg|_{q=\varphi^{-1}(s)}.$$

From (L) we get that the function $w(\xi, t)$ obeys the following conditions:

$$L_1 w \equiv w_t - \sigma(w)w_{\xi\xi} - \theta(w)w_\xi = 0 \quad \text{in } Q, \tag{24.1}$$

$$w = 0 \quad \text{on } \partial Q, \tag{24.2}$$

$$w(\xi, 0) = w_0(\xi) \equiv (\varphi(v_0(\xi))) \quad \text{in } [0, b], \tag{24.3}$$

where w_0 satisfies conditions (F.1)–(F.4) formulated in Subsection 18.1.

From now on this problem will be termed *problem* (L_1).

As above, a solution to problem (L_1) is obtained as the pointwise limit of a sequence $\{w_n\}$ of solutions of problems ($L_{1,n}$), regularized in the initial and boundary conditions. To define w_n, we pose the conditions

$$L_1 w_n = 0 \quad \text{in } Q, \tag{24.4}$$

$$w_n = \frac{1}{n} \quad \text{on } \partial Q, \tag{24.5}$$

$$w_n(\xi, 0) = w_{0n}(\xi) \quad \text{in } [0, b]. \tag{24.6}$$

The sequence of initial data is chosen in the following way:

(H.1) $w_{0n} \in C^\infty[0, b]$, $w_{0n} \searrow w_0$ as $\to \infty$,
$\frac{1}{n} \le w_{0n} \le 1 + M_0$, $M_0 = \max u_0$;
in a neighborhood of the point $\xi = 0$,
$\xi = b$, w_{0n} is a solution of the problem

$$\begin{cases} \sigma(w_{0n})w''_{0n} + \theta(w_{0n})w'_{0n} = 0, \\ w_{0n}(b) = \frac{1}{n};\ w'_{0n}(b) = -C \le -C_0;\ (w_{0n}(0) = \frac{1}{n};\ w'_{0n}(0) = C;) \end{cases}$$

(H.2) w_{0n} are uniformly Lipschitz continuous with Lipschitz constant L;

(H.3) there exist $\varepsilon_0 > 0$ and an integer $n(\varepsilon_0)$ such that for all $n > n(\varepsilon)$ and $\varepsilon \in (0, \varepsilon_0)$ each set $\{\xi \in (0, b) : w_{0n}(\xi) > \varepsilon\}$ is an interval in $(0, b)$.

The constant C_0 in (H.2) is taken from condition (B.3).

Such a sequence can be constructed by the same procedure that was utilized in Section 19 to approximate the data of problem ($E_{m,\lambda}$).

Under the conditions of Theorem 17.3, σ and θ are members of $C^{2+\tau}(\mathbb{R}^+)$ for some $\tau \in (0, 1]$, and the data of problems ($L_{1,n}$) are taken so that at the corner points of ∂Q the first-order compatibility conditions are fulfilled. Let us consider problem ($L_{1,n}$). Making use of the same trick as in the proof of solvability of the problems regularizing ($E_{m,\lambda}$), and referring to [97, Th. 5.6.1,Th. 4.5.2], we get: given an integer

n large enough, there exists a solution of $(L_{1,n})$ $w_n(\xi, t)$ taking values in the interval $\left[\frac{1}{n}, M_0 + \frac{1}{n}\right]$ and such that

$$w_n \in C^{3+\theta,(3+\theta)/2}_{\xi,t}\left(\overline{Q}\right) \quad \text{with some } \theta \in (0, 1].$$

To prove the solvability of problem (L_1) one has to establish three facts (cf. Sections 19–20):

1) w_n converge uniformly in $\overline{Q}$ to a function $w \in C\left(\overline{Q}\right)$;

2) $w_{n\xi}$ are equibounded in $\overline{Q}$;

3) the limiting function w is strictly positive in $(0, b) \times [0, T]$.

Lemma 24.1. *The functions*

$$V_n(\xi, t) \equiv -w_{n\xi} - \frac{\psi\left(\varphi^{-1}(w_n)\right)}{\varphi^{-1}(w_n)}$$

are equibounded in $\overline{Q}$.

Proof. The functions V_n satisfy in Q the uniformly parabolic equations

$$\mathcal{M}(w_n)V_n \equiv V_{nt} - \sigma(w_n)V_{n\xi\xi} - \left[\sigma'(w_n)w_{n\xi} + \theta(w_n)\right]V_{n\xi} = 0, \tag{24.7}$$

the boundary conditions following from (24.4)–(24.5)

$$V_{n\xi}\big|_{\partial Q} = -\left.\frac{w_{nt}}{\sigma(w_n)}\right|_{\partial Q} = 0, \tag{24.8}$$

and the initial condition

$$V_n\big|_{t=0} = -w'_{0n} - \frac{\psi\left(\varphi^{-1}(w_{0n})\right)}{\varphi^{-1}(w_{0n})}.$$

Conditions (24.8) mean that the lateral boundaries of Q do not contain a point of strict positive maximum or negative minimum of V_n. Further, the shape of equation (24.7) implies that neither the positive maximum nor negative minimum of V_n can be assumed inside Q. Thus, all local extreme points of V_n belong to B_0, where V_n is already bounded by virtue of (H.1)–(H.2). □

Corollary 24.1. *By* (17.12), *the boundedness of* V_n *automatically means the boundedness of* $w_{n\xi}$.

Corollary 24.2. *The maximum principle implies that sequence* $\{w_n\}$ *is monotone decreasing at each point of* $\overline{Q}$ *and nonnegative in* $\overline{Q}$. *Thus, there exists the limit function* w *which is at least Hölder continuous in* $\overline{Q}$ *(see* [62, 93]*).*

Lemma 24.2. *$w(\xi, t)$ is strictly positive in Q.*

Proof. Let us argue by contradiction. Assume that v vanishes at a point $(\xi', t') \in Q$.

1) It is asserted: there exist $\delta \in (0, \varepsilon_0)$ and an integer $n(\delta)$ such that each interval $(0, \xi')$, (ξ', b) for each $n > n(\delta)$ contains points $\xi_1(n)$ and $\xi_2(n)$, where

$$w_n(\xi_1(n), t') > \delta, \quad w_n(\xi_2(n), t') > \delta. \tag{24.9}$$

Let us rewrite equation (24.4) in the form

$$\left(\frac{1}{v_n}\right)_t = V_{n\xi}, \quad v_n = \varphi^{-1}(w_n), \tag{24.10}$$

and integrate it over the rectangle $(0, \xi') \times (0, t')$. We get

$$\int_0^{\xi'} \frac{ds}{v_n(s, t')} = \int_0^{\xi'} \frac{ds}{\varphi^{-1}(w_{0n})(s)} + \int_0^{t'} \left[V_n(\xi', \theta) - V_n(0, \theta)\right] d\theta < \infty. \tag{24.11}$$

Assume that (24.9) is false. The negation is

$$\forall \delta \in (0, \varepsilon_0),\ n' \in \mathbb{N}\ \exists n > n' : \ w_n(\xi, t') \le \delta \quad \text{everywhere in } (0, \xi'),$$

whence we immediately get the following contradiction to (24.11):

$$\int_0^{\xi'} \frac{ds}{v_n(s, t')} \ge \frac{\xi'}{\varphi^{-1}(\delta)} \to \infty \quad \text{as } \delta \to 0.$$

In the same way we check the validity of the second inequality of (24.9).

2) Let us take into consideration the sets

$$Q' \equiv Q \cap \{t \le t'\},$$

$$P_{n,\delta} \equiv \{(\xi, t) \in Q \cap \{t \le t'\} : \ w_n(\xi, t) > \delta\}.$$

Take $\delta \in (0, \varepsilon_0)$ and a number $n > n(\delta)$ so as to satisfy (24.9).

Assuming (H.3), we have: for each $\delta \in (0, \varepsilon_0)$, $n > n(\delta)$ the set $P_{n,\delta}$ contains at most one connected component.

Let us prove this. As a preliminary we observe that $P_{n,\delta}$ is open in the topology of Q' and that $(\xi_1(n), t')$, $(\xi_2(n), t') \in P_{n,\delta}$.

First, let us show that $P_{n,\delta}$ contains at least one curve l_0 connecting $P_{n,\delta} \cap B_{t'}$ to $\overline{P_{n,\delta}} \cap B_0$. We prove the existence of such a curve by arguing by contradiction. Unless it exists, there is a curve l_0 connecting two different points of the set $(Q' \setminus P_{n,\delta}) \cap B_{t'}$ and, besides, such that $w_n \le \delta' < \delta$ along l_0. By the continuity of w_n, the curve l_1 can be always taken piecewise-linear. Next, without loss of generality we may suppose that the ends $\mu' \in (\xi', b)$ and $\mu'' \in (0, \xi')$ of the loop l_1 on $B_{t'}$ are located so that

$$w_n(\xi, t') \le \delta' \quad \text{as } \xi \in [0, \mu'] \cup [\mu'', b].$$

By the assumption, l_1 always contains at least two adjoining piecewise-linear parts λ_1 and λ_2 with different directions of monotonicity with respect to t. Let us adopt the convention that traversing the loop l_1 clockwise, we first traverse the part λ_1 and then λ_2. Set

$$t^* = \sup\left\{t \in (0,t') : \lambda_1 \cap B_t \neq \emptyset \text{ and } \lambda_2 \cap B_t \neq \emptyset\right\}.$$

Applying to w_n the maximum principle in the domain with the parabolic boundary $\lambda_1 \cup \lambda_2$ for $t \leq t^*$ and the upper face $t = t^*$, we get that $w_n \leq \delta' < \delta$ everywhere in the closure of this domain and, hence, the considered nonmonotone part of the loop l_1 can be replaced by a segment of the straight line B_{t^*}. Repeating this procedure as often as is needed we will traverse all the loop l_1 and come to the conclusion that l_1 is merely the interval $[\mu', \mu'']$ within $B_{t'}$, which is impossible due to (24.9).

Proceeding in the same way and replacing, if needed, the nonmonotone parts by segments of horizontal lines, one can verify that the curve l_0 is monotone in t.

Assume now that $P_{n,\delta}$ contains two connected components. Taking within each of them some curves l_0' and l_0'' connecting $B_{t'}$ to B_0, and then applying to w_n the maximum principle in the domain H with the lateral boundaries l_0', l_0'' and the lower face

$$\left\{\xi \in (0,b) : l_0' \cap \{t=0\}\right\} < \xi < \left\{\xi \in (0,b) : l_0'' \cap \{t=0\}\right\},$$

we get, recalling (H.3): $w_n > \delta$ in $\overline{H}$.

Thus, the initial assumption has led us to a contradiction. □

Remark 24.1. Lemma 24.2 can be given the following alternative proof. Assume that $w_{nt} \geq -Lt^{-1}w_n$, $L = \text{const.} > 0$, $n = 1, 2, \ldots$ (see Lemma 24.4 below). Fix an arbitrary point $(\xi_0, t_0) \in Q$. Since $w_n \to w$ in $C\left(\overline{Q}\right)$ and $w_0(\xi_0) \geq \delta > 0$, there exists $t_* \in (0, t_0)$ such that $w_n(\xi_0, t_*) \geq \delta/2$. Now, integrating the inequality $\left(\ln\left(t^L w_n(\xi_0, t)\right)_t\right) \geq 0$ between the limits (t_*, t_0), we get the inequality

$$w_n(\xi_0, t_0) \geq w_n(\xi_0, t_*)\left(\frac{t_0}{t_*}\right)^L \geq \frac{\delta}{2}\left(\frac{t_0}{t_*}\right)^L,$$

and then passing to the limit as $n \to \infty$ we obtain the estimate

$$w(\xi_0, t_0) \geq \frac{\delta}{2}\left(\frac{t_0}{t_*}\right)^L > 0.$$

Proof of Theorem 18.2 (Existence of a solution to (L)).

Let $Q' \subset Q$ be an arbitrary subdomain of Q bounded away from Γ and $f(\xi, t) \in C_0^\infty(Q')$ be an arbitrary function. Fix an integer n, multiply equation (24.10) by f, and integrate the resulting equality by parts in Q:

$$\int_{Q'}\left\{\frac{f_t}{v_n} - f_\xi V_n\right\} d\xi\, dt = 0. \tag{24.12}$$

By Lemma 24.1, one may extract from the sequence $\{V_n\}$ a subsequence (which we assume to coincide with the whole sequence), converging weakly in $L^2(Q)$ to some function V. The uniform convergence of $\{w_n\}$ to w implies that

$$V \equiv -w_\xi - \frac{\psi\left(\varphi^{-1}(w)\right)}{\varphi^{-1}(w)}.$$

The legitimacy of the passage to the limit in the first term of (24.12) follows from the fact that, by Lemma 24.2, the functions $1/v_n$ already have in Q' an integrable majorant. As a result of the limiting procedure we obtain: given an arbitrary domain $Q' \subset Q$, bounded away from the parabolic boundary of Q, w satisfies in Q' equation (24.1) in the weak sense. Now we proceed in the standard way viewing w in Q' as a solution of the uniformly parabolic equation (24.1) and then applying [97, Th. 5.6.1]: in each subdomain $Q'' \subset Q'$ separated from the parabolic boundary of Q'

$$w \in C^{3+\delta,(3+\delta)/2}_{\xi,t}(Q'') \quad \text{with some } \delta \in (0,1],$$

e.g. inside Q w satisfies equation (24.1) in the classical sense. It remains to observe that the inclusion $w \in C\left(\overline{Q}\right)$ and the initial and boundary conditions for w are fulfilled by the method of regularization, and then to set $v = \varphi^{-1}(w)$. □

24.2 Further estimates. Here lower bounds for the functions w and w_t are established which will allow one to justify the representations (18.11), (18.12) and to verify the invertibility of the change of variables. In particular, it will be shown that the function $V(\xi,t)$ for each $t > 0$ has finite limits as $\xi \to 0$, $\xi \to b$, and that the functions $\eta(t)$, $\zeta(t)$ have generalized derivatives in the interval $(0, T)$ and obey the differential equations

$$\eta'(t) + \lim_{\xi \to b} V(\xi,t) = 0, \tag{24.13}$$

$$\zeta'(t) + \lim_{\xi \to 0} V(\xi,t) = 0 \quad \text{in } \mathcal{D}'(0,T). \tag{24.14}$$

Lemma 24.3. *There exist constants $C_1, C_2 > 0$ such that uniformly with respect to n*

$$C_1(b-\xi) + \frac{1}{n} \le w_n \le C_2(b-\xi) + \frac{1}{n} \quad \textit{in } \overline{Q_1} \equiv \overline{Q \cap \{\xi > b/2\}}, \tag{24.15}$$

$$C_1\xi + \frac{1}{n} \le w_n \le C_2\xi + \frac{1}{n} \quad \textit{in } \overline{Q_2} \equiv \overline{Q \cap \{\xi < b/2\}}. \tag{24.16}$$

Proof. a) The upper estimates in (24.15)–(24.16) follow directly from the boundedness of $w_{n\xi}$ in $\overline{Q}$.

b) Let us prove the lower estimate in (24.15). Set

$$\mu = \inf_{(0,T)} w\,(b/2, t) > 0.$$

Let us consider the function $h_n(\xi, \tau)$ defined by the conditions

$$h_n'' + \left[\frac{\psi\left(\varphi^{-1}(h_n)\right)}{\varphi^{-1}(h_n)}\right]' = 0 \quad \text{in } (b/2, b),$$

$$h_n(b,\tau) = \frac{1}{n}, \quad h_n'(b,\tau) = -\tau, \ \tau > 0.$$

It admits the representation

$$\int_{\frac{1}{n}}^{h_n(\xi,\tau)} \left[\tau + \frac{\psi(\varphi^{-1}(s))}{\varphi^{-1}(s)} - \frac{\psi(\varphi^{-1}(\frac{1}{n}))}{\varphi^{-1}(\frac{1}{n})}\right]^{-1} ds = b - \xi,$$

showing that h_n monotonically depends on τ and, also, that $h_n \to \frac{1}{n}$ as $\tau \to 0$. With the use of (17.13) we also obtain that

$$h_n \geq \frac{1}{n} + \tau(b - \xi).$$

Fix a number n' so large that $\mu > \frac{1}{n}$ and pick τ so that uniformly in n

$$h_n \geq w_n \quad \text{on the parabolic boundary of } Q_1.$$

Then it follows at once from the maximum principle that

$$w_n \geq h_n \geq \frac{1}{n} + \tau(b - \xi) \quad \text{in } Q_1.$$

c) To obtain the lower estimate in (24.16) we use the barrier function of the form

$$\omega_n(\xi) \equiv \frac{1}{n} + C\xi, \quad C = \text{const.} > 0.$$

Choose C so that

$$\omega_n \leq w_n \quad \text{on the parabolic boundary of } Q_2.$$

Next,

$$L_1\omega_n \equiv -\theta\left(\frac{1}{n} + C\xi\right),$$

whence the desired inequality. □

Remark 24.2. It is easy to verify that in Q_1 the limit function w admits a lower barrier of the form

$$h(\xi, t) \equiv C\frac{b - \xi}{1 + t}$$

with an appropriately chosen constant C.

We proceed to derive lower estimates for the functions w_{nt} and w_t.

Lemma 24.4. *There exists a constant $C_0 > 0$ such that uniformly in n*

$$w_{nt} \geq -C_0 \frac{w_n}{t} \quad \text{in } \overline{Q}. \tag{24.17}$$

Proof. Let us consider the function

$$s(\xi, t) \equiv \frac{w_{nt}}{w_n}.$$

It satisfies in Q the equation

$$\begin{aligned} s_t &- \sigma(w_n)s_{\xi\xi} - \left[\theta(w_n) + 2\frac{w_{n\xi}}{w_n}\right] s_\xi \\ &- \theta(w_n)\left[\frac{\theta'(w_n)}{\theta(w_n)} - \frac{\sigma'(w_n)}{\sigma(w_n)}\right] w_{n\xi} s - w_n \frac{\sigma'(w_n)}{\sigma w_n} s^2 = 0, \end{aligned}$$

vanishes at the lateral boundaries of Q, and is bounded as $t = 0$. With the use of (17.11) we find that

$$\frac{\alpha}{\beta} \leq s\frac{\sigma'(s)}{\sigma(s)} \leq \frac{\beta + 2}{\alpha}.$$

As follows from (17.12), the coefficient of s is uniformly bounded in $\overline{Q}$. It is easy to check now that the function $-C_0/t$ with an appropriate constant C_0 (independent of n), is a lower barrier for s in $\overline{Q}$. □

Corollary 24.3. *It follows from* (24.17) *as $n \to \infty$ that:*

$$w_t \geq -C_0 \frac{w}{t} \quad \text{in } Q. \tag{24.18}$$

Corollary 24.4. *Relying on* (24.18), *it is easy to check that $V(\xi, t)$ has finite limits as $\xi \to 0$ and $\xi \to b$ for each $t > 0$.*

Proceeding in the familiar way (cf. Section 20), we first check that $X(\xi, t)$ has finite limits at the lateral boundaries of Q for each $t > 0$, and then consider the function

$$\Phi(\xi) \equiv V(\xi, t) + \delta X(\xi, t)$$

depending on t as a parameter. As was already proven, Φ is bounded in $[0, b]$ for each $t > 0$ fixed, and its monotonicity (provided δ is taken sufficiently large) follows from (24.18).

Corollary 24.5. *There exist rectangles*

$$R_1 \equiv Q \cap \{\xi > \xi'\}, \quad R_0 \equiv Q \cap \{\xi < \xi'\}$$

such that

$$2C_2 \geq w_{n\xi} \geq \frac{C_1}{2} \quad \text{in } \overline{R_0}, \quad -\frac{C_1}{2} \geq w_{n\xi} \geq -2C_2 \quad \text{in } \overline{R_1}.$$

uniformly in n and with the constants C_1, C_2 from the conditions of Lemma 24.3.

For the proof we merely observe that for each n

$$- C_2 \leq w_{n\xi}(b,t) \leq -C_1, \quad C_1 \leq w_{n\xi}(0,t) \leq C_2, \tag{24.19}$$

by virtue of (24.15)–(24.16), and then make use of inequality (24.17) in the following representation yielded by (24.4): for all $\xi_1 \leq \xi_2$

$$w_{n\xi}(\xi_2,t) - w_{n\xi}(\xi_1,t) = \int_{\xi_1}^{\xi_2} \frac{\psi\left(\varphi^{-1}(w_n)\right)}{\varphi^{-1}(w_n)}\, ds + \int_{\xi_1}^{\xi_2} \frac{w_{nt}}{\sigma(w_n)}\, ds.$$

The convergence of these integrals is provided by conditions (17.11)–(17.13) and inequalities (24.15)–(24.16).

Corollary 24.6. *From Corollaries* 24.4–24.5 *we conclude that* w_ξ *is defined everywhere in* $\overline{Q}$ *and satisfies conditions* (24.19).

24.3 The inverse change of variables. It follows from (18.9) and the derived properties of w that $X(\xi,t) \in C(Q \cup B_0)$ and is strictly monotone in ξ for each $t \in [0,T]$. Also,

$$X_t = V(\xi,t) \quad \text{in } Q.$$

Applying Corollary 24.4, from this we derive (24.13)–(24.14). Due to (24.19), we also have: for a.e. $t \in (0,T)$

$$C_2 \geq \eta'(t) \geq C_1, \quad -C_2 \leq \zeta'(t) \leq -C_1.$$

Thus, the boundaries of $\Omega(t)$ are strictly monotone Lipschitz continuous curves. Let us now consider the function $\overline{u}(x,t)$ defined by equalities (18.10). It follows directly from the continuity of w and X in $\overline{Q}$ that $\overline{u}$ is continuous in $\overline{S}$. The following relations hold:

$$\begin{aligned}
v_t(\xi,t) &\equiv \overline{u}_t(X(\xi,t),t) + \overline{u}_x(X(\xi,t),t)X_t(\xi,t)\\
&= v^2\left[\varphi_\xi(v) + \frac{\psi(v)}{v}\right]_\xi\\
&\equiv \varphi_{xx}(\overline{u}) + \psi_x(\overline{u}) - \overline{u}_x\left(\frac{\varphi_x(\overline{u})}{\overline{u}} + \frac{\psi(\overline{u})}{\overline{u}}\right),
\end{aligned}$$

showing that $\overline{u}$ satisfies in Q equation (17.2) in the classical sense. Also, for each $t > 0$ fixed

$$\lim_{x\to\partial\Omega(t)} \overline{u}(x,t)\left[\frac{\varphi_x(\overline{u})}{\overline{u}} + \frac{\psi(\overline{u})}{\overline{u}}\right] = 0,$$

providing the integral identity (17.10). □

24.4 Estimating the time derivative. We have already introduced the sets R_0 and R_1 where the functions w_n and w were strictly monotone in ξ. Fix an integer n (large enough), and consider the rectangle R_1. Let us introduce in R_1 the new independent space variable η and the new sought function $y_n(\eta, t)$ by the formulas

$$\eta + \frac{1}{n} = w_n(b - y_n(\eta, t), t), \quad y_n(\eta, t) = b - \xi.$$

Adopting the notation

$$\sigma_n(\eta) = \sigma(\eta + \frac{1}{n}), \quad \theta_n(\eta) = \theta(\eta + \frac{1}{n}),$$

from (24.4) we derive the following equation for the function $y_n(\eta, t)$:

$$y_{nt} - \sigma_n(\eta)\frac{\partial}{\partial\eta}\left(-\frac{1}{y_{n\eta}}\right) = \theta_n(\eta). \tag{24.20}$$

Equations (24.20) are considered in the domain $H \equiv (0, \eta_0) \times (0, T]$, where η_0 is still defined by the sole condition: H is contained in the image of R_1 on the plane of variables (η, t) .

Lemma 24.5. *There exists a constant $C > 0$ such that*

$$\frac{\partial}{\partial\eta}(y_{nt}) \leq \frac{CC_1^{-2}}{t} \quad \text{in } \overline{H}. \tag{24.21}$$

(Here and elsewhere the constants C_1 and C_2 are the same as in (24.19)*).*

Proof. Let us consider the function $s = y_{nt}$. By (24.20), it satisfies the equation

$$s_t - \sigma_n(\eta)\frac{\partial}{\partial\eta}\left[\frac{1}{y_{n\eta}^2}\frac{\partial s}{\partial\eta}\right] = 0 \quad \text{in } H,$$

and the boundary condition $s|_{\eta=0} = 0$. Thus,

$$\frac{\partial}{\partial\eta}\left[\frac{1}{y_{n\eta}^2}\frac{\partial s}{\partial\eta}\right]\Bigg|_{\eta=0} = 0.$$

Denote

$$\rho = \frac{s_\eta}{y_{n\eta}^2}.$$

It follows from the equation for s that

$$(s_\eta)_t - \frac{\partial}{\partial\eta}\left[\sigma_n(\eta)\frac{\partial\rho}{\partial\eta}\right] = 0 \quad \text{in } H,$$

whence, with the use of the relation

$$\left[\frac{s_\eta}{y_{n\eta}^2}\right]_t = \frac{(s_\eta)_t}{y_{n\eta}^2} - 2y_{n\eta}\left[\frac{s_\eta}{y_{n\eta}^2}\right]^2,$$

we derive the following equation:

$$\mathcal{K}\rho \equiv \rho_t - \frac{1}{y_{n\eta}^2}\frac{\partial}{\partial\eta}\left[\sigma_n(\eta)\frac{\partial\rho}{\partial\eta}\right] + 2y_{n\eta}\rho^2 = 0 \quad \text{in } H.$$

Consider the function $z(t) \equiv C/t$ with a constant $C > 0$ chosen so that $z > \rho$ as $\eta = \eta_0$ and $t \in [0, T]$. Increasing C (if needed), we derive the inequality

$$\mathcal{K}z > 0 \quad \text{in } H.$$

Set $q(\eta, t) \equiv \rho - z$. The function q possesses the following properties:

$$q > 0 \quad \text{as } \{\eta = \eta_0,\ t \in [0, T]\} \quad \text{and as } t = 0,$$

$$\frac{\partial q}{\partial\eta} = 0 \quad \text{as } \eta = 0,$$

and

$$0 > \mathcal{K}\rho - \mathcal{K}q \equiv q_t - \frac{1}{y_{n\eta}^2}\frac{\partial}{\partial\eta}\left[\sigma_n(\eta)\frac{\partial q}{\partial\eta}\right] + 2y_{n\eta}(\rho + z)q \quad \text{in } H.$$

Treating this equation as a linear parabolic one and evoking the maximum principle we conclude that $q \geq 0$ in $\overline{H}$. □

Corollary 24.7. *One can introduce in the same way the variable η connected with the limit function w. The corresponding function $y(\eta, t)$ satisfies equation* (24.20) *with the coefficients $\sigma(\eta)$ and $\theta(\eta)$. It follows from* (24.21) *that*

$$\frac{\partial}{\partial\eta}(y_t) \leq \frac{C}{C_1^2}\frac{1}{t} \quad \textit{in } H. \tag{24.22}$$

Lemma 24.6.

$$\forall t > 0 \quad \lim_{\eta\to 0}\frac{1}{\varphi^{-1}(\eta)}y_t(\eta, t) = 0. \tag{24.23}$$

Proof. By (24.20) we have: for each $s < \eta_0/2$

$$\begin{aligned}
-\left[\frac{1}{y_\eta(2s, t)} - \frac{1}{y_\eta(s, t)}\right] &\equiv \int_s^{2s}\frac{\partial}{\partial\eta}\left[-\frac{1}{y_\eta}\right]d\eta \\
&= \int_s^{2s}\frac{1}{\sigma(\eta)}\left[\sigma(\eta)\frac{\partial}{\partial\eta}\left[-\frac{1}{y_\eta}\right] - \theta(\eta)\right]d\eta + \int_s^{2s}\frac{\theta(\eta)}{\sigma(\eta)}\,d\eta \\
&= \int_s^{2s}\frac{y_t}{\sigma(\eta)}\,d\eta + \int_s^{2s}\frac{\theta(\eta)}{\sigma(\eta)}\,d\eta \qquad (24.24)\\
&\equiv \int_s^{2s}\frac{1}{\sigma(\eta)}\left[y_t - \frac{C'\eta}{t}\right]d\eta + \frac{C'}{t}\int_s^{2s}\frac{\eta}{\sigma(\eta)}\,d\eta + \int_s^{2s}\frac{\theta(\eta)}{\sigma(\eta)}\,d\eta.
\end{aligned}$$

(Here C' is the maximum of the constants in (24.18) and (24.22).) The function $y_t - \frac{C'}{t}\eta$ is nonpositive and monotone in η; therefore

$$\int_s^{2s} \frac{1}{\sigma(\eta)}\left[y_t - \frac{C'}{t}\eta\right]d\eta \le \left[y_t - \frac{C'}{t}\eta\right]_{|\eta=s} \int_s^{2s} \frac{d\eta}{\sigma(\eta)} \tag{24.25}$$

$$\le \left(1 - \frac{\varphi^{-1}(s)}{\varphi^{-1}(2s)}\right)\frac{1}{\varphi^{-1}(s)}\left(y_t(s,t) - C'\frac{s}{t}\right) \equiv I.$$

Taking into account the condition of "quasi-homogeneity" of φ (see Subsection 17.1), we finally get from (24.25) that

$$I \le (1-\delta_*)\frac{1}{\varphi^{-1}(s)}y_t(s,t).$$

The integrals remaining in the right-hand side of (24.24) converge and tend to zero as $s \to 0$, while the left-hand side tends to zero because of the Cauchy criterion and the existence of a finite limit of the function y_η as $\eta \to 0$ (see Corollary 24.2). Combining the derived relation with (24.18), we arrive at (24.23). □

Now we strengthen (24.23) applying a special barrier construction proposed in [17]. Adopt the notation

$$S(\eta,t) \equiv \frac{y_t}{\eta}.$$

This function satisfies in H the equation

$$\mathcal{L}S \equiv S_t - \frac{\sigma(\eta)}{\eta}\frac{1}{y_\eta^2}\left(\eta S_{\eta\eta} + 2S_\eta\right) + \frac{2}{y_\eta}\left(S - \frac{\theta(\eta)}{\eta}\right)\left(\eta S_\eta + S\right) = 0.$$

Lemma 24.7. *For each $t_0 \in [0,T)$ there exist constants η_1^*, N_1 such that*

$$|y_t(\eta,t)| \le N_1\frac{\eta}{t-t_0} \quad \text{in } \overline{H_0}, \tag{24.26}$$

where $H_0 = (0,\eta_1^*) \times (t_0, t_0+1]$.

Proof. For the proof we construct a local lower barrier for S. The upper barrier was already obtained (see (24.18)). Consider the function

$$z(\eta,t) = -\frac{\gamma}{\Phi(\eta)} - \frac{\varepsilon}{\delta(t-t_0)+\Phi(\eta)}, \quad \Phi(\eta) \equiv \int_0^\eta \frac{ds}{\varphi^{-1}(s)},$$

depending on positive parameters $\gamma, \varepsilon, \delta$. Let us calculate $\mathcal{L}z$. Since always

$$z_\eta = \frac{\gamma}{\varphi^{-1}(\eta)\Phi(\eta)^2} + \frac{\varepsilon}{\varphi^{-1}(\eta)\left(\delta(t-t_0)+\Phi(\eta)\right)^2} > 0,$$

we obtain

$$\frac{2}{y_\eta}(\eta z_\eta + z) \le \frac{2}{y_\eta}\left(z^2 - \frac{\theta(\eta)}{\eta}z\right)$$
$$\le \frac{4}{y_\eta}\frac{\gamma^2}{(\Phi(\eta))^2} + \frac{4}{y_\eta}\frac{\varepsilon^2}{(\delta(t-t_0)+\Phi(\eta))^2} - \frac{2}{y_\eta}\frac{\theta(\eta)}{\eta}z.$$

With the use of these inequalities we calculate:

$$\mathcal{L}z \equiv \frac{\gamma}{y_\eta^2}\frac{1}{\Phi(\eta)}\left[\frac{2\varphi'(\varphi^{-1}(\eta))}{\Phi(\eta)} + 1 - \frac{2\varphi'(\varphi^{-1}(\eta))\varphi^{-1}(\eta)}{\eta}\right] + \frac{\varepsilon}{y_\eta^2}\frac{1}{(\delta(t-t_0)+\Phi(\eta))^2}$$
$$\times\left[\delta y_\eta^2 + \frac{2\varphi'(\varphi^{-1}(\eta))}{\delta(t-t_0)+\Phi(\eta)} + 1 - \frac{2\varphi'(\varphi^{-1}(\eta))\varphi^{-1}(\eta)}{\eta}\right]$$
$$+ \frac{2}{y_\eta}\left(z - \frac{\theta(\eta)}{\eta}\right)(\eta z_\eta + z)$$
$$\le \frac{\gamma}{y_\eta^2(\Phi(\eta))^2} \times \left\{4\gamma y_\eta + 2y_\eta\frac{\theta(\eta)}{\eta}\Phi(\eta)\right.$$
$$\left. + \left[\frac{2\varphi'(\varphi^{-1}(\eta))}{\Phi(\eta)} + 1 - \frac{2\varphi'(\varphi^{-1}(\eta))\varphi^{-1}(\eta)}{\eta}\right]\right\}$$
$$+ \frac{\varepsilon^2}{(\delta(t-t_0)+\Phi(\eta))^2}\left\{\delta y_\eta^2 + 4\varepsilon y_\eta + 2\frac{2y_\eta\theta(\eta)}{\eta}(\delta(t-t_0)+\Phi(\eta))\right.$$
$$\left. + \left[\frac{2\varphi'(\varphi^{-1}(\eta))}{\Phi(\eta)} + 1 - \frac{2\varphi'(\varphi^{-1}(\eta))\varphi^{-1}(\eta)}{\eta}\right]\right\}.$$

Condition (17.11) implies the inequalities

$$\frac{\beta+1}{\beta}\frac{\eta}{\varphi^{-1}(\eta)} \le \int_0^\eta \frac{ds}{\varphi^{-1}(s)} \equiv \int_0^{\varphi^{-1}(\eta)} \frac{\varphi'(s)}{s}\,ds \le \frac{\alpha+1}{\alpha}\frac{\eta}{\varphi^{-1}(\eta)}$$

which, in turn, give:

$$\mathcal{L}z \le \frac{\gamma}{y_\eta^2(\Phi(\eta))^2}\left[4\gamma y_\eta + 2y_\eta\frac{\theta(\eta)}{\eta}\Phi(\eta) + 1 - 2\frac{\alpha+1}{\beta+1}\right]$$
$$+ \frac{\varepsilon}{y_\eta^2(\delta(t-t_0)+\Phi(\eta))^2} \tag{24.27}$$
$$\times\left[4\varepsilon y_\eta + 2y_\eta\frac{\theta(\eta)}{\eta}\delta(t-t_0) + \Phi(\eta) + \delta y_\eta^2 + 1 - 2\frac{\alpha+1}{\beta+1}\right].$$

Recall that $2\frac{\alpha+1}{\beta+1} > 1$ (see (17.11)). Then it follows from (24.27) that $\mathcal{L}z < 0$ if only $\gamma, \delta, \varepsilon, (t-t_0), \eta_1^*$ are small enough. Let us fix some auxiliary values $\delta_0, \varepsilon_0, \gamma_0, \eta_1^*$ so as to satisfy the inequality

$$\mathcal{L}z < 0$$

$$\forall(\eta, t) \in (0, \eta_1^*) \times (t_0, t_0 + 1], \quad (\delta, \varepsilon, \gamma) \in (0, \delta_0) \times (0, \varepsilon_0) \times (0, \gamma_0)$$

and then specify their choice in order to fulfill the inequality $S - z > 0$ at the parabolic boundary of H_0.

There holds the inequality

$$S - z \geq \frac{\varepsilon}{\delta(t - t_0) + \Phi(\eta)} + \frac{1}{\Phi(\eta)} \left(\gamma - \frac{\beta + 1}{\beta} \frac{|y_t|}{\varphi^{-1}(\eta)} \right) \tag{24.28}$$

whence, with the use of (24.22),

$$\forall \gamma \in (0, \gamma_0) \quad S - z > 0 \quad \text{as } \eta \to 0. \tag{24.29}$$

For $t = t_0$ (24.28) can be continued in the following way:

$$(S - z)|_{t=t_0} \geq \frac{1}{\Phi(\eta)} \left(\varepsilon - \frac{\beta + 1}{\beta} \frac{|y_t|}{\varphi^{-1}(s)} \right). \tag{24.30}$$

Let us choose $\varepsilon = \varepsilon_0/2$ and fix η_1^* so as to make the right-hand side of (24.30) nonnegative. By virtue of (24.22), such a choice of η_1^* is always possible. Decreasing η_1^* if needed, we derive the inequality

$$\frac{\beta + 1}{\beta} \sup_{t \in (t_0, t_0+1)} \frac{|y_t|}{\varphi^{-1}(s)} \bigg|_{\eta=\eta_1^*} \leq \frac{\varepsilon_0}{4} \tag{24.31}$$

and then pick δ_0 such that

$$\delta_0 \leq \frac{1}{2} \int_0^{\eta_1^*} \frac{ds}{\varphi^{-1}(s)} \equiv \Phi(\eta_1^*).$$

Having chosen the set of the parameters ε, η_1^*, δ_0, it follows from (24.28) that

$$(S - z)|_{\eta=\eta_1^*} \geq \frac{\varepsilon_0}{2\delta_0 + 2\Phi(\eta_1^*)} - \frac{\varepsilon_0}{4} \frac{1}{\Phi(\eta_1^*)} > 0 \tag{24.32}$$

regardless of the value of γ_0.

By (24.29)–(24.30), (24.32), and the inequality $\mathcal{L}z < 0$ we conclude that $z \leq S$ in $\overline{H_0}$. Moreover, this is true *for every* $\gamma \in (0, \gamma_0)$. Letting $\gamma \to 0$, we finally get:

$$y_t \geq -\frac{\varepsilon\eta}{\delta(t - t_0) + \Phi(\eta)} \geq -\frac{\varepsilon}{\delta} \frac{\eta}{t - t_0} \quad \text{in } H_0.$$

□

Corollary 24.8. *In the variables* (ξ, t), *the estimate* (24.26) *takes the following form: for each* $t_0 > 0$ *there exist constants* $R \in (0, b)$ *and* $K_1 > 0$ *such that*

$$|w_t| \leq K_1 \frac{w}{t - t_0} \quad \text{in } G_R(t_0), \tag{24.33}$$

where $G_R(t_0) \equiv (R, b) \times (t_0, t_0 + 1]$.

24.5 Continuous differentiability of the right-side interface. As in Section 19, we specify now the barrier construction earlier proposed in [93, 62] to prove that the function V, viewed as a solution of a parabolic equation, preserves a certain regularity at the lateral boundary of the problem domain.

Fix arbitrary $t_0 \in (0, T)$, $\rho \in (R, b)$ and consider the functions

$$U^{\pm}(\xi, t) \equiv \pm[V(\xi, t) - V(b, t_1)] + \theta_1 \int_0^{\theta_5 \rho} \frac{d\sigma}{\varphi^{-1}(\sigma)} + \frac{\theta_2}{\rho^2}\left[\theta_3(t - t_1) + (b - \xi)^2\right] - \theta_4 \int_0^{\theta_5 (b-\xi)} \frac{d\sigma}{\varphi^{-1}(\sigma)}$$

in the domain $G_{b-\rho}(t_0)$, where $t_1 \in (t_0, t_0 + 1)$ is arbitrary and fixed, and the constants $\theta_1, \dots, \theta_5$ are to be defined.

Our aim is to prove that an appropriate choice of the constants $\theta_1, \dots, \theta_5$ makes both the functions $U^{\pm}$ nonnegative in $\overline{G_{b-R}(t - t_0)}$. Then, letting $\xi = b$ in the definition of $U^{\pm}$, and choosing ρ in a special way, we will be able to show the continuity of $V(b, t)$ at the point $t = t_1$, and to indicate its modulus of continuity.

a) *Continuity of* $V(\xi, t)$. With the use of (24.33) and (17.11) we find:

$$\begin{aligned} |V(\xi, t) - V(\mu, t)| &\le \int_{\xi}^{\mu} \left.\frac{|w_t|}{\sigma(\varphi^{-1}(w))}\right|_{t=t_1} d\sigma \\ &\le \frac{C_2}{C_1} \frac{K_1}{t_1 - t_0} \frac{1}{\alpha + 1} \int_{C_1(b-\mu)}^{C_1(b-\xi)} \frac{d\sigma}{\varphi^{-1}(\sigma)} \end{aligned} \tag{24.34}$$

with the constants C_1, C_2 from inequalities (24.15).

A byproduct of (24.34) is the following inequality: as $t = t_1$

$$\begin{aligned} U^{\pm}(\xi, t_1) \ge & -\frac{C_2 K_1}{C_1(\alpha + 1)} \frac{1}{t_1 - t_0} \int_0^{C_1(b-\xi)} \frac{d\sigma}{\varphi^{-1}(\sigma)} \\ & + \theta_1 \int_0^{\theta_5 \rho} \frac{d\sigma}{\varphi^{-1}(\sigma)} - \theta_4 \int_0^{\theta_5(b-\xi)} \frac{d\sigma}{\varphi^{-1}(\sigma)}. \end{aligned} \tag{24.35}$$

b) *Positivity of* $U^{\pm}(\xi, t)$. Take $\theta_5 = C_1$ and connect θ_1 to θ_4 by the relation

$$\theta_1 = 1 + \theta_4 + \frac{K_1 C_2}{C_1} \frac{1}{\alpha + 1} \frac{1}{t_1 - t_0}.$$

Then it follows at once from (24.35) that $U^{\pm}(\xi, t_1) > 0$ as $\xi \in (b - \rho, b)$.

c) *Positivity of* $U^{\pm}(b - \rho, t)$. Choose $\theta_2 = 1 + 2\max_{\overline{Q}} |V|$ and take into account the relation between the parameters θ_1 and θ_4.

d) *The sign of the space derivative of* $V^{\pm}$ *at the axis* $\xi = b$. With the use of (24.33) we find:

$$\frac{\partial U^{\pm}}{\partial \xi} = \pm \frac{\partial V}{\partial \xi} - \frac{2\theta_2(b - \xi)}{\rho^2} + \frac{C_1 \theta_4}{\varphi^{-1}(C_1(b - \xi))} \ge \tag{24.36}$$

$$\geq -\frac{K_1}{t-t_0}\frac{w}{\varphi^{-1}\varphi'(\varphi^{-1}(w))}\frac{1}{\varphi^{-1}(w)} + \frac{C_1\theta_4}{\varphi^{-1}(C_1(b-\xi))} - \frac{2\theta_2(b-\xi)}{\rho^2}$$
$$\geq \frac{1}{\varphi^{-1}(C_1(b-\xi))}\left[C_1\theta_4 - \frac{K_1}{(1+\alpha)(t-t_0)}\right] - \frac{2\theta_2(b-\xi)}{\rho^2}.$$

Choosing

$$\theta_4 > \frac{2K_1}{(1+\alpha)C_1(t-t_0)},$$

we get:

$$\frac{\partial U^\pm}{\partial\xi}(\xi,t) \to \infty \quad \text{as } \xi \to b.$$

e) The parameters $\theta_1, \theta_2, \theta_4, \theta_5$ are chosen so that for each $\theta_3, \rho \in (0, b-\rho)$ the functions $U^\pm$ cannot assume their negative minima at the parabolic boundary of $G_{b-\rho}(t_1)$. Let us now choose θ_3 so as to eliminate the possibility of occurrence of points of its negative minimum inside $G_{b-\rho}(t_1)$. Let $\mathcal{M}(\cdot)$ be the differential operator from (24.7). We have:

$$\mathcal{M}(w)U^\pm \equiv \frac{\theta_2}{\rho^2}\left[\theta_3 - 2\sigma(w) - 2\left(\sigma'(w)w_\xi + \theta(w)\right)(b-\xi)\right]$$
$$+ C_1\theta_4\frac{\sigma(w)}{\sigma(C_1(b-\xi))} - \frac{C_1\theta_4}{\varphi^{-1}(C_1(b-\xi))}\left[\sigma'(w)w_\xi + \theta(w)\right].$$

Making use of the relations

$$\frac{\sigma'(w)}{\varphi^{-1}(w)} = 2 + \left[s\frac{\varphi''(s)}{\varphi'(s)}\right]\Bigg|_{s=\varphi^{-1}(w)}, \qquad \frac{\theta(w)}{\varphi^{-1}(w)} = \left[s\left(\frac{\psi(s)}{s}\right)'\right]\Bigg|_{s=\varphi^{-1}(w)}$$

we find that all terms of $\mathcal{M}(w)U^\pm$ are bounded. Increasing θ_3, we arrive at the inequality

$$\mathcal{M}(w)U^\pm > 0 \quad \text{as } \xi \in (b, b-\rho), t \in (t_1, t_0+1]$$

and then evoke the maximum principle to conclude that

$$U^\pm \geq 0 \quad \text{in } \overline{G_{b-\rho}(t_1)} \cap \{t \leq t_0 + 1\}.$$

Letting $\xi = b$ and $t = t_2 > t_1$, we arrive at the following inequality: $\forall\rho \in (0, b-\rho)$, $t \in (t_1, t_0+1]$

$$|V(b,t_2) - V(b,t_1)| \leq \frac{\theta_2\theta_3}{\rho^2} + \theta_1\int_0^{C_1\rho}\frac{d\sigma}{\varphi^{-1}(\sigma)}.$$

Given t_1, t_2, let us define ρ as the solution of the equation

$$\Phi(\rho) \equiv \rho^2\int_0^{C_1\rho}\frac{d\sigma}{\varphi^{-1}(\sigma)} - (t_2 - t_1) = 0. \qquad (24.37)$$

Clearly, this equation always has a unique solution because

$$\Phi(0) < 0, \quad \Phi'(\rho) > 0, \quad \text{and } \Phi(\rho) \to \infty \text{ as } \rho \to \infty.$$

To ensure that this solution belongs to the interval $(0, R)$, one has to require that the difference $t_2 - t_1$ is small enough.

Having fixed t_2 and assuming that the lower face of the domain $G_\rho(t_1)$ is variable, we deduce inequality (24.36) with $t_2 \in (t_0, t_1)$, which completes the proof of continuity of $V(b, t)$ at the point $t = t_1$. Moreover, combining (24.36)–(24.37) we obtain the *modulus of continuity* of $V(b, t)$,

$$|V(b, t_2) - V(b, t_1)| \le (\theta_1 + \theta_2\theta_3) \int_0^{C_1\rho^*} \frac{d\sigma}{\varphi^{-1}(\sigma)},$$

where ρ^* is the minimal root of equation (24.37) with the given parameters t_1 and t_2. The constants $\theta_1, \ldots, \theta_3$ are defined as is demonstrated above.

24.6 Left-side interface. All arguments of this section remain true in the analogous treatment of the curve $x = \zeta(t)$. We introduce in the domain R_0 the variable η and the function $y(\eta, t)$ by the relations

$$\eta = w(y(\eta, t), t), \quad \xi = y(\eta, t),$$

whence

$$0 = \frac{d\eta}{dt} = w_t + w_\xi y_t, \quad w_\xi y_\eta = 1.$$

Substituting these relations, replacing (22.2), (22.3), into equation (24.1), we get

$$y_t - \sigma(\eta)\frac{\partial}{\partial\eta}\left(-\frac{1}{y_\eta}\right) = -\theta(\eta). \tag{24.38}$$

Equation (24.38) is different from (24.20) only in the sign of the right-hand side but this leads to a prominent simplification: condition (17.13) connecting the functions φ and ψ is no longer needed. Evidently, this condition was used only twice: in the proof of Lemma 24.3 to construct a lower barrier for $w(\xi, t)$, and in the proof of Lemma 24.7 to estimate the last term of the right-hand side of (24.27). Equation (24.38) shows that now all terms which earlier required a special estimation already have "good" signs and can in fact be omitted.

24.7 Infinite differentiability of the interfaces. Let us consider now the Cauchy problem for the equation

$$u_t = \left(u^m\right)_{xx} + \left(u^\lambda\right)_x$$

with the parameters $m > 1$, $\lambda > 1$. It is clear that under the conditions of Theorem 17.4 all the preceding arguments of this section remain true, and the occurring interfaces satisfy equations (24.13)–(24.14). For the sake of definiteness, let us confine ourselves to the case of the right-side interface and prove part a) of Theorem 17.4. The proof of part b) does not involve any major change and may be therefore omitted.

Let $\lambda \geq m$. Following the proof of C^{∞}-regularity of the interfaces in the case $\lambda \in (0,1)$ (Section 23), one has to check that given an arbitrary natural number k, the functions

$$D_t^k V(\xi,t) = \frac{\partial}{\partial \xi}\left(D_t^k w\right) + D_t^k\left(w^{(\lambda-1)/m}\right)$$

have finite limits as $\xi \to b$ for each t. Let us pass to the variables (η, t). The boundedness of $D_t^k V(\xi,t)$ is equivalent to the boundedness of the functions

$$\sigma_k = \frac{\partial}{\partial \eta}\left[D_t^{k-1}\left(\frac{y_t}{y_\eta}\right)\right] + D_t^{k-1}\left[\frac{\lambda-1}{m} w^{(\lambda)/m}\frac{y_t}{y_\eta}\right]$$

as $\eta \to 0$. It is easy to verify that σ_k are indeed bounded if, say,

$$\begin{cases} |y_t| \leq L_1\eta, \\ |D_t^k y| \leq L_2\eta^{1/m}, \\ \left\{\left|D_\eta\left(D_t^k y\right)\right|;\ \left|D_\eta^2\left(D_t^k y\right)\right|\right\} \leq L_3. \end{cases} \tag{24.39}$$

Similarly to (22.32), inequalities (24.39) can be proven by induction. Let us derive the equations for the functions $q^{(1)} = y_t$ and $g = q_\eta^{(1)}$. Differentiating (24.20) in t, we obtain

$$Rq^{(1)} \equiv q_t^{(1)} - m\frac{\eta^{(m+1)/m}}{y_\eta^2} q_{\eta\eta}^{(1)} + \frac{2}{y_\eta}\left[q^{(1)} - (\lambda-1)\eta^{\lambda/m}\right] q_\eta^{(1)} = 0. \tag{24.40}$$

For g we get from (24.40):

$$\begin{aligned} g_t &- m\frac{\eta^{(m+1)/m}}{y_\eta^2} g_{\eta\eta} + \left[\frac{4}{y_\eta}\left(q^{(1)} - (\lambda-1)\eta^{\lambda/m}\right) - (m+1)\frac{\eta^{1/m}}{y_\eta^2}\right] g_\eta \\ &- \left[\frac{2}{m\eta^{(m+1)/m}}\left(q^{(1)} - (\lambda-1)\eta^{\lambda/m}\right)^2 - \frac{2}{y_\eta}\left(g - (\lambda-1)\frac{\lambda}{m}\eta^{(\lambda-m)/m}\right)\right] = 0. \end{aligned} \tag{24.41}$$

Following the proofs of Lemmas 24.4, 24.5, we fix a point $(\overline{\eta}, \overline{t})$, define the corresponding rectangle G, and introduce the variables (ζ, τ) by the relations

$$\eta = \overline{\eta} + \zeta\left(\frac{\overline{\eta}}{2}\right)^s, \quad t = \overline{t} + \tau\left(\frac{\overline{\eta}}{2}\right)^p.$$

Thereafter we set $q^{(1)} = Q(\zeta,\tau)$, $s = 1$, $p = (m-1)/m$. The result of this rescaling is that equation (24.40) yields a uniformly parabolic equation for Q, whence, referring to [97, Th. 5.3.1], we get the estimate

$$|Q_\zeta(0,0)| \leq L_4 \max_{\overline{J}} |Q|, \quad J = (-1,1)\times(-1,0).$$

Returning to the variables (η, t) and applying inequality (24.33), we find

$$|q_\eta^{(1)}(\overline{\eta}, \overline{t})| \le \frac{L_5}{(\overline{\eta}/2)} \max_{\overline{G}} |q^{(1)}| \equiv L_6.$$

Relying on the last inequality and treating in the same way the solution of equation (24.41), we find that

$$|q_{\eta\eta}^{(1)}| \le \frac{L_7}{\overline{\eta}}.$$

The last estimate allows one to estimate $|q_t^{(1)}|$ directly from equation (24.40):

$$|q^2| \equiv |q_t^{(1)}| \le \frac{L_8}{\overline{\eta}^{1/m}}.$$

To estimate the higher-order derivatives we derive the equations

$$Rq^{(s)} = \Phi_s(\eta, q^{(1)}), \dots, q^{(s-1)}), \quad s = 1, 2, 3, \dots,$$

where the parabolic operator $R(\cdot)$ is defined in (24.40) and the right-hand sides Φ_s are polynomials. The desired estimates on $g^{(s)}$ then follow by induction in s.

24.8 A nonlocal equation in population dynamics. Let us extend the above-developed techniques to the study of regularity properties of the interfaces occurring in solutions of the Cauchy problem for nonlocal equations of a special class. Consider the Cauchy problem for the equation

$$u_t = \left(u^m\right)_{xx} + \left[u\left(\int_{-\infty}^{x} u(y,t)\,dy - \int_{x}^{\infty} u(y,t)\,dy\right)\right]_x \tag{24.42}$$

$$\text{in } S \equiv \mathbb{R} \times (0, T]$$

with the parameter of nonlinearity $m > 1$.

This equation represents a simplified model of the process of diffusion of a biological population when a nonlocal interaction of the individuals is taken into account. (See [114, 115] and the references therein.) The sought function $u(x, t)$ stands for the density of the biological population and is assumed nonnegative in $\overline{S}$. The initial distribution of the population density

$$u(x, 0) = u_0(x) \quad \text{in } \mathbb{R} \tag{24.43}$$

is subject to the following conditions:

(A.1) $u_0(x) > 0$ in $(0, a)$, $a < \infty$; $u_0(x) \equiv 0$ outside $(0, a)$;

(A.2) $\left\|\left(u_0^{m-1}\right)_x\right\|_{L^\infty(0,a)} \le M.$

(A.3) $\left(u_0^{m-1}\right)' \in C[0, a]$ and

$$\lim_{x\to a-0} \left(u_0^{m-1}\right)' \le -q, \quad \lim_{x\to 0+0} \left(u_0^{m-1}\right)' \ge q, \quad q = \text{const.} > 0.$$

We use the following concept of the solution.

Definition 24.1. A function $u(x,t)$ is said to be a weak solution to problem (24.42)–(24.43) if it is bounded and nonnegative in $\overline{S}$ and satisfies the following conditions:

(i) $u \in C(S) \cap L^{\infty}_{\mathrm{loc}}([0,T]; L^1(\mathbb{R}))$;

(ii) $(u^m)_x \in L^{\infty}(\mathbb{R} \times [\tau, T])$ for all $0 < \tau < T$;

(iii)
$$\int_{-\infty}^{x} u(y,t)\,dy \in C(S), \quad \int_{-\infty}^{\infty} u(y,t)\,dy \in C([0,T]),$$

$$\lim_{t\to 0} \int_{-\infty}^{x} u(y,t)\,dy = \int_{-\infty}^{x} u_0(y)\,dy \quad \forall\, x \in (-\infty, \infty];$$

(iv) for each $f \in C(S)$, compactly supported in S, the following identity holds:

$$\int_S \left\{ u f_t - \left[\left(u^m\right)_x + \left(\int_{-\infty}^{x} u(y,t)\,dy - \int_{x}^{\infty} u(y,t)\,dy \right) \right] f_x \right\} dx\,dt = 0.$$

As is shown in [114], there exists a unique weak solution of problem (24.42)–(24.43) $u(x,t)$. This solution displays the conservation property

$$\forall t \in (0,T] \quad \int_R u(x,t)\,dx = \int_R u_0(x)\,dx = b, \tag{24.44}$$

which suggests that Lagrangian coordinate transformation might be of benefit in the study of the regularity properties of solutions and interfaces.

Lagrangian coordinates. Let us view problem (24.42)–(24.43) as the mathematical model of some liquid flow with the density $\rho = u(x,t)$ and the velocity

$$V(x,t) = -\frac{m}{m-1}\left(u^m\right)_x - \left\{ \int_{-\infty}^{x} u(y,t)\,dy - \int_{x}^{\infty} u(y,t)\,dy \right\}.$$

Thereafter, following the routine scheme of the Lagrangian coordinates introduction, let us declare the quantity

$$\xi = \int_{-\infty}^{x} u_0(s)ds : [0,a] \mapsto [0,b]$$

the new independent space variable. Relying on (24.24) and literally repeating the arguments of Section 18.1, we find that the velocity of this flow is now given by the expression

$$W(\xi,t) \equiv V[X(\xi,t),t] = -\left(v^m\right)_\xi - 2\xi + b.$$

We arrive then at the following conditions for the density $v(\xi,t) \equiv u\left[X(\xi,t),t\right]$:

$$\left(\frac{1}{v}\right)_t + \left(v^m\right)_{\xi\xi} + 2 = 0 \quad \text{in } Q, \tag{24.45}$$

$$v(0,t) = v(b,t) = 0 \quad \text{as } t \in [0,T], \tag{24.46}$$

$$v(\xi,0) = v_0(\xi) \equiv u_0[X(\xi,0)] \quad \text{in } \xi \in [0,b]. \tag{24.47}$$

The formal basis for the employment of the Lagrangian coordinates is similar to that of the previous sections. Having proved the existence of a classical solution to problem (24.45)–(24.47), we then recover the solution of the original problem. The interfaces acquire the representations

$$\eta(t) = a - bt - \lim_{\xi\to b} \int_0^t \left(v^m\right)_\xi (\xi,\theta)\, d\theta, \quad t > 0,$$

$$\zeta(t) = bt - \lim_{\xi\to 0} \int_0^t \left(v^m\right)_\xi (\xi,\theta)\, d\theta, \quad t > 0,$$

which can be treated as in Sections 24.3–24.7.

Theorem 24.1. *Let assumptions* (A.1)–(A.3) *hold, and let* $u(x,t)$ *be a weak solution of the Cauchy problem* (24.42)–(24.43). *Then*

$$\eta(t), \zeta(t) \in C^\infty_{\text{loc}}(0,T).$$

Note that now the interfaces need not be monotone and can exhibit the turning points but, nonetheless, possess the smoothness properties identical to those known in the case of PME.

Let us point out here the key steps in the treatment of problem (24.45)–(24.47) which are required for the proof of Theorem 24.1[2]

a) *Solvability of problem* (24.45)–(24.47). A classical solution can be constructed by the standard regularizing procedure. We take a sequence of the initial and boundary data, approximating the data of the degenerate problem (24.45)–(24.47), so as to satisfy the infinite-order compatibility conditions at the corner points of the rectangle Q. To get the needed set of basic estimates we study the problems

$$Lw_n \equiv w_{nt} - mw_n^{(m+1)/m} w_{n\xi\xi} - 2mw_n^{(m+1)/m} = 0 \quad \text{in } Q, \tag{24.48}$$

$$w_n(0,t) = w_n(b,t) = \frac{1}{n} \quad \text{as } t \in [0,T],$$

$$w_n(\xi,0) \equiv w_{0n}(\xi) \quad \text{in } (0,b),\ n = 1,2,3,\ldots.$$

Instead of w_n, it will be convenient to work with the function

$$S_n = w_n + \xi^2 - b\xi,$$

which satisfies the equation

$$S_{nt} - mw_n^{(m+1/m)} S_{n\xi\xi} = 0 \quad \text{in } Q$$

[2]This presentation follows the paper by J. I. Diaz, T. Nagai, S. I. Shmarev, On the interfaces in a nonlocal quasilinear degenerate equation arising in population dynamics, *The Japan Journal of Industrial and Applied Mathematics* 13 (1996), 1–31.

following from (24.45), the same boundary conditions as w_n, and the initial condition

$$S_n(\xi, 0) = w_{0n}(\xi) + \xi^2 - b\xi \quad \text{in } [0, b].$$

The advantage of this change is that the functions S_n already obey the maximum principle for solutions of parabolic equations.

b) *Estimating the velocity.* To estimate the velocity amounts to proving the uniform boundedness of the functions $S_{n\xi}$ which satisfy the equations

$$\frac{\partial}{\partial t}\left(S_{n\xi}\right) - m\frac{\partial}{\partial \xi}\left[w_n^{(m+1)/m}\frac{\partial}{\partial \xi}\left(S_{n\xi}\right)\right] = 0.$$

Thus, S_n are permitted to assume their extreme values only on Γ, that is only at $t = 0$ (cf. (24.8)).

c) *Estimating the time derivative.* The estimates for the function $(v^m)_t$ are crucial for the study of the interface regularity. The very first estimate is analogous to (24.18) and has the form

$$\left(v^m\right)_t \geq -C\frac{v^m}{t}, \qquad C = \text{const.} > 0.$$

To derive it, we construct a lower barrier of the form $-C/t$ for the function $\sigma = (v^m)_t/v^m$, which satisfies the equation

$$\begin{aligned} \sigma_t \; &- \; mw^{(m+1)/m}\sigma_{\xi\xi} - 2mw^{1/m}w_\xi\sigma_\xi \\ &- \; 2(m+1)w^{1/m}(w-1)\sigma - \frac{m+1}{m}\sigma^2 = 0. \end{aligned}$$

Assumption (A.3) provides the inequalities

$$w_{n\xi}(b,t) \leq -L, \quad w_{n\xi}(0,t) \geq L, \qquad t > 0,\; L = \text{const.} > 0,$$

which allow one to introduce the new space variable η and the new sought function $y_n(\eta, t)$ (see Section 24.2). In some neighborhood of the lateral boundary $\eta = 0$ each function y_n satisfies the equation

$$y_{nt} = m\frac{\left(\eta + \frac{1}{n}\right)^{(m+1)/m}}{y_{n\eta}^2}y_{n\eta\eta} + 2my_{n\eta}\left(\eta + \frac{1}{n}\right)^{(m+1)/m}. \tag{24.49}$$

Treating these equations, which are very much like (24.20), we consequently establish for the limit function y_t the estimates

$$\eta^{-1/m}y_t(\eta, t) \to 0 \quad \text{as } \eta \to 0,$$

$$y_t \geq -C\frac{y}{t}, \tag{24.50}$$

completing the lower bound for $(v^m)_t$. The former of these two estimates is derived by the literal repetition of the proofs of (22.6) and (24.23). To get the latter one, we follow the proof of Lemma 24.7. Let us denote $\sigma \equiv y_t$. The equation for σ follows from (24.49) and has the form

$$\mathcal{N}\sigma \equiv \sigma_t - m\eta^{(m+1)/m}\frac{1}{y_\eta^2}\sigma_{\eta\eta} - 2m\eta^{(m+1)/m}\sigma_\eta + \frac{2}{y_\eta}\sigma\sigma_\eta = 0. \qquad (24.51)$$

It is easy to check that the function

$$z(\eta,t) \equiv -\gamma\eta^{1/m} - \varepsilon\frac{\eta}{\delta(t-t_0)+\eta^{(m-1)/m}},$$

depending on the positive parameters ε, δ, γ, satisfies the inequality

$$\begin{aligned}\mathcal{N}z \le{}& \gamma\eta^{-1+\frac{2}{m}}\left[-\frac{m-1}{m}\frac{1}{y_\eta^2} + \frac{2\varepsilon(m+1)}{m}\frac{1}{y_\eta}\right]\\ &+ \frac{\eta}{\left(\delta(t-t_0)+\eta^{(m-1)/m}\right)^2}\\ &\times\left[\varepsilon\delta + \frac{2\varepsilon^2}{y_\eta} - 2\varepsilon(m-1)\eta - \varepsilon\frac{m-1}{m}\frac{1}{y_\eta^2} + 2m\varepsilon\delta\eta^{1/m} + 2m\varepsilon\eta\right]\end{aligned}$$

For ε, δ, γ, η small enough we get that $\mathcal{N}z < 0$ and, hence, z may be taken as a lower barrier for σ. The choice of the parameters has to be carried out so as to provide the positive sign of the difference $\sigma - z$ at the parabolic boundary of the rectangle (small) where the sign of $\mathcal{N}z$ is controlled (cf. the proof of Lemma 24.7).

The further arguments needed for the proof of $C^{1,\alpha}$-regularity of the interfaces do not require any changes in comparison with those of Sections 24.5–24.6.

C^∞-regularity of the interfaces. Having proved $C^{1,\alpha}$-regularity of the interface, we can improve this result, reducing the problem to the one already solved for PME.

Let us consider the Cauchy problem for PME with the initial function u_0 subject to conditions (A.1)–(A.3). It is easy to see that the same arguments, i.e. the passage to Lagrangian coordinates, the proof of the unique solvability of the corresponding problem in the time-independent domain, the proof of the equivalence between the original and Lagrangian statements, and the study of the interface regularity are true. (See also [162] for the direct proof of this equivalence.) The interface in the Cauchy problem for PME has the representation formula

$$\eta(t) = a - \int_0^t (v^m)_\xi(b,\theta)\,d\theta,$$

where v is the solution of the equation

$$\left(\frac{1}{v}\right)_t + (v^m)_{\xi\xi} = 0 \quad \text{in } Q$$

satisfying boundary and initial conditions (24.46), (24.47).

It is shown in [17] that in the case of PME $\eta \in C^\infty(\tau, T)$, $\tau > 0$. This result holds for the monotone moving part of the interface, i.e. under condition (A.3), for the whole interface. The result of [17], written in the variables (ξ, t), is the following. Let

(i) $\begin{cases} \frac{1}{C}(b-\xi) \le v^m(\xi,t) \le C(b-\xi), \ C = \text{const.} > 0, \\ \text{in some rectangle } E(\tau), \tau > 0; \end{cases}$

(ii) $|(v^m)_t(\xi,t)| \le C_1 v^m(\xi,t)$ in $\overline{E(\tau)}$.

Then the functions $D_t^s((v^m)_\xi)$, $s = 1, 2, \dots$, have finite limits as $\xi \to b$. This conclusion is arrived at by consideration of the parabolic equations obtained for $D_t^s((v^m)_\xi)$ from PME. The structure of their coefficients and the estimates of (i) allow one to choose a special rescaling transformation rendering (locally) these equations uniform parabolic ones; thereafter one has to have recourse to the results of the classical parabolic theory [97, Th. 5.3.1], which provide the required estimates.

Let us now write down another Lagrangian counterpart of the Cauchy problem (24.42), (24.43). Eliminating from the trajectory equation and the mass balance law of the fictitious flow the velocity $v(\xi, t)$, for the trajectory $X(\xi, t)$ we obtain the following conditions:

$$X_t = m\frac{1}{X_\xi^{m+1}}X_{\xi\xi} - 2\xi + b \quad \text{in } Q.$$

A solution to this problem is given by the constructed solution to (L). Formally, the function $\sigma = X_t$ satisfies the equation

$$\sigma_t = m\frac{1}{X_\xi^{m+1}}\sigma_{\xi\xi} - m(m+1)\frac{X_{\xi\xi}}{X_\xi^{m+2}}\sigma_\xi.$$

The same equation with the coefficients possessing the same properties holds for the function $\overline{X}_t$, if we denote by $\overline{X}$ the function standing for the trajectory of the particle in the Lagrangian counterpart of the Cauchy problem for PME. This coincidence proves that the differential properties of the functions X and $\overline{X}$ must be the same.

A generalization of equation (24.42). Here we point out how all the preceding analysis may be extended to the case of the Cauchy problem for the general equation

$$u_t = \left(u^m\right)_{xx} + \left[u\Phi\left(\int_{-\infty}^x u(y,t)\,dy\right)\right]_x.$$

Assume that

$$\Phi'(s) > 0 \quad \text{as } s \ge 0.$$

Following the above-described scheme of the introduction of Lagrangian coordinates, we arrive at the equation

$$\left(\frac{1}{v}\right)_t + \left(v^m\right)_{\xi\xi} + \Phi'(\xi) = 0 \quad \text{in } Q.$$

Thus, the only difference between the above-considered case of the linear function Φ and the general case is that now one has to deal with equation (24.45), having the coefficient of the lower-order term not constant but bounded and positive. This provides the validity of all principal steps of the previous work that relied only on the maximum principle. Passing to a sequence of solutions v_n of the regularized problems, corresponding to (24.45), and denoting $w_n := v_n^m$ we then introduce the functions

$$S_n := w_n + \int_0^{\xi} \Phi(s)\, ds.$$

It is easy to see that if Φ is smooth enough, one can literally repeat the proofs of all propositions of this section. As a conclusion we obtain

Theorem 24.2. *Let conditions* (A.1)–(A.3) *hold. Assume that* $\Phi \in C^\infty$ *and* $\Phi'(s) > 0$ *for* $s \geq 0$. *Then*

$$\eta \in C^\infty_{\text{loc}}(0, T).$$

25 Axially symmetrical solutions of the porous medium equation: long-time asymptotic behavior

Let us start by recalling the main purpose of this section. We are interested in the long-time behavior of solutions of the following problem:

$$(\mathrm{P}_1) \quad \begin{cases} u_t = r^{1-N}\left(r^{N-1}\varphi_r(u)\right)_r & \text{in } H = (0,\infty)\times(0,T], \\ \left[r^{N-1}\varphi_r(u)\right]\big|_{r=0} = 0 & \text{as } t \in [0,T], \\ u(r,0) = u_0(r) & \text{for } r \geq 0. \end{cases}$$

The initial function u_0 and the function $\varphi(\cdot)$ are always assumed to satisfy conditions (C.1)–(C.3) of Section 17 and (17.23).

25.1 Lagrangian statement. Let us consider problem (P_1) as the mathematical problem governing a one-dimensional liquid flow with the density

$$\rho = R^{N-1}u(R,t)$$

and the velocity

$$V(R,t) = -\frac{\partial}{\partial R}\left(\int_0^{u(R,t)} \frac{\varphi'(s)}{s}\, ds\right).$$

Equation (17.24) can then be viewed as the mass balance law of this flow and condition (17.25) means that the liquid is motionless at the axis $R = 0$. To describe this one-dimensional motion we employ the mass Lagrangian coordinate

$$\xi = \int_0^R \sigma^{N-1} u_0(\sigma) d\sigma : [0, a] \mapsto [0, b].$$

Let us denote by $r(\xi, t)$ the trajectory of the liquid particle located at the point ξ at the instant $t = 0$, and adopt the notation $v(\xi, t)$ for the density at this particle:

$$\rho = r(\xi, t) u[r(\xi, t), t] \equiv r(\xi, t) v(\xi, t).$$

To define $r(\xi, t)$ and $v(\xi, t)$, we have the trajectory equation (18.1) and the continuity equation, now having the form

$$\left(r^N(\xi, t)\right)_\xi = \frac{N}{v(\xi, t)} \quad \text{in } Q. \tag{25.1}$$

(Wherever possible we keep the notation of the previous sections.) Condition (17.25) can be rewritten in the form

$$r(0, t) = 0 \quad \text{as } t \in [0, T], \tag{25.2}$$

and from (25.1)–(25.2) we infer that

$$r^N(\xi, t) = N \int_0^\xi \frac{d\sigma}{v(\sigma, t)} \quad \text{in } Q. \tag{25.3}$$

Substituting (25.3) into (18.1), we arrive at the equation for $v(\xi, t)$:

$$\left(\frac{1}{v}\right)_t + \left(r^{2(N-1)} \varphi_\xi(v)\right)_\xi = 0 \quad \text{in } Q. \tag{25.4}$$

The image of the interface is now the straight line $\xi = b$, where

$$v(b, t) = 0 \quad \text{as } t \in [0, T]. \tag{25.5}$$

Finally, the initial conditions for $v(\xi, t)$ are given by condition (17.26):

$$v(\xi, t) = v_0(\xi) \quad (\equiv u_0(r(\xi, 0))) \qquad \text{as } \xi \in [0, b]. \tag{25.6}$$

As above, problem (25.4)–(25.26), (25.2) is called *problem* (L).

Equation (25.4) degenerates (fails to be parabolic) at the lateral boundary of Q $\{\xi = b\}$ due to (25.5), and at $\xi = 0$ by virtue of (25.2). Moreover, it contains the nonlocal coefficient $r^{2(N-1)}(\xi, t)$. The last feature is a complication both for the proof of solvability of problem (L) and for the investigation of the qualitative properties of solutions. Had we derived the equation for the function $r(\xi, t)$ (equations of the types (25.15), (25.16) below), we would see that these equations do not contain any nonlocal term. However, the boundary condition for r^N (corresponding to (25.5)) has the form

$$\left(r^N\right)_\xi (\xi, t) \to \infty \quad \text{as } \xi \to b,$$

and so is far from being classical.

To circumvent both these difficulties (a "good" equation with a "bad" boundary condition or, conversely, a "bad" equation supplied with the usual boundary conditions), we combine these two possible Lagrangian counterparts of problem (P_1). The nonlocal problem (L) is used to prove the existence theorem for Lagrangian counterpart of the original problem (P_1). Having done this, we automatically get the corresponding function $r(\xi, t)$ which, at the same time, is a solution of the initial-boundary value problem for the degenerate parabolic equation (25.16). The solutions of the latter obey the usual maximum principle, and the required estimates for $r(\xi, t)$ are then derived in the standard way via comparison with some barriers (sub/super solutions).

25.2 Solvability of problem (L). Regularization. Invertibility of the coordinate transformation.

Theorem 25.1. *Let the conditions* (C.1)–(C.3) *and* (17.23) *hold. Then problem* (L) *has at least one classical solution*

$$v(\xi, t) \in C\left(\overline{Q}\right) \cap C^{\infty}(Q)$$

strictly positive in Q *and such that*

$$0 \geq r^{N-1}(\xi, t)\varphi_{\xi}(v) \geq -C, \quad C = \text{const.} > 0. \tag{25.7}$$

Under the assumptions of Theorem 25.1 the function v_0 satisfies the conditions:

(N.1) $v_0 > 0$ as $\xi \in [0, b)$, $v_0(b) = 0$, $v_0 \in C[0, b]$;

(N.2) v_0 is monotone decreasing and Lipschitz continuous in $[0, b]$;

(N.3) there exist constants $M_0 > 0$ and $\xi_0 \in [0, b)$ such that

$$\varphi_{\xi}(v_0) \leq -M_0 \quad \text{in } [\xi_0, b].$$

Denote

$$r_n^N(\xi, t) = \frac{1}{n} + N \int_0^{\xi} \frac{d\sigma}{\varphi^{-1}(w_n(\sigma, t))}, \quad \sigma(s) = \varphi'\left(\varphi^{-1}(s)\right)\left(\varphi^{-1}(s)\right)^2$$

and consider the sequence $\{w_n\}$ of solutions of the regularized problems (L_n)

$$w_{nt} = \sigma(w_n)\frac{\partial}{\partial \xi}\left[r_n^{2(N-1)}(\xi, t)\frac{\partial w_n}{\partial \xi}\right] \quad \text{in } Q, \tag{25.8}$$

$$w_n(b, t) = \frac{1}{n} \quad \text{as } t \in [0, T], \tag{25.9}$$

$$w_n(\xi, 0) = w_{0n}(\xi) \quad \text{as } \xi \in [0, b], \tag{25.10}$$

$$\left[\frac{\partial w_n}{\partial \xi}\right]\Bigg|_{\xi=0} = 0. \tag{25.11}$$

The sequence of initial data $\{w_{0n}\}$ approximating w_0 we choose as follows:

(K.1) $w_{0n} \in C^{\infty}[0, b]$, $w_{0n} \to \varphi(v_0)$ as $n \to \infty$,
$w_{0n} \equiv \varphi\left(v_0\left(\frac{1}{n}\right)\right)$ as $\xi \in [0, \frac{1}{n}]$,
$w_{0n} \equiv \frac{1}{n}$ as $\xi \in \left[b - \frac{1}{n}, b\right]$;

(K.2) $\frac{1}{n} \le w_{0n} \le M = \text{const.}$ in $[0, b]$,
$w_{0n} \ge M_2(b - \xi) + \frac{1}{n}$ in $[0, b]$,
$0 \ge r^{N-1}(\xi, 0) w'_{0n}(\xi) \ge -M_3$ in $[0, b]$.

Equation (25.8) has the functional (or nonlocal) coefficient $r(\xi, t)$ and thus its solvability does not follow from any of the standard references. Nevertheless, the existence of a solution w_n to problem (L_n) can be established, for example, by the direct application of the Schauder Fixed Point Theorem, as is done in [97, Th. 6.5.1].

Let us assume first that the existence of a classical solution to problem (L_n) is already established.

Lemma 25.1. *Given a classical solution $w_{n\xi}$ to problem* (L_n), *the function*

$$q_n \equiv r^N w_{n\xi}$$

is nonpositive in $\overline{Q}$, and is bounded from below in $\overline{Q}$ by a constant not depending on n.

Proof. Denote $v_n = \varphi^{-1}(w_n)$ and consider the equations for q_n. In Q

$$\begin{aligned}\mathcal{L}_n q_n &\equiv q_{nt} - v_n^2 \varphi'(v_n) r_n^{2(N-1)} q_{n\xi\xi} - \\ &- \left[2 v_n r_n^{N-1} q_n - 2(N-1) v_n \varphi'(v_n) r_n^{N-2} v_n \frac{v_n \varphi''(v_n)}{\varphi'(v_n)} q_n\right] q_{n\xi} \\ &+ (N-1)\frac{\varphi'(v_n)}{r_n^2} q_n - (N-1)\frac{v_n \varphi''(v_n)}{\varphi'(v_n)} \frac{1}{r_n} q_n^2 = 0 \end{aligned} \tag{25.12}$$

By (25.11)

$$q_n(0, t) = 0 \tag{25.13}$$

and, due to (25.8)–(25.9),

$$\left[r_n^{N-1} q_{n\xi} + \frac{N-1}{v_n r_n} q_n\right]\Bigg|_{\xi=b} = 0. \tag{25.14}$$

(25.13)–(25.14) imply that q_n can assume its extreme values only in $Q \cup B_0$. The form of equation (25.12) shows that q has no points of local negative minimum in the interior of Q. At last, $q_n(\xi, t) \ge -M_3$ by the choice of w_{0n}.

Let us estimate q_n from above. For each constant $C > 0$

$$\mathcal{L}_n C \equiv C\frac{N-1}{r_n}\left[\frac{\varphi'(v_n)}{r_n} - v_n C \frac{\varphi''(v_n)}{\varphi'(v_n)}\right].$$

Given an arbitrary n, the constant C may be chosen so small that $\mathcal{L}_n C \geq 0$. For $t = 0$, $q_n \leq 0$ by the assumption. At $\xi = 0$, $q_n = 0$ by (25.13). Next, since each point of the vertical line $\xi = b$ gives to the function w_n its minimal value, $C > 0 \geq q_n$ as $\xi = b$. By the maximum principle we conclude now that $q_n < C$ in $\overline{Q}$ and can pass to the limit as $C \to 0$. □

Corollary 25.1. *$r_n(\xi, t)$ are equibounded in $\overline{Q}$.*

For the proof, let us integrate equation (25.8) with respect to ξ and make use of (25.11). Then

$$\left(r_n^N\right)_t = -N r_n^{2(N-1)} \varphi_\xi(v_n) \quad \text{in } Q \tag{25.15}$$

or, equivalently,

$$r_{nt} = -r_n^{N-1} \varphi_\xi(v_n) \quad \text{in } Q. \tag{25.16}$$

Integrating (25.16) with respect to t and applying Lemma 25.1, we get:

$$0 \leq r_n \leq r_n(\xi, 0) + M_3 t.$$

Corollary 25.2. *The sequence $\{w_n\}$ converges uniformly in $C\left(\overline{Q}\right)$.*

It follows from Lemma 25.1 and the obvious inequality

$$r_n^N \geq \frac{N\xi}{\varphi^{-1}(M)}$$

that for all $\xi, \eta \in [0, b]$, $\xi < \eta$, and $t \in [0, T]$

$$\begin{aligned} 0 \leq w_n(\xi, t) - w_n(\eta, t) &= -\int_\xi^\eta w_{n\xi} r_n^{N-1} \frac{ds}{r_n^{N-1}} \\ &\leq M_3 \left[\frac{\varphi^{-1}(M)}{N}\right]^{(N-1)/N} (\eta - \xi)^{1/N}. \end{aligned}$$

The Hölder equicontinuity of w_n in ξ implies its Hölder equicontinuity in t [62, 93]. By the Ascoli–Arzelà theorem one can extract from the sequence $\{w_n\}$ a subsequence (which we assume coinciding with the original) that converges in $C\left(\overline{Q}\right)$ to some function w, and then set $v = \varphi^{-1}(w)$.

Later on we will check that the function v defined in this way is a classical solution of problem (L).

Corollary 25.3. *$v(\xi, t)$ is strictly positive in $[0, b) \times [0, T]$.*

This assertion follows immediately from the monotonicity of v_n in ξ and the following relation obtained from (25.8): for each $(\xi_0, t_0) \in Q$

$$\int_{\xi_0}^b \frac{ds}{v_n(s, t_0)} = \int_{\xi_0}^b \frac{ds}{\varphi^{-1}(w_{0n}(s))} - N \int_0^{t_0} \left[r_n^{2(N-1)} \frac{\partial w_n}{\partial \xi}\right]\Bigg|_{\xi_0}^b d\tau \leq M_4$$

where M_4 is an absolute constant depending only on $a, t_0, M_0, \dots, M_3$.

Let us introduce now the following family of operators depending on the parameter $\tau \in [0, 1]$:

$$L_\tau(w^{(\tau)}) \equiv w_t - \sigma(w^{(\tau)})\frac{\partial}{\partial \xi}\left[r_\tau^{2(N-1)}\frac{\partial w^{(\tau)}}{\partial \xi}\right] = 0 \quad \text{in } Q, \tag{25.8$_\tau$}$$

where

$$r_\tau^N = (1-\tau) + \tau r_n^N.$$

Consider the problem of finding a solution $w^{(\tau)}(\xi, t)$ of equation $(25.8)_\tau$ satisfying conditions (25.9)–(25.11). We introduce the operator $w^{(\tau)} = \Phi(w_n; \tau)$ that assigns to the pair (w_n, τ) the solution $w^{(\tau)}$ of the problem just formulated. We will view it as an operator from a ball $B_R \subset C_{\xi,t}^{1+\delta,(1+\delta)/2}$ of radius R into itself (the values of R and $\delta \in (0, 1)$ will be defined only through the data in (25.9)–(25.10)). Our objective is to prove that the operator $\Phi(w_n; \tau)$ has a fixed point for $\tau = 1$, which will be the sought solution to problem (L_n). A checking of the proof in [97, Th. 6.5.1] shows that actually we need only the following estimates:

1) for $\tau = 0$ problem $(25.8)_\tau$, (25.9)–(25.11) has at least one non-trivial solution;
2) there exist constants ε, M independent of τ such that every possible fixed point $w^{(\tau)}$ of the mapping Φ satisfies the inequalities
$$\varepsilon \le w^{(\tau)} \le M;$$
3) the fixed point satisfies the estimate $|w_\xi^{(\tau)}| \le M_1$ with a constant M_1 not depending on τ;
4) the Hölder norm of $w_\xi^{(\tau)}$ in Q is bounded by a constant M_2 independent of τ.

Assertion 1) needs no special verification. The problem for $\tau = 0$ is easy to reduce to problem (E), already solved in Section 24. Let us come back to the estimate of Lemma 25.1 and its consequences. It is easy to verify that every fixed point of the mapping Φ satisfies the estimate of Lemma 25.1, whence the validity of assertions 2)–3). The last assertion is a byproduct of the standard regularity theory [97] once 2) and 3) are established.

This proves that for each $n \in \mathbb{N}$ problem (L_n) has a solution w_n and we now proceed to show that the sequence $\{w_n\}$ converges to a solution of the degenerate problem (L).

Proof of Theorem 25.1. Let $f(\xi, t) \in C^\infty(Q)$ be compactly supported in Q. By virtue of (25.8), for each integer n large enough

$$\int_Q \left\{\frac{f_t}{v_n} + f_\xi r_n^{2(N-1)} w_{n\xi}\right\} d\xi\, dt = 0. \tag{25.17}$$

It follows from the pointwise convergence $v_n \to v$, the strict positivity of v, and the uniform boundedness of r_n and q_n that

$$r_n \to r \quad \text{pointwise in } \overline{Q}, \qquad q_n \to \sigma \quad \text{weakly in } L^2(Q).$$

Moreover, since the functions r_n are bounded away from zero and infinity uniformly in n in each domain, bounded away from the parabolic boundary of Q, then

$$\sigma \equiv r^{N-1}\varphi_{\xi}(v) \in L^2(Q).$$

Passing in (25.17) to the limit as $n \to \infty$ we conclude that $v(\xi, t)$ is a weak solution of equation (25.4) in Q.

Taking into account Corollary 25.3, one can then use the standard trick of viewing equation (25.4) as a uniformly parabolic one in each subdomain $Q' \subset Q$, bounded away from the parabolic boundary of Q. This allows one to improve the smoothness of the weak solution v within the confines imposed by the assumptions on the regularity of the function $\varphi(\cdot)$.

The inclusion into $C\left(\overline{Q}\right)$, and the fulfillment of the initial and boundary condition for v at $\xi = b$, follow just from the construction. Condition (25.2) is an implication of convergence of the integral representing $r(\xi, t)$. □

To complete the justification of the Lagrangian approach amounts to indicating the connection between the solutions of problems (L) and (P_1).

Denote

$$\Omega(t) = \{R > 0 : \ R = r(\xi, t), \ \xi \in [0, b)\}$$

and set

$$\overline{u}(R, t) = \begin{cases} v(\xi, t) & \text{as } R \in \Omega(t), \\ 0 \quad \text{as } R \in \mathbb{R}^+ \setminus \Omega(t), & t \in [0, T]. \end{cases}$$

Theorem 25.2. *Under the conditions of Theorem* 25.1 *the function* $\overline{u}$ *is a weak solution of problem* (P_1).

The proof merely follows the corresponding arguments of Sections 20 and 24. Given an arbitrary function $f(R, t)$ satisfying the conditions of Definition 17.2, we check the validity of the integral identity (17.28) by passing to the Lagrangian coordinate in the expression

$$\int_0^a f(R, 0) R^{N-1} u_0(R)\, dR.$$

The requirements on the smoothness of $\overline{u}$ are automatically fulfilled.

Note that

$$\eta(t) = r(b, t) \quad \text{as } t \in [0, T],$$

by the monotonicity of $r(\xi, t)$ in ξ. Also,

$$r_t = -r^{N-1}\varphi_{\xi}(v) \quad \text{in } Q. \tag{25.18}$$

25.3 Similarity solutions. Let us assume that $\varphi(s) \equiv s^m$ with $m > 1$. In this case, problem (P) admits the class of self-similar solutions $U(x, t; M)$ (see Section 17). Let us construct the counterpart of this class in the variables (ξ, t).

Consider the function

$$W(\xi, t; \tau) \equiv \frac{Q(\xi)}{(\tau + t)^\alpha},$$

$$\alpha = \frac{N}{2 + N(m-1)}, \qquad \tau = \text{const.} > 0.$$

Substituting W into (25.4) we get the equation for defining $Q(\xi)$. In the one-dimensional case, it was established in [23] that the function W defined in this way was the image of the Barenblatt solution U after the passage to the Lagrangian coordinate. In what below, we will fix the correspondence between the given solutions W and U of problems (L) and (P), respectively, by proving the uniqueness of a solution to problem (L) in an appropriately chosen function class.

It follows from the conditions of problem (L) that $Q(\xi)$ has to satisfy the conditions

$$\frac{\alpha}{Q} + \left\{ \left(N \int_0^\xi \frac{d\sigma}{Q(\sigma)} \right)^{2(N-1)/N} \left(Q^m\right)' \right\}' = 0 \quad \text{in } (0, b), \tag{25.21}$$

$$Q(b) = 0; \qquad \int_0^\xi \frac{ds}{Q(s)} \to 0 \quad \text{as } \xi \to 0.$$

To these conditions we also add the following one

$$\left[N \int_0^\xi \frac{d\sigma}{Q(\sigma)} \right]^{(N-1)/N} \left(Q^m\right)'(\xi) \to 0 \quad \text{as } \xi \to 0. \tag{25.22}$$

Condition (25.22) is not restrictive in the sense that the uniqueness result for problem (L) will be proven in the class of functions satisfying (25.22).

Let us pass to the construction of the function $Q(\xi)$.

1) $N = 1$. Let us introduce an auxiliary constant

$$\left(Q^m\right)'(b) \in (-\infty, 0).$$

Denote $q(\xi) = Q^m(\xi)$. Integration of equation (25.21) in the interval (ξ, b) yields

$$-\frac{\alpha m}{m-1} q^{(m-1)/m} + \frac{1}{2}\left\{ \left(q'(b)\right)^2 - \left(q'\right)^2 \right\} = 0. \tag{25.23}$$

Observe that

$$\frac{1}{2}\left(q'(b)\right)^2 = \frac{\alpha m}{m-1} q^{(m-1)/m}(b). \tag{25.24}$$

Having integrated (25.23), we get the following representation for $q(\xi)$:

$$\int_0^{q(\xi)} \left\{ \left(q'(b)\right)^2 - \frac{2\alpha m}{m-1} s^{(m-1)/m} \right\}^{-1/2} ds = b - \xi$$

with the constant $q'(b)$ given, due to (25.24), by the equality

$$q'(b) = \left(B\left(\frac{m+1}{m}; \frac{3}{2}\right)\right)^{-(m-1)/(m+1)} (2\alpha)^{m/(m+1)} \left(\frac{m}{m-1}\right)^{1/(m+1)},$$

$B(\cdot, \cdot)$ is Euler's function.

2) $N = 2$. The direct integration of (25.21) leads to the equality

$$\frac{\alpha}{2} + (Q^m)'(\xi) = 0 \quad \text{in } (0, b),$$

whence

$$Q(\xi) = \left[\frac{\alpha}{2}(b - \xi)\right]^{1/m} \quad \text{in } (0, b).$$

3) $N \geq 3$. Having integrated equation (25.21), let us write the equation for the function

$$\rho(\xi) = N \int_0^\xi \frac{ds}{Q(s)}.$$

We get

$$\frac{\alpha}{N} + mN^m \rho^{(N-2)/N} \frac{\rho''}{(\rho')^{m+1}} = 0 \quad \text{in } (0, b), \tag{25.25}$$

$$\rho(0) = 0, \quad \rho'(0) = C,$$

where C is also an auxiliary constant to be defined. It follows from (25.25) that

$$\left(\frac{1}{\rho'}\right)^{m-1} = \frac{1}{C^{m-1}} - \frac{\alpha(m-1)}{2mN^m}\rho^{2/N}. \tag{25.26}$$

Equation (25.26) leads to the following relation for $\rho(\xi)$:

$$\frac{1}{C}\int_0^{\rho(\xi)} \left\{1 - (Ls)^{2/N}\right\}^{1/(m-1)} ds = \xi, \tag{25.27}$$

$$L = \left[\frac{\alpha(m-1)}{2nN^m}C^{m-1}\right]^{N/2}.$$

Let us calculate the constant C. The condition

$$\rho'(\xi) = \frac{N}{Q(\xi)} \to \infty \quad \text{as } \xi \to b$$

and equation (25.26) give:

$$C = \left[\frac{2mN^m}{\alpha(m-1)}(\rho(b))^{-2/N}\right]^{1/(m-1)}. \tag{25.28}$$

Let us substitute (25.28) into (25.27) and set $\xi = b$. As a result we get

$$C = \left\{\frac{N}{2b} B\left(\frac{N}{2}; \frac{m}{m-1}\right)\right\}^{2/N\alpha} \left\{\frac{2nN^m}{\alpha(m-1)}\right\}^{1/\alpha}.$$

The function $Q(\xi)$ is now given by the relation

$$Q(\xi) = \frac{N}{\rho'(\xi)}.$$

Let us mention two properties of the constructed self-similar solutions to problem (L).

$$1) \qquad \left[N \int_0^\xi \frac{d\sigma}{W(\sigma, t; \tau)}\right]^{(N-2)/N} (W^m)_\xi = -\frac{\alpha}{t+\tau} < 0; \qquad (25.29)$$

$$2) \qquad (W^m)_{\xi\xi} = -\frac{N-2}{N} \frac{(W^m)_\xi}{W} \left[\int_0^\xi \frac{d\sigma}{W(\sigma, t; \tau)}\right]^{-1}.$$

Thus, W is always monotone in ξ, is convex when $N = 1$, concave when $N \geq 3$, and is represented by a segment of a straight line if $N = 2$.

25.4 Power nonlinearity. Differentiable asymptotics. Until the end of this subsection we assume that

$$\varphi(s) \equiv s^m, \quad m > 1.$$

We begin with establishing the uniqueness of the solution to problem (L).

Lemma 25.2. *Let* v_0 *satisfy* (N.1)–(N.3) *and, additionally,*

$$r^{N-2}(\xi, 0) \left(v_0^m\right)_\xi (\xi) \geq -L \quad in\ (0, b) \qquad (25.30)$$

with some constant L. Then problem (L) *admits at most one classical solution.*

Proof. Let us first prove that property (25.30) is preserved for all $t > 0$. Let $\{v_n\}$ be a sequence of solutions of problems (L_n) converging to v. The corresponding sequence of functions $\{w_{0n}\}$ can be taken to satisfy conditions (K.1)–(K.2) and the additional condition

(K.3) $r_n^{N-2}(\xi, 0) \left(v_0^m\right)' (\xi) \geq -S, \quad$ in $[0, b]$, $S = \text{const.} > 0$.

Let us consider the functions $\omega_n = r_n^{N-2} \left(v_n^m\right)_\xi$ satisfying the equations

$$\omega_{nt} - a_n \omega_{n\xi\xi} + b_n \omega_{n\xi} = [(m-1)N + 2]\, \omega_n^2 \geq 0,$$

and the boundary conditions

$$\omega_n|_{\xi=0} = 0, \qquad \left.\frac{\partial \omega_n}{\partial \xi}\right|_{\xi=b} + \left.\frac{\omega_n}{v_n r_n^N}\right|_{\xi=b} = 0.$$

(The exact form of the coefficients a_n and b_n is not essential here.) These conditions imply that ω_n can assume their negative minima only on the lower face of Q where (K.3) holds. Passing to the limit as $n \to \infty$, we get the estimate

$$r^{N-2}\left(v^m\right)_\xi \geq -S \quad \text{in } Q. \tag{25.31}$$

Let v_1 and v_2 be two solutions of problem (L), satisfying (25.30). Consider the functions

$$\rho_i(\xi, t) = N \int_0^\xi \frac{ds}{v_i(s,t)}, \qquad i = 1, 2.$$

By Theorem 25.2 each of the functions v_1 and v_2 generates a solution of problem (P_1). However, by virtue of Proposition 17.4 these solutions must coincide, whence

$$\eta^N(t) = \rho_1(b,t) = \rho_2(b,t) \quad \text{as } t \in [0, T].$$

Further, $\rho_1(0,t) = \rho_2(0,t) = 0$ by the definition. As follows from (25.4), each of ρ_i satisfies in Q the equation

$$\mathcal{M}\rho_i \equiv \rho_{it} - mN^{m+1}\rho_i^{2(N-1)/N} \frac{\rho_{i\xi\xi}}{\left(\rho_{i\xi}\right)^{m+1}} = 0.$$

Let us denote

$$A(s,p,q) \equiv mN^{m+1} s^{2(N-1)/N} \frac{q}{p^{m+1}}$$

and derive the equation for the function

$$R(\xi,t) \equiv (\rho_1 - \rho_2)\, e^{-\lambda t}, \qquad \lambda = \text{const.} > 0.$$

We have:

$$\begin{aligned} R_t &- \left(\int_0^1 \frac{\partial}{\partial q}[A(s,p,q)]\,d\theta\right) R_{\xi\xi} - \left(\int_0^1 \frac{\partial}{\partial p}[A(s,p,q)]\,d\theta\right) R_\xi \\ &+ \left[\lambda - \int_0^1 \frac{\partial}{\partial s}[A(s,p,q)]\,d\theta\right] R = 0, \end{aligned} \tag{25.32}$$

where

$$s = \theta\rho_2 + (1-\theta)\rho_1, \qquad p = s_\xi, \qquad q = s_{\xi\xi}.$$

Assume, for instance, that R attains its positive maximum at a point $P \in Q$. By (25.32)

$$\left[\lambda - \int_0^1 \frac{\partial}{\partial s}[A(s,p,q)]\,d\theta\right] R \leq 0 \quad \text{at } P. \tag{25.33}$$

By Lemma 25.1 and due to the definition of ρ_i

$$\rho_{i\xi\xi} \geq 0 \quad \text{in } \overline{Q},\ i = 1, 2. \tag{25.34}$$

Also, $\rho_{1\xi\xi}(P) \geq \rho_{2\xi\xi}(P)$ by assumption. It follows now from (25.33)–(25.34) that

$$0 \geq \lambda - 2mN^m(N-1)\left[\rho_1^{(N-2)/N}\frac{\rho_1}{(\rho_1\xi)^{m+1}}\right]\Bigg|_{(\xi,t)=P}$$
$$\equiv \lambda + \frac{2(N-1)}{m}\omega^{(1)}(P) \geq \lambda - \frac{2(N-1)}{m}S,$$

where $\omega^{(1)} = r_1^{N-2}\left(v_1^m\right)_\xi$, and the constant S is taken from (25.31). It remains to choose

$$\lambda = 1 + \frac{2(N-1)}{m}S$$

to obtain the contradiction with the assumption $v_1 \neq v_2$. □

Corollary 25.4. *Property* (25.29) *means that the self-similar solutions* $W(\xi, t; \tau)$ *can also be obtained as the pointwise limits of sequences of solutions to problems* (L_n) *with the initial functions* w_{0n} *approximating the function* $\tau^{-\alpha}Q^m(\xi)$ *and satisfying* (K.1)–(K.3).

Lemma 25.3.

$$v_t \geq -\alpha\frac{v}{t} \quad in\ Q, \quad \frac{1}{\alpha} = m - 1 + \frac{2}{N}.$$

Proof. Fix an arbitrary integer n. Let $U(\xi, t)$ be the even continuation of w_n on the rectangle $G = (-b, b) \times (0, T]$. Accept the notation

$$R^N(\xi, t) = \frac{1}{n} + N\int_0^{|\xi|}\frac{d\sigma}{U^{1/m}(\sigma, t)}, \quad W(\xi, t) = U^{1/m}\left(R^{2(N-1)}U_\xi\right)_\xi \equiv \frac{\left(U^{1/m}\right)_t}{U^{1/m}}.$$

The function W satisfies in G the following equation

$$\mathcal{A}W \equiv W_t - mR^{2(N-1)}UW_{\xi\xi} - a_nW_\xi - (1+m)W^2$$
$$+ 4(N-1)\left[R^{N-2}\operatorname{sign}\xi U_\xi\right]W - 2N(N-1)\left(R^{N-2}U_\xi\right)^2 = 0,$$

where a_n is bounded and continuous in $\overline{G}$. For each finite n

$$W = 0 \quad \text{on } \partial G, \quad \text{and} \quad W|_{t=0} \text{ is finite.}$$

Consider the function $\sigma(t) = -C/t$ with a constant $C > 0$ to be defined. Clearly, $\sigma(t) < W$ on the parabolic boundary of G. Next, for $C \in (0, \alpha)$

$$\mathcal{A}\sigma(t) \equiv \frac{C}{t^2}(1 - C(1+m)) - \frac{4C(N-1)}{t}\left(R^{N-2}\operatorname{sign}\xi U_\xi\right)$$
$$- 2N(N-1)\left(R^{N-2}U_\xi\right)^2 \equiv$$

$$\equiv \frac{C}{t^2}(1 - C(1+m)) - 2(N-1)\left[\sqrt{N}\left(R^{N-2}\operatorname{sign}\xi U_\xi\right) + \frac{C}{t\sqrt{N}}\right]^2$$
$$+ \frac{2(N-1)}{N}\cdot\frac{C^2}{t^2}$$
$$\le \frac{C}{t^2}\left[1 - C\left(1 + m - \frac{2(N-1)}{N}\right)\right] \equiv \frac{C}{t^2}\left(1 - \frac{C}{\alpha}\right) < 0.$$

The statement of the lemma follows now by the maximum principle applied to W and the consequent passage to the limit as $n \to \infty$. □

Corollary 25.5. *The functions*

$$\left(r^N\right)_t \quad \text{and} \quad r^{2(N-1)}\left(v^m\right)_\xi$$

have finite limits as $\xi \to 0$ *and* $\xi \to b$ *for each* $t \in (0, T)$.

The proof is analogous to those presented in the previous sections: we consider the bounded function

$$\Phi(\xi) = \delta r^N + r^{2(N-1)}\left(v^m\right)_\xi$$

and check that it is monotone in ξ if δ is chosen in an appropriate way.

Corollary 25.6. *Making recourse to* (25.18), *we obtain the interface equation*

$$\left(\eta^N\right)'(t) + N\lim_{\xi\to b} r^{2(N-1)}\left(v^m\right)_\xi = 0 \quad \text{in } \mathcal{D}'(0, T). \tag{25.35}$$

Lemma 25.4. *There exist positive constants* θ, τ *such that in* $\overline{Q}$

$$N\int_0^\xi \frac{d\sigma}{W(\sigma, t; \tau)} \ge r^N \ge N\int_0^\xi \frac{d\sigma}{W(\sigma, t; \theta)}. \tag{25.36}$$

Proof. Recall condition (N.3) and take the constants τ, θ so that

$$\tau^{-\alpha}Q(\xi) \le v_0(\xi) \le \theta^{-\alpha}Q(\xi) \quad \text{in } [0, b]. \tag{25.37}$$

Let us choose the sequences $\left\{v_{0n}^{(i)}\right\}$ so that $v_{0n}^{(i)}$ satisfies (K.1)–(K.3),

$$v_{0n}^{(1)} \to W(\xi, 0; \tau), \quad v_{0n}^{(2)} \to v_0, \quad v_{0n}^{(3)} \to W(\xi, 0; \theta),$$

and consider the corresponding solutions of problems (L_n). Set

$$\rho_n^{(i)} = \frac{1}{n} + N\int_0^\xi \frac{ds}{v_n^{(i)}(s, t)}, \quad i = 1, 2, 3.$$

The functions $\rho_n^{(i)}$ satisfy in Q the equations $\mathcal{M}\rho_n^{(i)} = 0$, and the conditions

$$\rho_n^{(i)}(0,t) = \frac{1}{n}, \quad \left.\frac{\partial \rho_n^{(i)}}{\partial \xi}\right|_{\xi=b} = nN, \quad \rho_n^{(1)}(\xi,0) \geq \rho_n^{(2)}(\xi,0) \geq \rho_n^{(3)}(\xi,0).$$

Following the proof of Lemma 25.2, it is easy to check that for each integer n the functions $\rho_n^{(2)} - \rho_n^{(1)}$ and $\rho_n^{(3)} - \rho_n^{(2)}$ are nonpositive in $\overline{Q}$. Letting $n \to \infty$, we arrive at (25.36). □

Proof of Theorem 17.5. 1) Set $\xi = b$ in (25.36) and rewrite this relation in the equivalent form

$$A\left[\left(\frac{\theta+t}{\lambda+t}\right)^{\alpha} - 1\right] \leq \frac{\eta^N(t)}{(\lambda+t)^{\alpha}} - A \leq A\left[\left(\frac{\theta+t}{\lambda+t}\right)^{\alpha} - 1\right], \tag{25.38}$$

where $\lambda \in [\theta, \tau]$ is arbitrary and

$$A = N\int_0^{\xi} \frac{ds}{Q(s)}.$$

Consider the function

$$\Phi(s) = \left(\frac{s+t}{\lambda+t}\right)^{\alpha} - 1, \quad s \in [\theta, \lambda],$$

viewing the variable t as a parameter. By the mean value theorem,

$$\Phi(s) = -\alpha\left(\frac{s+t}{\lambda+t}\right)^{\lambda-1}\frac{\lambda-s}{\lambda+t} \geq -\alpha\frac{\tau-\theta}{\lambda+t},$$

$$\Phi(s) \equiv \Phi(s) - \Phi(\lambda) \leq \alpha\frac{\tau-\theta}{\lambda+t}.$$

Combining these inequalities with (25.38), we get at once (17.30):

$$\forall t \geq 0 \quad \left|\frac{\eta^N(t)}{(\lambda+t)^{\alpha}} - A\right| \leq \alpha A\frac{\tau-\theta}{\lambda+t}$$

with an arbitrary $\lambda \in (\theta, \tau)$.

2) *Asymptotics of solutions to problem* (L) *in cases* $N = 1, 2$. Choose the constants θ, τ so that (25.37) holds. First, the functions $\rho^{(i)}$ satisfy inequalities (25.36). On the other hand, $\rho^{(i)}(0,t) = 0$. This means that at each point of the line $\xi = 0$ the functions $\rho^{(1)} - \rho^{(2)}$ and $\rho^{(2)} - \rho^{(3)}$ assume their minimal values in $\overline{Q}$, i.e.,

$$0 \leq \left.\frac{\partial}{\partial \xi}\left(\rho^{(1)} - \rho^{(2)}\right)\right|_{\xi=0} = N\left(\frac{1}{W(0,t;\tau)} - \frac{1}{v(0,t)}\right),$$

$$0 \le \frac{\partial}{\partial \xi}\left(\rho^{(2)} - \rho^{(3)}\right)\Big|_{\xi=0} = N\left(\frac{1}{v(0,t)} - \frac{1}{W(0,t;\theta)}\right).$$

Finally, $v(b,t) = W(b,t;\tau) = W(b,t;\theta)$. Thus, both the positive maximum of $v^m - W^m(\xi,t;\theta)$ and the negative minimum of $v^m - W^m(\xi,t;\tau)$ can be attained only inside Q. Let us consider the function

$$S(\xi,t) = \left(v^m - W^m(\xi,t;\theta)\right)e^{\lambda t}, \quad \lambda = \text{const.} > 0.$$

Adopt the notation

$$B(s,p,q,) = ms^{(m+1)/m}p^{(N-2)/N}q, \quad C(s,p,q) = 2m(N-1)sp^{(N-2)/N}q$$

and write the equation for the function S:

$$S_t - \left(\int_0^1 B_q d\mu\right) S_{\xi\xi} - \left(\int_0^1 C_r d\mu\right) S_\xi + \left[\lambda - \left(\int_0^1 (B_s + C_s)\, d\mu\right)\right] S$$
$$= \left[\int_0^1 (B_p + C_p)\, d\mu\right]\left(\rho^{(3)} - \rho^{(2)}\right), \tag{25.39}$$

where

$$s = \mu v^m + (1-\mu)W^m, \quad s_\xi = r, \quad s_{\xi\xi} = q, \quad \mu\rho^{(2)} + (1-\mu)\rho^{(3)} = p.$$

Let us assume that S has in Q a point of local positive maximum. At this point

$$S > 0, \quad S_t > 0, \quad S_\xi = 0, \quad S_{\xi\xi} \le 0.$$

By Lemma 25.3, $\rho^{(2)} \ge \rho^{(3)}$ in $\overline{Q}$. Moreover, for $N = 1, 2$ $(W^m)_{\xi\xi} \le 0$. At last, $(v^m)_\xi$ and $(W^m)_\xi$ are nonpositive in $\overline{Q}$. Substituting these relations into (25.39), we get that at the point of local positive maximum

$$\left[\lambda - 2m(N-1)\left(W^m\right)_\xi \left(\rho^{(3)}\right)^{(N-2)/N}\right]\left(v^m - W^m\right) \le 0.$$

Choosing now

$$\lambda = 1 + 2m\alpha\frac{N-1}{\theta},$$

we arrive at a contradiction.

Likewise we estimate the difference $v^m - W^m(\xi,t;\tau)$ to get

$$W^m(\xi,t;\theta) \ge v^m(\xi,t) \ge W^m(\xi,t;\tau) \quad \text{in } Q. \tag{25.40}$$

Having rewritten (25.40) in the form

$$W(\xi,t;\lambda)\left[\left(\frac{\lambda+t}{\theta+t}\right)^\alpha - 1\right] \ge v(\xi,t) - W(\xi,t;\lambda)$$
$$\ge W(\xi,t;\lambda)\left[\left(\frac{\lambda+t}{\tau+t}\right)^\alpha - 1\right], \tag{25.41}$$

and applying the arguments already employed, we arrive at the inequality

$$|v(\xi,t) - W(\xi,t;\lambda)| \le B\frac{W(\xi,t;\lambda)}{\lambda+t}, \qquad \lambda = \text{const.} > 0.$$

Now, let us rewrite (25.40) as

$$\begin{aligned} W^m(b,t;\theta) - W^m(\xi,t;\theta) &\le v^m(b,t) - v^m(\xi,t) \\ &\le W^m(b,t;\tau) - W^m(\xi,t;\tau) \quad \text{in } \overline{Q} \end{aligned} \tag{25.42}$$

and make use of the mean value theorem. In the limit as $\xi \to b$ we have:

$$\left(Q^m\right)'(b)(\theta+t)^{-\alpha m} \le \left(v^m\right)_\xi(b,t) \le \left(Q^m\right)'(b)(\tau+t)^{-\alpha m}. \tag{25.43}$$

Denote

$$D = -N\left(Q^m\right)'(b)A^{2(N-1)/N}.$$

Applying the asymptotics found for $\eta^N(t)$, we get from (25.41) and (25.35)

$$D\left[\left(\frac{\tau+t}{\lambda+t}\right)^{\alpha-1} - 1\right] \le \frac{\left(\eta^N(t)\right)'}{(\lambda+t)^{\alpha-1}} - D \le D\left[\left(\frac{\theta+t}{\lambda+t}\right)^{\alpha-1} - 1\right],$$

whence (17.31). □

25.5 Equations with arbitrary nonlinearity. Let us denote by $y(t)$ the solution of the Cauchy problem

$$y' = Ny^{(N-2)/N}\varphi'\left(\frac{N}{y}\right) \quad \text{as } t > 0,\ y(0) = y_0 > 0, \tag{25.44}$$

and let $R_\theta(\xi)$ be the solution of the boundary value problem

$$\begin{cases} \left(R'_\theta\right)^{\theta+1} = R_\theta^{(N-2)/N}R''_\theta \quad \text{as } \xi \in (0,b), \\ R_\theta(0) = \varepsilon; \quad R'_\theta(b) = 1/\gamma, \quad \gamma,\varepsilon > 0,\ \theta > 1. \end{cases} \tag{25.45}$$

The solutions of (25.44) are given by the formula

$$N^2 t = \int_{N/y(0)}^{N/y(t)} \frac{d\mu}{\mu^{1+\frac{2}{N}}\varphi'(\mu)},$$

whence we infer that $y(t)$ is monotone, $y(t) \to \infty$ as $t \to \infty$.

Lemma 25.5.

$$\forall K, q = \text{const.} > 0 \quad \frac{y(t)}{y(Kt+q)} \to 1 \quad \textit{as } t \to \infty.$$

Proof. Let us assume first that $K > 1$. We have:

$$\begin{aligned} 1 \geq \frac{y(t)}{y(Kt+q)} &\equiv 1 - \frac{y(Kt+q) - y(t)}{y(Kt+q)} \\ &= 1 - N\frac{y(s)}{y(Kt+q)} y^{-N/2}(s)\varphi'\left(\frac{N}{y(s)}\right) \equiv \sigma \quad \text{for some } s \in (t, Kt+q). \end{aligned}$$

Observe that for all $s > t > 0$

$$\left(\frac{s}{t}\right)^{\alpha} \leq \frac{\varphi'(s)}{\varphi'(t)} \leq \left(\frac{s}{t}\right)^{\beta}, \quad \left(\frac{s}{t}\right)^{\alpha+1} \leq \frac{\varphi(s)}{\varphi(t)} \leq \left(\frac{s}{t}\right)^{\beta+1}. \tag{25.46}$$

Then

$$\begin{aligned} \sigma &\geq 1 - Ny^{-N/2}(s)\frac{\varphi'(N/y_0)}{(N/y_0)^{\alpha-1}}\left(\frac{N}{y(s)}\right)^{\alpha-1} \\ &= 1 - \text{const.}\,(y(s))^{-\left(\alpha+\frac{2}{N}\right)} \to 1 - 0 \quad \text{as } t \to \infty. \end{aligned}$$

Let $K < 1$. Then, since $Kt + q < t$ as $t \to \infty$, we get by the analogy with the previous case

$$\begin{aligned} 1 \leq \frac{y(t)}{y(Kt+q)} &\leq 1 + Ny^{-N/2}(s)\varphi'\left(\frac{N}{y(s)}\right) \\ &= 1 + \text{const.}\,(y(s))^{-\left(\beta+\frac{2}{N}\right)} \to 1 \quad \text{as } t \to \infty,\ s \in (Kt+q, t). \end{aligned}$$

□

The solution of (25.45) is given by the formula

$$\left(N\frac{\theta-1}{2}\right)^{1/(\theta-1)} \int_{\varepsilon}^{R_\theta(\varepsilon)} \left[C - s^{2/N}\right]^{1/(\theta-1)} ds = \xi, \tag{25.47}$$

where

$$C = \frac{2}{N(\theta-1)}\left[\gamma^{\theta-1} + \frac{N(\theta-1)}{2} R_\theta^{2/N}(b)\right],$$

$$\left(\frac{N(\theta-1)}{2}\right)^{1/(\theta-1)} \int_{\varepsilon}^{R_\theta(b)} \left[C - s^{2/N}\right]^{1/(\theta-1)} ds = b.$$

Clearly, $R_\theta(b)$ is always finite and bounded away from zero. Proceeding to the limit as $\varepsilon, \gamma \to 0$, we get from (25.47) the solution of the limit problem with the parameters $\varepsilon = \gamma = 0$.

An upper barrier for $\rho(\xi, t) = N\int_0^\xi \frac{ds}{v(s,t)}$. Let us consider the sequence of functions

$$\rho_n(\xi, t) = \frac{1}{n} + N\int_0^\xi \frac{ds}{v_n(s,t)},$$

approximating $\rho(\xi, t)$. Due to (25.4), for each integer n

$$\mathcal{M}\rho_n \equiv \rho_{nt} - N\rho_n^{2(N-2)/N}\varphi'\left(\frac{N}{\rho_{n\xi}}\right)\frac{\rho_{n\xi\xi}}{\left(\rho_{n\xi}^2\right)} = 0 \quad \text{in } Q,$$

$$\rho_n(0,t) = \frac{1}{n}; \qquad \rho_{n\xi}(b,t) = nN.$$

Let us introduce the function

$$\overline{z}(\xi, t) = y(\delta t)R_{\alpha+1}(\xi),$$

depending on the parameters $\delta > 0, y(0), \varepsilon$, and γ. Let δ be an arbitrary positive number; we choose ε and $y(0)$ satisfying the following conditions:

1) $\overline{z}(0,t) - \rho_n(0,t) \geq \varepsilon y(0) - \frac{1}{n} > 0$,
2) $\overline{z}(b,t) - \rho_{n\xi}(b,t) \geq \frac{y(0)}{\gamma} - nN > 0$,
3) $\overline{z}(\xi, 0) - \rho_n(\xi, 0) \geq 0$.

The last condition is fulfilled if condition 1) holds and the inequality $(\overline{z}_\xi - \rho_{n\xi})_{|_{t=0}} \geq 0$ is true which, in essence, means the following:

$$\frac{N}{y(0)R'_{\alpha+1}(\xi)} \leq v_{0n}(\xi) \quad \text{as } \xi \in [0,b].$$

Let us make use of (25.46): we have to check that

$$\left(\frac{N}{y(0)}R'_{\alpha+1}(\xi)\right)^{\alpha+1} \leq (\alpha+1)\frac{M^{\alpha+1}}{\varphi(M)}\varphi(v_{0n}) \quad \left(\leq v_{0n}^{\alpha+1}(\xi)\right).$$

By (K.2), there exists a constant $M_2 > 0$ such that

$$\varphi(v_{0n}) \geq M_2(b-\xi) + \frac{1}{n} \quad \text{as } \xi \in [0,b].$$

Thus, the required inequality will be fulfilled if, given an arbitrary M_2, we are able to choose $y(0)$ so that

$$\frac{1}{y^{\alpha+1}(0)}\left(\frac{N}{M}\right)^{\alpha+1}\frac{\varphi(M)}{\alpha+1}\frac{1}{\left(R'_{\alpha+1}\right)^{\alpha+1}} \leq M_2(b-\xi)$$

in a neighborhood of the point $\xi = b$. By the mean value theorem, the previous inequality can be rewritten in the form

$$-\frac{1}{(\alpha+1)y(0)^{\alpha+1}}\left(\frac{N}{M}\right)^{\alpha+1}\varphi(M)\frac{R''_{\alpha+1}}{\left(R'_{\alpha+1}\right)^{\alpha+2}}\Bigg|_{\xi=s} \geq -M_2 \quad \text{as } s \in (\xi, b),$$

that is,

$$\frac{1}{y(0)^{\alpha+1}}\left(\frac{N}{M}\right)^{\alpha+1}\varphi(M)\frac{1}{(R_{\alpha+1})^{(N-2)/N}}\bigg|_{\xi=s}\leq M_2.$$

Since $R_{\alpha+1}$ is bounded as $\xi\to b$ and bounded away from zero, it is easy to satisfy this inequality just by increasing $y(0)$.

Thus, $\bar{z}(\xi,t)\geq\rho_n$ on Γ. Let us calculate $\mathcal{M}\bar{z}$. We have:

$$\mathcal{M}\bar{z}\equiv\delta y'(\delta t)R_{\alpha+1}-Ny^{(N-2)/N}R_{\alpha+1}^{2(N-1)/N}\times\frac{R''_{\alpha+1}}{\left(R'_{\alpha+1}\right)^2}\frac{\varphi'\left(\frac{N}{yR'_{\alpha+1}}\right)}{\varphi'\left(\frac{N}{y}\right)}. \tag{25.48}$$

We consider two possibilities.

1) Assume that

$$\mu\equiv\min R'_{\alpha+1}=R'_{\alpha+1}(0)\geq 1.$$

In this case, it follows at once from (25.46) that

$$-\frac{\varphi'\left(\frac{N}{yR'_{\alpha+1}}\right)}{\varphi'\left(\frac{N}{y}\right)}\geq-\frac{1}{\left(R'_{\alpha+1}\right)^{\alpha}}$$

and, by virtue of (25.47),

$$\mathcal{M}\bar{z}\geq y'(\delta t)R_{\alpha+1}(\delta-1)\quad\text{in } Q \text{ as } \mu\geq 1.$$

2) Let $\mu<1$. Then

$$\begin{aligned}\mathcal{M}\bar{z}&\equiv y'(\delta t)R_{\alpha+1}(\xi)\left[\delta-R_{\alpha+1}^{(N-2)/N}\frac{R''_{\alpha+1}}{\left(R'_{\alpha+1}\right)^2}\cdot\frac{\varphi'\left(\frac{N}{y'_{\alpha+1}}\right)}{\varphi'\left(\frac{N}{y\mu}\right)}\right]\\&\geq y'(\delta t)R_{\alpha+1}(\xi)\left[\delta-\mu^{\alpha-\beta}\right].\end{aligned} \tag{25.49}$$

Thus, setting $\delta=\left(\min\left(1;\,R'_{\alpha+1}(0)\right)\right)^{\alpha-\beta}$, $\gamma=0$, we get the inequality

$$\bar{z}\geq\rho_n\quad\text{in }\overline{Q}.$$

Observe that the parameter $y(0)$ here can be taken not depending on n, and that $\varepsilon\to 0$ as $n\to\infty$. In the limit as $n\to\infty$ and $\varepsilon\to 0$ we get:

$$\bar{z}\geq\rho\quad\text{in }\overline{Q}.$$

A lower barrier for ρ. For a lower barrier for ρ we take the function

$$\underline{z}\equiv y(\delta' t)R_{\beta+1}(\xi)$$

with the parameters $y(0)$, $\varepsilon = 0$, $\gamma > 0$, and some $\delta' > 0$ still undefined. Repeating the preceding arguments, we obtain that

$$\rho \geq \underline{z} \quad \text{in } \overline{Q},$$

provided $y(0)$ and γ are taken small enough, and with

$$\delta' = \left(\min\left(1;\, R'_\beta(0)\right)\right)^{\alpha-\beta}.$$

The desired estimate follows then by passage to the limit as $\gamma \to 0$.

Now, let $y_*(t)$ be the solution of (25.44) with the initial condition $y(0) = 1$. Collecting the estimates for $\rho(\xi, t)$ as $\xi = b$ and applying Lemma 25.5, we have that

$$R_{\beta+1}(b)\frac{y_*(\delta' t)}{y_*(t+\nu)} \leq \frac{\eta^N(t)}{y_*(t)} \leq R_{\alpha+1}(b)\frac{y_*(\delta t)}{y_*(t)} \leq R_{\alpha+1}(b)\frac{y_*(\delta t+\mu)}{y_*(t)}$$

with some positive constants ν, μ. Letting $t \to \infty$, we get from this the inequality

$$R_{\beta+1}(b) \leq \underline{\lim}_{t\to\infty}\frac{\eta^N(t)}{y_*(t)} \leq \overline{\lim_{t\to\infty}}\frac{\eta^N(t)}{y_*(t)} \leq R_{\alpha+1}(b),$$

whence (17.32).

The proof of Theorem 17.6 is completed. □

26 A non-classical problem for a degenerate parabolic equation: uniqueness of unbounded solutions

In this section we study the following problem: to find a solution $\omega(\xi, t)$ of the equation

$$\omega_t = \left[\nu(\omega)\omega_\xi\right]_\xi \quad \text{in } \Pi_T = \{(\xi, t) : |\xi| < a,\ 0 < t < T\} \tag{26.1}$$

satisfying the initial and boundary conditions

$$\omega(\xi, 0) = \omega_0(\xi) \quad \text{as } |\xi| < a, \tag{26.2}$$

$$\omega(\xi, t) \to +\infty \quad \text{as } \xi \to \pm a,\ t \in (0, T). \tag{26.3}$$

The nonstandard nature of this problem consists in condition (26.3). It is clear that its statement makes sense only for those equations which degenerate as $\omega \to \infty$, and in the class of initial data growing unboundedly as $\xi \to a$ and $\xi \to -a$.

Let us rigorously formulate the requirements to the functions ν and ω_0 forming the problem. Assume that the function $\nu(r)$ satisfies the following conditions:

$$\nu(r) > 0 \quad \text{as } r > 0, \tag{26.4}$$

$$\nu(r) \in C^1(\mathbb{R}^+) \cap L^1(1, \infty), \tag{26.5}$$

and $\nu'(r)$ is locally Hölder continuous in $\mathbb{R}^+$. We claim that

$$\omega_0(\xi) \in C(-a, a), \tag{26.6}$$

$$\omega_0 > 0 \quad \text{for } |\xi| < a, \tag{26.7}$$

$$\omega_0 \to +\infty \quad \text{as } \xi \to \pm a. \tag{26.8}$$

Among with (26.1), let us consider the equation reciprocal to it with respect to L-transformation:

$$u_t = [k(u)u_x]_x\,, \tag{26.9}$$

where $k(s) = s^{-2}\nu(s^{-1})$. The connection between the variables ξ, ω and x, u is given by the formulas

$$u(x,t) = [\omega(\xi,t)]^{-1}\,, \tag{26.10}$$

$$x = \int_0^\xi \omega(\eta, t)\, d\eta + \int_0^t \nu\,[\omega(0,s)]\, \omega_\xi(0,s)\, ds.$$

These formulas admit the inversion

$$\omega(\xi,t) = [u(x,t)]^{-1}\,, \tag{26.11}$$

$$\xi = \int_0^x u(y,t)\, dy + \int_0^t k\,[u(0,s)]\, u_x(0,s)\, ds.$$

Assume now that ω_0 satisfies the additional conditions

$$\int_{-a}^0 \omega_0(\xi)\, d\xi = \int_0^a \omega_0(\xi)\, d\xi = \infty. \tag{26.12}$$

Let us construct the function $u_0(x)$ by eliminating ξ from relations (26.10) taken at $t = 0$. According to (26.12), the function u_0 is defined for all $x \in \mathbb{R}$. Completing equation (26.9) by the initial condition

$$u(x,0) = u_0(x) \quad \text{as } x \in \mathbb{R}, \tag{26.13}$$

we get the Cauchy problem already given a detailed discussion in Section 17. More precisely, we have arrived at problem (E) (see Section 17) with $\varphi'(u) = k(u)$ and $\psi(u) = 0$.

Let $\varphi(s)$ be defined as the primitive to the function $k(s) = s^{-2}\nu(s^{-1})$ and normalized by the condition $\varphi(0) = 0$. Proceeding from (26.4)–(26.5), it is easy to check that this function satisfies all the restrictions imposed in the formulation of Proposition 17.2. With respect to the function $u_0(x)$ we conclude, due to (26.6) -(26.8), (26.10), that it is bounded, positive, and continuous in $\mathbb{R}$. Moreover, as follows from (26.11) with $t = 0$, and the properties of ω_0, the inclusion $\omega_0 \in L^1(\mathbb{R})$ and the equality $\|u\|_{L^1(\mathbb{R})} = 2a$ hold.

Relying on Proposition 17.2 one can now conclude that in each strip $S_T = \mathbb{R}\times(0,T)$ the Cauchy problem (26.9), (26.13) has a unique solution in the sense of Definition 17.1. It follows from results of [137, 109, 65, 136] that for $t > 0$ the function $u(x,t)$ possesses continuous derivatives u_t, u_x, u_{xx} and equation (26.9) is thus satisfied in the classical sense. Moreover, for all finite $t > 0$ and $x \in \mathbb{R}$ there holds the inequality $u(x,t) > 0$. Finally, the solutions of problem (26.9), (26.13) possess the property of conservation of "the total mass":

$$\int_{\mathbb{R}} u(x,t)\,dx = \int_{\mathbb{R}} u_0(x)\,dx = 2a. \tag{26.14}$$

The above-listed properties of the function $u(x,t)$ show that the second equality of (26.11) defines a bijective mapping of the strip S_T onto the rectangle Π_T. Let us now define a solution $\omega(\xi,t)$ of problem (26.1)–(26.3) by the first of equalities (26.11). The function ω defined in this way will be a classical solution of equation (26.1) in the open domain Π_T, and condition (26.2) will be fulfilled in the ordinary sense. The continuity and positivity of $u(x,t)$ in S_T, along with convergence of the first of integrals in (26.14), guarantee the fulfillment of condition (26.3). The correctness of this method of defining a solution to the non-classical problem (26.1)–(26.3) is confirmed by the following assertion.

Theorem 26.1. *Let conditions* (26.4)–(26.8), (26.12) *hold. Then the above-defined positive solution of problem* (26.1)–(26.3) *is unique.*

The assertion of this theorem follows immediately from the uniqueness of the solution for the Cauchy problem (26.9), (26.13) and the fact that formulas (26.10) realize a bijective mapping. Let us show now that Theorem 26.1 remains true if condition (26.12) is replaced by the condition $\omega_0(\xi) \in L^1(-a,a)$. In this case $u_0(x)$ will be compactly supported in $[b,c]$, where

$$b = -\int_{-a}^{0} \omega_0(\xi)\,d\xi, \quad c = \int_0^a \omega_0(\xi)\,d\xi.$$

Moreover, $u_0 > 0$ if $b < x < c$. It follows from (26.8), (26.10) with $t = 0$ that $u_0(x) \in C[b,c]$. Having continued u_0 by zero outside the segment $[b,c]$, we arrive once again at the Cauchy problem for equation (26.9). Generally speaking, this problem need not have a classical solution but it has a unique weak solution in the sense of Definition 17.1 [137, 109, 65, 136]. Here the support of the solution $u(\cdot,t)$ will be finite for each fixed t, $0 \le t < T < \infty$.

Let us define the subdomain Ω_T of S_T by the relations

$$\Omega_T = \{(x,t) : \; u(x,t) > 0, \; 0 < t < T\}.$$

Repeating the arguments analogous to those preceding the formulation of Theorem 26.1, we come to the conclusion that the second of formulas (26.11) realizes a bijection of the domain Ω_T (now bounded) onto the rectangle Π_T. Then the first of formulas (26.11) defines a solution of problem (26.1)–(26.3) in the case when $\omega_0(\xi) \in L^1(-a,a)$. With this definition, the solvability of the problem automatically follows. The uniqueness of its solution follows from the uniqueness of the weak solution for the Cauchy problem (26.9)–(26.13).

Similar arguments allow one to construct a solution of problem (26.1)–(26.3) in cases when $\omega_0(\xi)$ belongs to the classes $L^1(-a, 0)$ or $L^1(0, a)$, but does not belong to $L^1(-a, a)$. Let us note that in all the above-mentioned cases equalities (26.14) hold for the solutions of the corresponding Cauchy problems for equation (26.9), reciprocal to (26.1). Summarizing, we conclude that condition (26.12), or any other condition controlling the growth of the function ω_0 as $\xi \to -a$ and $\xi \to a$, is unessential for the fulfillment of Theorem 26.1. It holds only if assumptions (26.4)–(26.8) are true.

27 The Stefan problem with degeneracy at the free boundary: example of exact solution

Let us consider the degenerate parabolic equation

$$w_t = \left[\frac{w^m w_x}{(\lambda w + \mu)^{m+2}}\right]_x, \tag{27.1}$$

where m, μ, λ are positive constants. As shown in Section 10, equation (27.1) possesses the hidden symmetry property. This makes it possible to construct its nonlocal-invariant solutions. The most interesting among them are the solutions that have singularities at the lines of degeneracy $w(x, t) = 0$ stronger than those displayed by the well-studied generalized solutions of such equations (see Section 17 and the literature cited). To simplify the computations, let us restrict ourselves to the case $m = 1$. As for the parameters λ and μ, without any loss of generality we may take them equal to unity. Thus, let us consider the equation

$$w_t = \left[\frac{w}{(w+1)^3} w_x\right]_x. \tag{27.2}$$

Let us perform in (27.2) the transformation $w = u - 1$ and then apply to the equation obtained for $u(x, t)$ the L-transformation (8.4). The corresponding equation for the function $\omega(\xi, t)$ is of the following form:

$$\omega_t = \left[(1 - \omega)\omega_\xi\right]_\xi .$$

The substitution $\omega = 1 - \psi$ reduces the last equation to the form

$$\psi_t = \left(\psi\psi_\xi\right)_\xi . \tag{27.3}$$

In contrast to equation (27.2) for which the largest admissible Lie group is three-parameter, equation (27.3) possesses a four-parameter Lie group. The same assertion is true for all the family of equations (27.1) and the corresponding equations

$\psi_t = (\psi^m \psi_\xi)_\xi$. In particular, equation (27.3) admits the transformation $\xi = \tilde{\xi}$, $t = a\tilde{t} - 1$, $\psi = a^{-1}\tilde{\psi}$ ($a > 0$ is a constant) which does not correspond to any pointwise transformation preserving the original equation (27.2). One of the solutions of (27.3) invariant with respect to that transformation is given by the formula

$$\psi = \frac{\varphi'(\xi)}{t+1}, \tag{27.4}$$

where the function φ satisfies the equation

$$\varphi'\varphi'' + \varphi = 0.$$

(Here the prime denotes differentiation in ξ). Let us consider the even solutions of the last equation, assuming that $\varphi' > 0$ for ξ small enough. They are defined by the formula

$$\xi = \int_0^{\varphi} \frac{dz}{\left(A - 3z^2/2\right)^{1/3}}, \qquad A = \text{const.} > 0.$$

Set

$$\xi_* = \int_0^{\sqrt{2A/3}} \frac{dz}{\left(A - 3z^2/2\right)^{1/3}} \equiv 3\pi^{-1} \{\Gamma(2/3)\}^2 (A/2)^{1/6}.$$

Then $\varphi' > 0$ for $|\xi| < \xi_*$ and $\varphi' = 0$ as $\xi = \xi_*$. The function φ attains the maximum value $\sqrt{2A/3}$ at $\xi = \xi_*$. Solution (27.4) of equation (27.3) induces a connection between the variables of Euler and Lagrange:

$$x = \xi - \frac{\varphi(\xi)}{t+1}. \tag{27.5}$$

The requirement of smooth bijective correspondence between ξ and x in the interval $[-\xi_*, \xi_*]$ for each $t \geq 0$ leads to the restriction $A < 1$. The invariant solution (27.4) of equation (27.3), defined in the half-strip $|\xi| \leq \xi_*$, $t \geq 0$, corresponds to the nonnegative solution $w(\xi, t)$ of equation (27.2), parametrically defined by the formulas (27.5) and

$$w = \frac{\varphi'(\xi)}{t + 1 - \varphi'(\xi)}. \tag{27.6}$$

Continuing by zero the solution of (27.4) outside the lateral boundaries of the half-strip, we get $w(\xi, t) = 0$ for $|x| \geq \eta(t)$, where

$$\eta(t) = \xi_* - \frac{\varphi(\xi_*)}{t+1}$$

is the function defining the boundary of the support of $w(x, t)$.

The function w defined in this way is continuous in the half-plane $x \in \mathbb{R}$, $t \geq 0$. This function is a classical solution of equation (27.2) in the domain $|x| < \eta(t)$, $t > 0$ and in the completion of the half-plane to the closure of this domain. Also, the function

$w\, w_x$ has nonzero limits as $x \to \eta(t) + 0$ and $x \to \eta(t) - 0$, $t \geq 0$. The last feature leads this solution beyond the borders of the standard class of generalized solutions to the Cauchy problem for equation (27.2) and gives a reason to term it "a strange solution" of this equation.

Too strong a singularity of the function w at the curves $x = \pm\eta(t)$ leads to the violation of the conservation law,

$$\frac{d}{dt}\left(\int_{\mathbb{R}} w(x,t)\,dx\right) = 0,$$

which holds for the usual generalized solutions of equation (27.2). This is why one fails to attach to this solution a physical sense in trying to identify w with the density of some continuum (for instance, with the density of a gas filtrating through a porous medium). However, this is possible if w is taken not for the density, but for the temperature of a medium with phase transitions. This approach is quite exotic since once the temperature assumes the value of the transition, $w = 0$, the coefficient of heat conductivity vanishes.

Due to (27.5), (27.6), the "differential equation" for the function $\eta(t)$ can be given the form

$$\frac{d\eta}{dt} = - \lim_{x\to\eta(t)-0}\left[\frac{w w_x}{(w+1)^3}\right]. \tag{27.7}$$

Relation (27.7) is equivalent to the condition at the free boundary in the one-phase Stefan problem corresponding to the linear heat equation (27.2) and the unit value of the latent heat of melting (the mathematical statement of the one-phase Stefan problem is contained in Section 4). It is interesting to remark that when t varies from 0 to ∞, the melting front $x = \eta(t)$ advances to the finite distance $\varphi(\xi_*) = \sqrt{2A/3}$. This makes a qualitative difference between the behaviour of the boundary of support in the constructed "strange solution" from the behavior of supports of the generalized solutions to equation (27.2). Analogous "strange" solutions can be constructed for the general equation (27.1) with $m,\ \lambda,\ \mu > 0$.

Chapter IV

Parabolic equations in several space dimensions

28 Review of results

In the previous sections we were dealing with evolution equations of divergence form with a single space variable. In their analysis, the effectiveness of passage to the Lagrangian coordinates was to a prominent extent due to the fact that for such a class of equations this transition was realized by a reciprocity transformation. However, the possibility itself to introduce Lagrangian coordinates is not connected with the one-dimensional character of the process described by an evolution equation. The key point here is the possibility to attribute to this equation the sense of some balance law.

In this chapter, the method of Lagrangian coordinates is extended to multi-dimensional equations. The main objects of investigation here will be the one-phase Stefan problem for the homogeneous heat conduction equation and the Cauchy problem for the porous medium equation. The simplest formulation of the first of these problems is the following.

Let us denote by $\Omega(t) \subset \mathbb{R}^n$ a time-dependent domain with boundary $\Gamma(t)$. It is required to define the surface $\Gamma(t)$ and the temperature $\theta(x, t)$ by the conditions

$$D_t \theta = \Delta \theta, \quad x \in \Omega(t),\ t \in (0, T), \tag{28.1}$$

$$\theta = 0, \quad V_\nu = -\sum_{i=1}^{n} \nu_i D_i \theta, \quad x \in \Gamma(t),\ t \in (0, T), \tag{28.2}$$

$$\theta(x, 0) = \theta_0(x), \quad x \in \Omega(0). \tag{28.3}$$

Here we use the notation $D_t = \partial/\partial t$, $D_i = \partial/\partial x_i$, $i = 1, \dots, n$. In (28.2) V_ν is the velocity of displacement of the hypersurface $\Gamma(t)$ in the direction of the outer normal ν to this hypersurface. It is assumed that the domain $G = \Omega(0)$ is given, $\theta_0 > 0$ if $x \in G$, $\theta_0(x) = 0$ if $x \in \Gamma(0) = S = \partial G$.

The method presented below of treating the Stefan problem was proposed in [109] and is based on its analogy with the problem of the nonstationary motion of the ideal fluid with free boundary [116]. It turned out to be convenient to carry out the analysis of this problem in Lagrangian coordinates where the problem domain is rendered stationary and thus is known beforehand.

The next section is devoted to the passage to Lagrangian coordinates in problem (28.1)–(28.3). In the new coordinates the Stefan problem corresponds to an initial-boundary value problem in a fixed domain for a quasilinear system of n equations. This system does not belong to any of the types so far studied and may be conventionally termed a degenerate parabolic system. The degeneracy of the system is partially confirmed by the fact that on the lateral boundary of the cylinder $G_T = G \times (0, T)$ for n sought functions only one boundary condition is posed. To clarify the mathematical nature of the problem obtained, in Section 30 its linearization is undertaken. If the problem is linearized on an exact solution, the linear model proves to be well-posed. The proof of this fact is given in Section 31.

In Section 32 the self-similar solutions of problem (28.1)–(28.3) are studied in which G is a cone in $\mathbb{R}^n$ and $\theta_0(x)$ is a homogeneous function of the variables $x_1, \ldots, x_n$ of degree of homogeneity zero. It is to be noted that while the one-dimensional similarity solutions of the one-dimensional Stefan problem were already well studied (see, for instance, [169]), their multi-dimensional analogs were discovered only recently [107, 158, 138, 139, 6]. In paper [107] a self-similar solution of the planar problem was constructed such that the domain $\Omega(t)$ was close to a circular ring. The solvability of the multi-dimensional Stefan problem in the self-similar statement corresponding to the case when G is a cone is proven in [158]. This paper was preceded by paper [139] which dealt with the linearized version of this problem. Finally, in [6] the solvability of the planar self-similar problem was established where the domain $\Omega(t)$ was included between the free boundary and the prescribed boundary of a sector with opening angle less than π. Also, at the given boundary the temperature was maintained constant.

Another problem where the Lagrangian coordinates method proved to be fruitful was the Cauchy problem for the so-called *porous medium equation* (PME)

$$D_t u = \Delta (u^m), \quad x \in \mathbb{R}^n, \ t \in (0, T), \tag{28.4}$$

where $m = \text{const.} > 1$. Equation (28.4) was already discussed in the previous chapter. It serves to model the process of isentropic filtration of the polytropic gas in a homogeneous isotropic porous medium. Here the function u is proportional to the gas density and $m-1$ is the polytropic exponent. Equation (28.4) also describes the processes of nonlinear heat conduction or diffusion under the power dependence of the heat conductivity or the diffusion coefficients on the temperature or concentration. In all these applications, the sought function $u(x, t)$ takes on nonnegative values. That is why in the study of the Cauchy problem

$$u(x, 0) = u_0(x) \quad x \in \mathbb{R}^n \tag{28.5}$$

for the porous medium equation it is usually assumed that u_0 is nonnegative.

We will be interested in the situations when the support of u_0 does not coincide with the whole space. As was already pointed out in Section 17, in this case the problem need not have classical solutions. The zero-level surfaces of the function $u(x, t)$ are, generally speaking, the surfaces where the space derivatives are discontinuous. Since their location is a priori unknown, we have every reason to term the boundary of support of $u(x, t)$ the free boundary.

In paper [109] the formulation of problem (28.4)–(28.5) was proposed in which the free boundary "stopped": it coincided with the boundary of support of u_0. Since in [109] the geometrical Lagrangian coordinates were used, the equivalent to (28.4) system of n scalar equations contained some information about the initial distribution u_0 in (28.5). It is shown in Section 33 how one may introduce in the obtained problem an analog of the mass Lagrangian coordinates so that the resulting system will take on a canonical form. There we also point out a wide class of exact solutions of the Cauchy problem (28.4)–(28.5) with the initial datum a measure. These solutions can be interpreted as generalizing the famous Barenblatt solution. Apart from the latter they are not self-similar, though.

The simplest consistent model of the situation under discussion is the Cauchy problem for equation (28.4) with $m = n = 2$. In this case, it is called the Boussinesq equation. Apart from the above-mentioned applications, it serves for an approximate description of the process of the planar filtration of the incompressible fluid in homogeneous soil over a horizontal bed [132]. The function $u(x, t)$ defines the level of the soil water.

Section 34 is devoted to the study of the Boussinesq equation in the so-called canonical Lagrangian variables. In particular, it is shown that an equivalent to (28.4) system of two quasi-linear equations possesses broader group properties than the original equation. This feature makes it possible to simplify the statement of the boundary value problem for that system. A detailed analysis is given of the self-similar solutions of the Boussinesq equation. We present an example of calculation of self-similar solutions in Lagrangian coordinates. Finally, a rational method is proposed to linearize the Boussinesq equation (it is far from evident that in the original variables this equation can be linearized near the boundary of the support).

The concluding section of the chapter contains the study of the regularity properties of interfaces displayed by solutions of the Cauchy problem for the multi-dimensional PME. For this we employ the same idea as in Chapter III. Having rendered the interface a stationary hypersurface (the support of the initial data), we thus reduce the matter to the study of the function representing the particles' trajectories. Since the interface is constituted by the same particles (in our fictitious model), it amounts to verifying that the "trajectories" maintain a certain regularity at the lateral boundary of the initial support. To be precise, we prove that the free boundary admits (locally in time) a parametrization which has the same regularity in the tangential directions in each plane $\{t = \text{const.}\}$ as had the initial interface, but becomes C^∞ in time.

29 Lagrangian coordinates in the one-phase Stefan problem

29.1 Formulation of the problem. In this section we propose the Lagrangian approach for the analysis of the multi-dimensional one-phase Stefan problem for the homogeneous heat equation. The statement of this problem was already given in the previous section: this is problem (28.1)–(28.3).

To formulate this problem in the Lagrange variables, let us consider (28.1) as the continuity equation for a compressible liquid with the density ρ and the velocity $v = (v_1, \dots, v_n)$:

$$D_t \rho + \sum_{i=1}^{n} D_i (\rho v_i) = 0. \tag{29.1}$$

We shall consider (28.2) as the kinematic condition on the surface $\Gamma(t)$ (the zero level surface of θ):

$$D_t \theta + \sum_{i=1}^{n} v_i D_i \theta = 0 \quad \text{as } \theta = 0. \tag{29.2}$$

Equation (29.1) will coincide with (28.1), provided that

$$\rho = \theta + \psi, \quad v_i = -\frac{1}{\rho} D_i \theta, \tag{29.3}$$

where ψ is an arbitrary function independent of t. Due to the equalities

$$V_\nu = \frac{D_t \theta}{|D\theta|}, \quad \sum_{i=1}^{n} v_i D_i \theta = -|D\theta| \quad \text{as } \theta = 0,$$

the second condition of (28.2) reduces to

$$D_t \theta - |D\theta|^2 = 0 \quad \text{for } \theta = 0.$$

By (29.3), this equality is equivalent to (29.2) if $\psi \equiv 1$.

Let us write the equation of the trajectory of the "liquid particle" with the coordinates x and the velocity vector v at time t as

$$\frac{dx}{dt} = v = -\frac{1}{1+\theta} D\theta. \tag{29.4}$$

This equality follows from (29.3). Treating (29.4) as a system of ordinary differential equations with respect to the components of the vector x, we shall complement this system by the initial conditions

$$x = \xi \quad \text{for } t = 0. \tag{29.5}$$

By the *Lagrangian coordinates* we shall mean the initial coordinates ξ of the "liquid particle".

The next objective is to reformulate relations (28.1)–(28.3) in the new independent variables (ξ, t). Set

$$\theta(x(\xi, t), t) = u(\xi, t)$$

and note that $D\theta = (M^*)^{-1} \nabla u$, where $M^* = \nabla_\xi x$ is the matrix obtained by transposing the Jacobian M with the entries $M_{ij} = D_j x_i, i, j = 1, \dots, n$. (Throughout Chapters 29 and 30, ∇ denotes the gradient in the variables $\xi_1, \ldots, \xi_n$). Inserting the last equality into (29.4), we obtain

$$M^* D_t x = \nabla \left\{ \log \frac{1}{1+u} \right\}. \tag{29.6}$$

It remains to express u in terms of M and ξ. To this end we apply the Cauchy identity

$$D_t\,(\tilde{\rho} \cdot J) = 0, \quad \tilde{\rho}(\xi, t) = \rho(x, t), \tag{29.7}$$

which is well known in continuum mechanics, where the notation $J = \det M$ is used. Equation (29.7) is a consequence of the continuity equation (29.1). Prescribing the initial conditions (28.3), (29.5) and using the equality $\tilde{\rho} = u + 1$, we arrive at the relations

$$(u+1)\,J = \rho_0(\xi) = 1 + \theta_0(\xi). \tag{29.8}$$

The substitution of (29.8) into (29.6) leads to the sought system of equations for the functions $x_i(\xi, t)$, $i = 1, \ldots, n$:

$$M^*\, D_t x = \nabla \left\{ \log \frac{\det M}{1 + \theta_0(\xi)} \right\}. \tag{29.9}$$

Equalities (29.5) impose initial conditions for system (29.9). The appropriate boundary condition can be derived from the obvious identity

$$D_t u = D_t \theta + \sum_{i=1}^{n} v_i\, D_i \theta.$$

By (29.2), this identity implies that the equality $D_t u = 0$ holds on the free boundary whose equation in Lagrangian coordinates is $u = 0$. The zero-level surface of $u(\xi, t)$ is thus time-independent. By (28.3), (29.5) and the definition of $\Gamma(t)$, the equation of the free surface may be given the form $\theta_0(\xi) = 0$. Therefore, with $\Gamma(t)$ treated as the image of $\Gamma(0)$ under the transformation $\xi \mapsto x$, condition (28.2) will be automatically satisfied. Recall now condition (29.8) and take into account the equalities $u = \theta_0 = 0$, to conclude that

$$J = 1 \quad \text{for } \xi \in \Gamma(0). \tag{29.10}$$

We now give the complete formulation of the Stefan problem in Lagrangian variables. In the cylinder $G_T = G \times (0, T)$, it is required to find a solution $x = (x_1, \ldots, x_n)$ of system (29.9) that satisfies condition (29.5) on the lower face of the cylinder and condition (29.10) on its lateral boundary $S_T = S \times (0, T)$, $S = \Gamma(0)$.

In the sequel we shall refer to the above problem as *problem* (A).

29.2 Linearization. Let $x(\xi, t)$ and $X(\xi, t)$ be smooth vector functions, with x not necessarily satisfying system (29.9). Set

$$\tilde{x} = x + \varepsilon X, \quad \tilde{M}^* = \nabla_\xi \tilde{x}, \quad M^* = \nabla_\xi x,$$

where ε is a formal small parameter.

We shall compute the terms of $\tilde{J} = \det \tilde{M}$ and $\tilde{M}^* D_t \tilde{x}$ which are linear in ε. We have

$$\tilde{M}^* D_t \tilde{x} = M^* D_t x + \varepsilon \left\{ M^* D_t X + N^* D_t x \right\} + \cdots, \tag{29.11}$$

where $N^* = \nabla_\xi X$. We shall assume that $J = \det M \neq 0$ and define the function Y by the relation

$$Y = M^{-1} X.$$

Applying the obvious identity

$$N^* D_t x = \nabla \{ x\, D_t x \} - D_t \left\{ M^* x \right\}$$

and the definition of Y, we get

$$\begin{aligned} M^* D_t X + N^* D_t x &= M^* M D_t Y + \nabla \left\{ Y\, M^* D_t x \right\} \\ &\quad + \left(M^*\, D_t M - D_t M^* M \right) Y. \end{aligned} \tag{29.12}$$

Further, as $\varepsilon \to 0$,

$$\det \tilde{M} = J \det \left(I + \varepsilon\, M^{-1} N \right) = J + \varepsilon J \operatorname{tr} \left[M^{-1} N \right] + O(\varepsilon^2).$$

In terms of Y, this expression can be written as

$$\det \tilde{M} = J \left\{ 1 + \varepsilon \left(\sum_{i=1}^{n} (D_i Y_i + b_i Y_i) + O(\varepsilon) \right) \right\}, \tag{29.13}$$

where $b = (b_1, \ldots, b_n)$,

$$b_i = \sum_{j,k=1}^{n} \frac{\partial \xi_i}{\partial x_k} \frac{\partial}{\partial \xi_j} \left(\frac{\partial x_k}{\partial \xi_i} \right). \tag{29.14}$$

Indeed,

$$\begin{aligned} \operatorname{tr} \left(M^{-1} N \right) &= \sum_{i,k=1}^{n} \frac{\partial \xi_i}{\partial x_k} \frac{\partial X_k}{\partial \xi_i} = \sum_{i,k=1}^{n} \frac{\partial \xi_i}{\partial x_k} \frac{\partial}{\partial \xi_i} \left(\sum_{j=1}^{n} Y_j \frac{\partial x_k}{\partial \xi_i} \right) \\ &= \sum_{i,j,k=1}^{n} \frac{\partial \xi_i}{\partial x_k} \frac{\partial x_k}{\partial \xi_j} \frac{\partial Y_j}{\partial \xi_i} + \sum_{i,j,k=1}^{n} \frac{\partial \xi_i}{\partial x_k} \frac{\partial}{\partial \xi_i} \left(\frac{\partial x_k}{\partial \xi_j} \right) Y_j = \sum_{i=1}^{n} D_i Y_i + \sum_{j=1}^{n} b_j Y_j. \end{aligned}$$

We transform the right-hand side of (29.14) using the formula

$$\sum_{i,k=1}^{n} \frac{\partial \xi_i}{\partial x_k} \frac{\partial}{\partial \xi_i} \left(\frac{\partial x_k}{\partial \xi_j} \right) = \sum_{k=1}^{n} \frac{\partial}{\partial x_k} \left(\frac{\partial x_k}{\partial \xi_j} \right) = \frac{1}{J} D_j J.$$

Hence, $b = \nabla \{ \log J \}$.

From (29.11)–(29.13), the linearized system (29.9), expressed in terms of $Y = (Y_1, \dots, Y_n)$, can be written in the form

$$A D_t Y = \nabla \{\operatorname{div} Y + a\, Y\} + BY + f, \tag{29.15}$$

where

$$A = M^* M, \quad B = M\, D_t M_* - M^* D_t M,$$

$$a = \nabla \{\log J\} - M^*\, D_t x, \quad \operatorname{div} Y = \sum_{i=1}^{n} D_i Y_i.$$

By (29.13), the linearization of the boundary condition (29.10) leads to the relation

$$\operatorname{div} Y + bY = \varphi \quad \text{for } \xi \in S. \tag{29.16}$$

Equalities (29.15), (29.16) are to be completed by the initial condition

$$Y(\xi, 0) = Y_0(\xi) \quad \text{as } \xi \in G. \tag{29.17}$$

The complete form of the linearized system (in the sequel referred to as *problem* (B)) is the following: to find a solution Y of system (29.15) in the cylinder G_T which satisfies conditions (29.15) on the lateral boundary S_T of G_T and condition (29.17) on the lower face G.

In (29.15)–(29.17) the functions f, φ and Y_0, defined on G_T, S_T and G, respectively, are assumed to be given.

It is worth noting that problem (B) can be reduced to a formulation with one unknown scalar function. Let $Y = (Y_1, \dots, Y_n)$; set

$$U = \operatorname{div} Y + a\, Y \tag{29.18}$$

and consider the following system of ordinary differential equations for Y, parametrized by ξ:

$$A\, D_t Y - BY = F, \quad F = \nabla U + f. \tag{29.19}$$

Let $\Psi(\xi, t)$ be the fundamental matrix of the homogeneous system

$$D_t Y = A^{-1} BY.$$

By the method of variation of constants, we get the following representation of the solution to the Cauchy problem (29.19), (29.17)):

$$\begin{aligned} Y(\xi, t) &= \int_0^t \Psi(\xi, t)\Psi^{-1}(\xi, \tau) A^{-1}(\xi, \tau)[F(\xi, \tau)]d\tau + \Psi(\xi, \tau)\Psi^{-1}(\xi, 0)[Y_0] \\ &\equiv \int_0^t Q_0(\xi, t, \tau)\nabla U(\xi, \tau) d\tau + g(\xi, t). \end{aligned} \tag{29.20}$$

Further, using equation (29.19) and the representation (29.20), we compute the derivative

$$D_t u = \operatorname{div}(D_t Y) + a D_t Y + D_t a Y$$

to obtain

$$D_tU = \operatorname{div}\left(A^{-1}\nabla U\right) + \nabla U\, A^{-1}a + \operatorname{div}\left(\int_0^t Q(\xi,t,\tau)\nabla U(\xi,\tau)\,d\tau\right)$$
$$+ \int_0^t \omega(\xi,t,\tau|a)\,\nabla U(\xi,\tau)d\tau + h(\xi,t), \tag{29.21}$$

where

$$Q(\xi,t,\tau) = A^{-1}B(\xi,t)Q_0(\xi,t,\tau),$$
$$\omega(\xi,t,\tau|a) = Q_0^*(\xi,t,\tau)[D_ta(\xi,t)] - A^{-1}(\xi,t)B(\xi,t)[a(\xi,t)],$$
$$h(\xi,t) = f\,A^{-1}a + \operatorname{div}\left(A^{-1}\,Bg\right) + \left(D_ta - A^{-1}Ba\right)g.$$

Combining (29.16) and (29.18) leads to the equality

$$U = (a-b)\,Y + \psi,$$

which by differentiation in time gives the boundary condition for the potential U on the lateral boundary S_T:

$$D_tU = \nabla U\, c + \int_0^t \omega(\xi,t,\tau|a-b)\,\nabla U(\xi,t)\,d\tau + \psi, \tag{29.22}$$

where $c = a - b$,

$$\psi = g(D_tc - A^{-1}B[c]) + f + A^{-1}[c]D_t\psi.$$

Moreover, the function U satisfies the initial condition

$$U(\xi,0) = U_0(\xi) \equiv \operatorname{div} Y_0(\xi) + a(\xi,0)Y_0(\xi). \tag{29.23}$$

We are now able to formulate the linearized problem (A) in terms of the function U: *to find a solution of equation* (29.21) *in the cylinder* G_T, *satisfying condition* (29.22) *on the lateral boundary of* G_T *and condition* (29.23) *on its base.*

30 Correctness of the linear model

In this section we shall prove the existence and uniqueness theorem for problem (B) linearized on the exact solution of problem (A).

Lemma 30.1. *Let* $u(x,t)$ *be the exact solution of problem* (A), *with the third derivatives in* ξ *and* t *continuous in the domain* G_T. *Then*

$$B \equiv M\,D_tM^* - M^*D_tM = 0 \quad \text{for } (\xi,t)\in G_T, \tag{30.1}$$
$$a - b = -q\,\nu, q \ge a_0 = \text{const.} > 0 \quad \text{for } (\xi,t)\in S_T, \tag{30.2}$$
$$y\,A[y] \ge a_0|y|^2 \quad \text{for } (\xi,t)\in G_T, \tag{30.3}$$

where ν *is the unit outward normal to* S.

Proof. The first relation follows from the equalities

$$B_{j,k} = \sum_{i=1}^{n} \left\{ D_j \left(D_t x_i \right) D_k x_i - D_j x_i D_k \left(D_t x_i \right) \right\} = \sum_{i=1}^{n} \left\{ D_j \left(D_t x_i D_k x_i \right) \right\}$$

$$+ \sum_{i=1}^{n} \left\{ D_t x_i \left(D_j D_k x_i - D_k D_j x_i \right) \right\}$$

$$= D_j D_k \left\{ \log \left(\frac{J}{\rho_0} \right) \right\} - D_k D_j \left\{ \log \left(\frac{J}{\rho_0} \right) \right\} = 0, \tag{30.4}$$

which we have obtained by using equation (30.1), and the equality of the second derivatives of the scalar functions x_i and $(\log J - \log \rho_0)$.

To derive (30.2), let us note that the function $w = -\log(1+u)$ satisfies the homogeneous parabolic equation

$$D_t w = \operatorname{div} \left\{ M^{-1} M^{*-1} [\nabla w] \right\} - M^{-1} [\operatorname{div}(M^{-1})^*] \nabla w \tag{30.5}$$

in G, it vanishes on the lateral boundary of S_T, is strictly negative on the lower face of G_T and

$$a - b = -M^*[D_t x] = -\nabla w \quad \text{for } \xi \in S.$$

S_T is the minimum level surface of w, while ∇w is directed along with the outward normal to the surface S and vanishes nowhere due to the maximum principle.

Equation (30.5) results from the relations

$$D_t \tilde{\rho} = D_t \rho + \sum_{i=1}^{n} v_i D_i \rho = -\rho \sum_{i=1}^{n} D_i v_i = -\rho \operatorname{div}_x v,$$

$$D_t w = D_t \{-\log \tilde{\rho}\} = \operatorname{div}_x v, \quad v = M^{*-1} [\nabla_\xi w]$$

by applying the operator div_x expressed in Lagrangian variables to the latter equation.

Finally, inequality (30.3) follows from the definition of the matrix A, the relation

$$w = \log J - \log (1 + \theta_0),$$

and the boundedness of $|w|$ and $\log(1+\theta_0)$. □

Let us denote

$$(X, Y) = \int_G X(\xi, t) Y(\xi, t)\, d\xi, \quad (X, Y)_S = \int_S X(\xi, t) Y(\xi, t)\, d\sigma,$$

where X, Y are either vector functions (then $XY = \sum_{i=1}^{n} X_i Y_i$) or scalar functions, $d\sigma$ is an element of the surface S. Denote by $L_2(G_T)$ ($L_2(S_T)$) the space of either vector or scalar functions with modulus square integrable in G_T (on S_T), without specifying whether a given function is vector or scalar valued.

Under the hypotheses of Lemma (30.1), problem (B) can be reformulated as follows: to determine a vector $Y(\xi, t)$ which in G_T satisfies the system of equations

$$A[D_t Y] = \nabla(\operatorname{div} Y + aY) + f, \tag{30.6}$$

the condition

$$\operatorname{div} Y + aY = -q\,(Y\nu) + \varphi \quad \text{on } S, \tag{30.7}$$

and the initial condition

$$Y(\xi, 0) = 0 \quad \text{for } \xi \in G. \tag{30.8}$$

We shall refer to problem (30.6)–(30.8) as *problem* (C).

Taking the scalar product of equation (30.6) with an arbitrary smooth vector function Ψ and integrating the result by parts over domain G, we arrive at the identity

$$\begin{aligned}(\Psi,\ A[D_t Y]) \;+\; (\operatorname{div} Y + aY,\ \operatorname{div}\Psi) + (q(Y\nu),\ (\Psi\nu))_S \\ = (f,\ \Psi + (\varphi, (\Psi\nu)))_S\,,\end{aligned} \tag{30.9}$$

which leads to the definition of the generalized solution.

Definition 30.1. By the *generalized solution* of problem (C) we shall mean a vector function $Y \in L_2(G_T)$ such that

$$Y\nu \in L_2(G_T), \qquad |D_t Y|,\ \operatorname{div} Y \in L_2(G_T),$$

which fulfills identity (30.9) with an arbitrary smooth function Ψ and initial condition (30.8).

Theorem 30.1. *Let conditions* (30.2), (30.3) *be satisfied and let*

$$K_0 = \max\left\{|a|^{(0)}_{G_T},\ |D_t A|^{(0)}_{G_T}\right\} < \infty.$$

Then the generalized solution of problem (C) *is unique.*

Proof. The difference of two possible solutions $Y_{(1)}$ and $Y_{(2)}$ of problem (C) satisfies identity (30.9) with $f = 0$ and $\varphi = 0$. Since there always exists a sequence of smooth functions Ψ_n strongly convergent in $L_2(G_T)$ to Y, such that $\operatorname{div}\Psi_n$ converges to $\operatorname{div} Y$ strongly in $L_2(G_T)$, and $\Psi_n\nu$ converges to $Y\nu$ strongly in $L_2(S_T)$, we can assume $\Psi = Y$ in (30.9) strongly in $L_2(S_T)$

$$(A[D_t Y], Y) + (\operatorname{div} Y + aY,\ \operatorname{div} Y) + (q(Y\nu), (Y\nu))_S = 0.$$

Using Hölder's and Young's inequalities, we get from this

$$\frac{1}{2}\|\operatorname{div} Y\|^2_{2,G} + \frac{1}{2}\frac{d}{dt}(Y, A[Y]) + a_0\|Y\nu\|^2_{2,S} \le \frac{1}{2}K_0\,\|Y\|^2_{2,G}. \tag{30.10}$$

Integrating (30.10) in time, using (30.2), (30.3) and the initial condition (30.8) gives the inequality

$$\|Y(t)\|^2_{2,G} \le K_0\, a_0^{-1} \int_0^t \|Y(\tau)\|^2_{2,G}\, d\tau,$$

which implies the assertion of the Theorem. □

Theorem 30.2. *Let hypothesis* (30.2), (30.3) *be fulfilled, the boundary S be of* C^2 *and*

$$K_1 = \max\left\{|A, D_t A|^{(0)}_{G_T},\ |a, D_t a|^{(0)}_{G_T},\ |q, D_t q|^{(0)}_{S_T}\right\} < \infty, \tag{30.11}$$

$$K_2 = \max\left\{\|f, D_t f\|_{2,G_T},\ \|\varphi, D_t\varphi\|_{2,S_T}\right\} < \infty. \tag{30.12}$$

Then there exists a generalized solution Y of problem (C) *such that*

$$\max_{t\in(0,T)}\left\{\|D_tY\|_{2,G},\ \|D_tU\|_{2,G},\ \|U(t)\|^{(1)}_{2,G},\ \|Y\nu\|_{2,S}\right\} < \infty, \tag{30.13}$$

where $U = \operatorname{div} Y + aY$.

Proof. Let $\Psi_n(\xi)$ be sufficiently regular vector functions which form an orthogonal basis in $L_2(G)$. We construct a generalized solution of problem (C) as the limit, as $N \to \infty$, of Galerkin approximations

$$Y^N = \sum_{k=1}^{N} l_k^N(t)\,\Psi_k(\xi)$$

with the functions $l_k^N(t)$ satisfying the Cauchy problem

$$\sum_{j=1}^{N} a_{kj}\frac{dl_j^N}{dt} = \sum_{j=1}^{N} b_{kj} l_j^N + \varphi_k, \quad l_k^N(0) = 0,\ k = 1,\dots,N. \tag{30.14}$$

Equations (30.14) represent the conditions ensuring the orthogonality of the expression

$$A[D_tY^N] - \nabla\left(\operatorname{div} Y^N + aY^N\right) - f$$

to the first N basis functions $\Psi_1,\dots,\Psi_N$, with the boundary conditions (30.7) taken into account.

From these conditions, the functions $a_{kj}, b_{kj}, \varphi_k$ can be represented by

$$a_{kj} = \left(\Psi_k,\ A[\Psi_j]\right),$$

$$-b_{kj} = \left(\operatorname{div}\Psi_j + a\Psi_j,\ \operatorname{div}\Psi_k\right) + \left(q(\Psi_j\nu),\ (\Psi_k\nu)\right)_S,$$

$$\psi_k = (f,\ \Psi_k) + (\varphi,\ \Psi_k\nu)_S.$$

The existence of a solution to the Cauchy problem (30.14) follows from the standard theory of ordinary differential equations. Indeed, by condition (30.3) the matrix $\{a_{kj}\}$ admits the bounded inverse

$$\sum_{k,j=1}^{N} a_{kj}y_ky_j = \left(\sum_{k=1}^{N} y_k\Psi_k,\ A\left[\sum_{j=1}^{N} y_j\Psi_j\right]\right) \geq a_0\left\|\sum_{k=1}^{N} y_k\Psi_k\right\|^2_{2,G} = a_0|y|^2, \tag{30.15}$$

hence system (30.14) can be solved with respect to the time derivatives.

Conditions (30.11) and (30.12) indicate that the derivatives of the functions a_{kj}, b_{kj} are bounded, and that the functions φ_k belong to $W_2^2(0,T)$. Consequently, there exists a unique solution of problem (30.14) in the space $W_2^2(0,T)$.

It follows easily that the functions Y^N fulfill identity (30.9) for an arbitrary function

$$\Psi = \sum_{k=1}^{N} c_k(t)\Psi_k(\xi).$$

To justify the passage to the limit as $N \to \infty$ in that identity, we obtain uniform estimates on the Galerkin approximations Y^N, independent of N:

$$\left\{\max_{t\in(0,T)} \|Y^N\|_{2,G},\ \|Y^N\nu\|_{2,S_T},\ \|\operatorname{div} Y^N|_{2,G_T}\right\} \le C_1, \tag{30.16}$$

$$\left\{\max_{t\in(0,T)} \|D_tY^N(t)\|_{2,G},\ \|D_tY^N\nu\|_{2,S_T},\ \|\operatorname{div} D_tY^N\|_{2,G_T}\right\} \le C_2, \tag{30.17}$$

$$C_i = C_i(K_1, K_2), \quad i = 1, 2.$$

Estimates (30.16), ((30.17) can be derived in a standard way. Since the proof of (30.17) is more difficult, we present it in detail. Let us differentiate (30.14) with respect to time, multiply by $D_tl_k^N$ and take the sum in k from 1 to N. We have

$$\begin{aligned}
&\frac{1}{2}\frac{d}{dt}\left(D_tY^N,\ A[D_tY^N]\right) + \|\operatorname{div} D_tY^N\|_{2,G}^2 + \left(q\left(D_tY^N\nu\right),\ -\left(D_tY^N\nu\right)\right)_S \\
&= -\frac{1}{2}\left(D_tY^N,\ D_tA\left[D_tY^N\right]\right) + \left(D_tf,\ D_tY^N\right) \\
&\quad - \left(aD_tY^N + D_taY^N,\ \operatorname{div} D_tY^N\right) - \left(D_tq\left(Y^N\nu\right),\ \left(D_tY^N\nu\right)\right)_S \\
&\quad + \left(D_t\varphi,\ \left(D_t\varphi,\ \left(D_tY^N\nu\right)\right)_S\right).
\end{aligned} \tag{30.18}$$

We integrate (30.18) in time over the interval $(0,t)$, bound the right-hand side from above using Hölder's and Young's inequalities, and the left-hand side with the use of (30.2), (30.3), to deduce the inequality

$$\begin{aligned}
&\|D_tY^N(t)\|_{2,G}^2 + \|\operatorname{div} D_\tau Y^N\|_{2,G_t}^2 + \|D_\tau Y^N\nu\|_{2,S_t}^2 \\
&\qquad \le c(K_1)\left\{\|D_\tau Y^N\|_{2,G_t}^2 + K_2^2\right\},
\end{aligned} \tag{30.19}$$

which, by Gronwall's inequality, yields (30.17).

Relations (30.16) and (30.17) imply in a standard way the existence of a generalized solution to problem (C) for which these estimates also hold.

Coming back to identity (30.9) we see that

$$(A[D_tY] - f,\ \Psi) + (U,\ \operatorname{div}\Psi) = 0,$$

for arbitrary finite functions Ψ, hence equivalently

$$A[D_t Y] - f = \nabla U. \tag{30.20}$$

The last relation, combined with (30.17) and (30.12), implies that

$$\max_{t\in(0,T)} \|\nabla U\|_{2,G} < C.$$

Thus we can speak of the trace of U on S for almost all $t \in (0, T)$.

We finally take an arbitrary function Ψ, not necessarily finite, to infer from (30.9) by (30.20) that the identity

$$(U - q\,(Y\nu) - \varphi,\ \Psi\nu)_S = 0$$

holds. Hence, the boundary condition (30.7) is satisfied almost everywhere on S. □

31 Similarity solutions of the Stefan problem

31.1 Self-similar statement. The Lagrangian counterpart of the one-phase Stefan problem was formulated in Section 29: to find a solution of the system of equations

$$M^* x_t = \nabla_\xi \ln\left(\frac{|M|}{1+\theta_0}\right) \quad \text{in } \Omega_0 \times (0, T), \tag{31.1}$$

satisfying the initial and boundary conditions

$$|M||_{\Gamma_0} = 1 \quad \text{as } t \in (0, T), \quad x|_{t=0} = \xi \in \Omega_0. \tag{31.2}$$

Here $M = \nabla_\xi^* x$ is the Jacobian matrix with elements $M_{ij} = \partial x_i/\partial \xi_j$, $i, j = 1, \ldots, N$; M^* is the matrix transpose to M.

Given a solution $x \equiv x(\xi, t)$ of problem (31.1)–(31.2), the solution of the original problem is recovered with the help of the continuity equation for the motion of the fictitious fluid written in the Lagrange variables:

$$\theta(x,t) = \frac{1+\theta_0(\xi)}{|M|} - 1, \quad \xi \in \Omega_0.$$

The equation of the surface Γ_t is then obtained as the image of Γ_0 under the mapping $\xi \to x(\xi, t)$.

Let us assume that Ω_0 is a cone in $\mathbb{R}^N$, and $1+\theta_0(x)$ is a homogeneous function of the variables $x_1, \ldots, x_N$ of zero homogeneity degree. Under these assumptions conditions (31.1), (31.2) and the domain Ω_0 admit the rescaling transformation

$$x' = k\,x, \quad \xi' = k\,\xi, \quad t' = k^2 t,$$

which allows one to search for self-similar solutions of the form

$$x = \sqrt{t}\, f(\zeta), \quad \zeta = \frac{\xi}{\sqrt{t}}. \tag{31.3}$$

Let $r = |\zeta|$, $\varphi = \{\varphi_1, \dots, \varphi_{N-1}\}$ be the spherical coordinates in the space of the variables ζ. Denote by K the image of the cone Ω_0, by S_{N-1} the unit sphere centered at the top of K, and by $\omega = K \cap S_{N-1}$ the simply-connected domain cut out by K on S_{N-1}. By the assumption, θ_0 does not depend on $|\zeta|$ and can be viewed therefore as a function of the angular variables φ: $\theta_0 = \theta_0(\varphi)$.

The substitution of formulas (31.3) into conditions (31.1), (31.2) leads to the self-similar statement of the one-phase Stefan problem in Lagrangian coordinates: to find a vector $h = f(\zeta) - \zeta$ satisfying the conditions

$$F(h) \equiv \{F_1,\ F_2\} = 0, \tag{31.4}$$

$$F_1(h) \equiv (I + \nabla h)\left(h - r\,\frac{\partial h}{\partial r}\right) - 2\,\nabla \ln\left[\frac{|M|}{1+\theta_0}\right] \quad \text{in } K,$$

$$F_2 \equiv \left[|I + \nabla^* h| - 1\right]\Big|_{\partial K}, \quad |h| \to 0 \text{ as } r \to \infty. \tag{31.5}$$

The "initial" condition in (31.5) follows from (31.2).

Unless specially indicated, here and elsewhere all differential operators are assumed to be acting in the coordinates (r, φ). By $(\nabla h)_{ij}$ we denote the elements of the matrix ∇h in the spherical coordinate system. The notation I is used for the identity matrix and the identity operator.

As we already pointed out in the introduction, the problem formulated is far from classical. First of all, the operator $F_1(\cdot)$ is not elliptic. Next, for N sought components of vector h only one boundary condition is posed. These peculiarities of the problem make it hardly possible to come at once to any conclusion about its solvability.

The main line of proceeding further is the following. To reduce the difficulties caused by the nonlinear character of the problem, we shall construct a solution on the basis of Newton's method. This is an iteration method enabling one to replace a complicated nonlinear problem by a sequence of linear ones. Roughly speaking, we shall have to study in detail the linear version of problem (31.4)–(31.5). This study suggests the choice of the function spaces to which a solution to the complete problem might belong. Having chosen them, we will be able to prove the convergence of the iteration procedure.

The linear version of problem (31.4)–(31.5) is not easy to investigate, however. Let us linearize this problem at zero. That is, let us calculate the formal limit

$$\lim_{\varepsilon \to 0} \frac{1}{\varepsilon}\,[F(\varepsilon\, h) - F(0)]\,.$$

Denote by G the resulting operator. After linearization, problem (31.4)–(31.5) becomes:

$$G(0)\, h \equiv \{G_1(0),\ G_2(0)\}\, h,$$

where

$$G_1(0)\,h = h - r\,h_r - 2\,\nabla \mathrm{div}\, h \quad \text{in } K, \quad G_2(0)\,h = (\mathrm{div}\, h)|_{\partial K}\,.$$

Thus, the linear problem need not be elliptic. Moreover, since the problem is considered in the cone K, the coefficient of the lower-order derivative grows to infinity. These are the chief difficulties we will face in proceeding further.

31.2 Definitions. The main result. We shall use the notation

$$\frac{\partial^{|\kappa|+\delta} f}{\partial r^{\delta} \partial \varphi^{\kappa}} = D_r^{\delta} D_{\varphi}^{\kappa} f, \quad \kappa = (\kappa_1, \dots, \kappa_{N-1}), \; |\kappa| = \sum_{i=1}^{N-1} \kappa_i,$$

$$D_{\varphi}^{\kappa} f = D_{\varphi_1}^{\kappa_1} \dots D_{\varphi_{N-1}}^{\kappa_{N-1}} f.$$

If a function $f(r, \varphi)$ is measurable for some $r > 0$ as a function of the variables φ, then

$$\|f\|_2(r) \equiv \|f\|_{L_2(\omega)}(r) = \left\{ \int_{\omega} f^2(r, \varphi)\, d\omega \right\}^{1/2},$$

where $d\omega$ is the squared element of the surface S_{N-1}.

Let us introduce the semi-norms

$$\|f|_{m,\beta}^{(p,\tau)} = \sum_{0 \le l \le m} \sum_{\delta + |\kappa| = l} \sup_{r>1} \left\{ r^{\beta - m + (N-1)/2} \rho(r, \delta, p, \tau) \, \|D_r^{\delta} D_{\varphi}^{\kappa} f\|_2(r) \right\},$$

$$\|f|_{m,\beta} = \sum_{0 \le l \le m} \sum_{\delta + |\kappa| = l} \sup_{r \in (0,1)} \left\{ r^{\beta - m + (N-1)/2} \|D_r^{\delta} D_{\varphi}^{\kappa} f\|_2(r) \right\}$$

with the weights

$$\rho(r, \delta, p, \tau) = \begin{cases} r^{p-\delta} & \text{as } \delta \ge p - \tau, \\ r^{\delta} & \text{as } \delta < p - \tau. \end{cases}$$

Let us define the Banach space $V_{m,\beta}^{(p,\tau)} \cap V_{m,\gamma}$ as the completion of $C^{\infty}(K)$ by the norm

$$\|f\| = \|f|_{m,\beta}^{(p,\tau)} + \|f|_{m,\gamma}.$$

Introduce the space of vector-valued functions

$$W_{m,\beta}^{(p,\tau)} \cap W_{m,\gamma} = \left\{ f = (f_1, \dots, f_N) : \; f_i \in V_{m,\beta}^{(p,\tau)} \cap V_{m,\gamma} \right\},$$

with the norm

$$\|f\| = \sum_{i=1}^{N} \|f_i\|_{V_{m,\beta}^{(p,\tau)} \cap V_{m,\gamma}}.$$

The notation Y_∂ will be used to denote the space of traces on ∂K of the functions from the space Y with the norm

$$\|\varphi\|_{Y_\partial} = \inf\{\|f\|_Y : \ f \in Y,\ f|_{\partial K} = \varphi\}.$$

$H^\sigma(\omega)$ is the Sobolev space with the norm

$$\|f\|_{H^\sigma(\omega)} = \left\{\int_\omega \sum_{|\gamma|\le\sigma} |D^\gamma_\varphi f|^2 \, d\omega\right\}^{1/2}.$$

Also, $H^0(\omega) = L_2(\omega)$. For $\sigma \ge 1$ we denote by $H^\sigma_0(\omega)$ the closure of the space $C^\infty_0(\omega)$ in the norm of $H^\sigma(\Omega)$.

We always use the notation $\alpha = m + 1 - (N-1)/2$ and

$$\begin{aligned}
X &= W^{(2,0)}_{m+2,\alpha+2} \cap W_{m+2,\alpha}, & R &= V^{(2,1)}_{m+1,\alpha} \cap V_{m+1,\alpha},\\
Z &= W^{(0,0)}_{m,\alpha} \cap W_{m,\alpha}, & Y &= V^{(0,0)}_{m,\alpha} \cap V_{m,\alpha},\\
Q &= W^{(0,-1)}_{m+1,\alpha+2} \cap W_{m+1,\alpha}, & S &= V^{(0,-1)}_{m+1,\alpha+2} \cap V_{m+1,\alpha},\\
U &= W^{(4,1)}_{m+3,\alpha+2} \cap W_{m+3,\alpha}, & T &= V^{(4,1)}_{m+3,\alpha+2} \cap V_{m+3,\alpha}.
\end{aligned}$$

By T^0, $S^0, \ldots$ we denote the subspaces of functions having zero trace on ∂K.

The notation ∇_S is reserved for the differential operator acting in the local coordinate system on S_{N-1}:

$$\nabla f = \vec{e}_r\, f_r + \frac{1}{r}\nabla_S f,$$

where $\vec{e}_r$ is the unit vector of the r axis in the spherical coordinate system.

Let $\{\psi_i\}$ and $\{\lambda_i\}$ be the sequences of eigenfunctions and eigenvalues of the first boundary value problem for the Laplace–Beltrami operator Δ_S :

$$-\Delta_S\psi_i = \lambda_i\psi_i, \quad \text{for } \varphi \in \omega; \quad \psi_i|_{\partial\omega} = 0, \qquad i = 1, 2, \ldots. \tag{31.6}$$

The system $\{\psi_i\}$ is complete in $L_2(\omega)$. Without loss of generality, one may suppose that it forms a basis of $L_2(\omega)$, and that the values λ_i grow monotonely in i. It is known that all eigenvalues of problem (31.6) are positive, finite-multiple, and that the sequence $\{\lambda_i\}$ may have only the accumulation point $\lambda = \infty$. $\psi_i \in C^\infty_0(\omega)$ if $N = 2$. For $N = 3$, the same is true if $\partial\omega \in C^\infty$.

We can formulate now the main result of the study.

Theorem 31.1. *Let $K \subset \mathbb{R}^N$ $(N = 2, 3)$, be such that ω is simply-connected, $\partial\omega \in C^\infty$, and the eigenvalues of problem* (31.6) *obey the inequality*

$$\min_{i=1,2,\ldots} \lambda_i > 2N. \tag{31.7}$$

Given an arbitrary integer $m \ge 3$, there exists a constant $q > 0$ depending only on m, N (and the differentiability properties of the curve $\partial\omega$ if $N = 3$), such that for every function

$$\theta_0(\varphi) \in H^{m+1}_0(\omega), \quad \|\theta_0\|_{H^{m+1}(\omega)} < q,$$

problem (31.4)-(31.5) *has a unique solution* $h \in X$. *This solution satisfies the inequality*

$$\|h\|_X \le C \|\theta_0\|_{H^{m+1}(\omega)}$$

with a constant $C > 0$ *depending only on* m, N *(and* $\partial\omega$ *for* $N = 3$*), but independent of* h *and* θ_0.

Theorem 31.1 asserts a "local" existence of self-similar solution of the Stefan problem, i.e. under the assumption that the initial distribution of the temperature θ_0 is close enough to zero in the metric of $H^{m+1}(\omega)$.

Actually, condition (31.7) on the eigenvalues of problem (31.6) is a restriction on ω. For example, let $N = 2$. Due to the variational property of the eigenvalues of problem (31.6), condition (31.7) is easy to satisfy by just requiring that the opening angle of the sector K is small enough. If $N = 3$ and K is a circular cone, (31.7) is provided by the smallness of the solid angle of K. However, in the general case the value of λ_1 depends on the differentiability properties of the boundary $\partial\omega$ of ω.

31.3 Auxiliary propositions. In this subsection, we collect a few assertions of small independent value but useful in the further proceeding.

Lemma 31.1. *Under condition* (31.7) *the cone* K *occupies in* $\mathbb{R}^N$ *a domain strictly smaller than the half-space.*

Proof. It is enough to check that condition (31.7) is violated even if K coincides with the half-space. The required assertion then follows from the variational principle for the eigenvalues of the Laplace–Beltrami operator: given two domains ω and ω' such that $\omega' \subset \omega$, the corresponding eigenvalues λ and λ' in problem (31.6) satisfy the inequality $\lambda' \ge \lambda$, see [110].

1) Let $N = 2$. Then problem (31.6) takes the form

$$\varphi''(\tau) + \lambda\,\varphi(\tau) = 0 \quad \text{as } \tau \in (0, \pi); \qquad \varphi(0) = \varphi(\pi) = 0.$$

The value of λ_1 may be calculated explicitly; $\lambda_1 = 1$, which contradicts the condition $\lambda_1 > 2N = 4$.

2) Let $N = 3$. Now K is the hemisphere $r = 1$, $\varphi_1 \in (0, \pi)$, $\varphi_2 \in (0, 2\pi)$ in the spherical coordinate system. Equation (31.6) takes the form

$$\frac{1}{\sin\varphi_1}\frac{\partial}{\partial\varphi_1}\left(\sin\varphi_1\frac{\partial\psi}{\partial\varphi_1}\right) + \frac{1}{\sin^2\varphi_1}\frac{\partial^2\psi}{\partial\varphi_1^2} + \lambda\psi = 0.$$

It admits separation of variables and can be explicitly solved (see [118, Ch. 4]). The bounded solutions of this problem vanishing as $\varphi_1 \to \pm 1$ have a representation via the associated Legendre functions of the first kind $P_\nu^m(\cos\varphi_1)$

$$\psi(\varphi_1, \varphi_2) = P_\nu^m(\cos\varphi_1)\cos m\varphi_2, \tag{31.8}$$

$$\psi(\varphi_1, \varphi_2) = P_\nu^m(\cos\varphi_1)\sin m\varphi_2,$$

where $m = 0, 1, 2, \ldots$. The eigenvalues λ are represented as

$$\lambda_{mn} = \nu_{mn}(1+\nu_{mn}), \quad m \le \nu_{m1} \le \cdots \le \nu_{mn} \le \cdots.$$

Also, all ν_{mn} must be positive integers.

The associated Legendre functions $P_\nu^m(x)$ are obtained from the Legendre polynomials $P_\nu(x)$ by the formula

$$P_\nu^m(x) = (-1)^m(1-x^2)^{m/2}\frac{d^m P_\nu(x)}{dx^m},$$

where the Legendre polynomials themselves are obtainable from Rodrigues' formula

$$P_\nu(x) = \frac{1}{2^\nu \nu!}\frac{d^\nu}{dx^\nu}(x^2-1)^\nu.$$

Since $P_0(x) = 1$, $P_1(x) = x$, and we are interested in the solutions that satisfy the condition $\psi(\pm 1, \varphi_2) = 0$ for all admissible φ_2, we have to conclude that $\min \nu_{mn} = 1$ and, correspondingly, $\lambda_1 = 2 < 2N = 6$. □

Now that $\omega \subset S_{N-1}$ occupies a domain strictly smaller than the hemisphere, one can always choose a local coordinate system on S_{N-1} so that the coefficients of the differential operators ∇_S and Δ_S have no singularities in $\overline{\omega}$. Then the operators ∇ and Δ admit the estimates

$$|\nabla f| \le C_1(\omega)\left\{\sum_{\delta+|\kappa|=1} r^{\delta-1}\left|D_r^\delta D_\varphi^\kappa f\right|\right\},$$

$$|\Delta f| \le C_2(\omega)\left\{|\nabla f| + \sum_{\delta+|\kappa|=2} r^{\delta-2}\left|D_r^\delta D_\varphi^\kappa f\right|\right\}, \tag{31.9}$$

where $C_i(\omega)$ are bounded constants.

Lemma 31.2. *Let* $m \ge 3$.

1) *If* $f \in V_{m,\beta}^{(p,\tau)} \cap V_{m,\gamma}$, *then*

$$r D_r f \in V_{m-1,\beta-1}^{(p,\tau+1)} \cap V_{m-1,\gamma-1} \subset V_{m-2,\beta-2}^{(p,\tau+1)} \cap V_{m-2,\gamma-2}.$$

2) *If* $f \in V_{m,\beta}^{(p,\tau)} \cap V_{m,\gamma}$, *then* $\nabla f \in W_{m-1,\beta}^{(p-2,\tau-1)} \cap W_{m-1,\gamma}$.

3) *Let the following inequalities hold:*

$$\begin{cases} \beta - m + (N-1)/2 \ge 0, \\ \delta - m + (N-1)/2 \le 0, \\ \alpha - m + (N-1)/2 \ge 0, \\ \gamma - m + (N-1)/2 \le 0. \end{cases} \tag{31.10}$$

If $f_1 \in V_{m,\beta}^{(p,\tau)} \cap V_{m,\gamma}$, $f_2 \in V_{m,\alpha}^{(q,t)} \cap V_{m,\delta}$ *and the parameters* θ, σ, μ, ν *satisfy the relations*

$$\nu = \min\{\alpha;\ \beta\}, \quad \theta \le \min\{p;\ q\}, \quad \mu = \max\{\gamma;\ \delta\},$$

$$\theta \ge 2\sigma \ge 2\{t + \tau - \max\{p;\ q\}\},$$

then $f_1 f_2 \in V_{m,\nu}^{(\theta,\sigma)} \cap V_{m,\mu}$.

4) *If* $f_1, \dots, f_k \in V_{m,\beta}^{(p,\tau)} \cap V_{m,\gamma}$ *with* $p \ge 2\tau$ *and conditions* (31.10) *hold, then*

$$\left(\prod_{i=1}^{k} f_i\right) \in V_{m,\beta}^{(p,\tau)} \cap V_{m,\beta}, \qquad \left\|\prod_{i=1}^{k} f_i\right\| \le C k \prod_{i=1}^{k} \|f_i\|.$$

Proof. The truth of properties 1)–2) directly follows from the definition of the metric in the function spaces used. Let us check 3). Fix an arbitrary $r > 0$ and consider any of the terms of the expression

$$\sum_{|\sigma_1|+|\sigma_2|=\gamma} \left\{ \sum_{\delta_1+|\kappa_1|=\sigma_1} \sum_{\delta_2+|\kappa_2|=\sigma_2} r^{-|\kappa_1|-|\kappa_2|} \left(D_r^{\delta_1} D_\varphi^{\kappa_1} f_1\right) \left(D_r^{\delta_2} D_\varphi^{\kappa_2} f_2\right) \right\}.$$

For the sake of definiteness, let us assume $|\sigma_1| = \max\{|\sigma_1|;\ |\sigma_2|\}$. Then, for sure, $\sigma_2 \le [m/2]$. (Here $[a]$ denotes "the integer part of a".) Consider now $D_r^{\delta_i} f_i$, $i = 1, 2$, as functions of the variables φ depending on r as a parameter. There holds the inequality

$$\left\|\prod_{i=1}^{2} D_r^{\delta_i} D_\varphi^{\kappa_i} f_i\right\|_2 (r) \le C \left\|D_r^{\delta_1} D_\varphi^{\kappa_1} f_1\right\|_2 (r) \left\|D_r^{\delta_2} D_\varphi^{\kappa_2} f_2\right\|_{C(\omega)} (r). \tag{31.11}$$

In our assumption $|\sigma_2| \le m/2$, which allows one to apply the Sobolev embedding theorem to the second factor of the right-hand side of (31.11):

$$\left\|\prod_{i=1}^{2} D_r^{\delta_i} D_\varphi^{\kappa_i} f_i\right\|_2 (r) \le C \left\|D_r^{\delta_1} D_\varphi^{\kappa_1} f_1\right\|_2 (r) \left\|D_r^{\delta_2} D_\varphi^{\kappa_2} f_2\right\|_{H^{1+[(N-1)/2]}(\omega)} (r). \tag{31.12}$$

Let $r \in (0, 1)$. Multiply inequality (31.12) by $r^{\mu-m+\delta_1+\delta_2+(N-1)/2}$. Taking into account (31.10) and the fact that for each integer $m \ge 3$ and $N = 2, 3$ $|\sigma_2| + 1 + [(N-1)/2] \le m$, we get

$$\left\|\prod_{i=1}^{2} f_i\right|_{m,\mu} \le 2 C_1 \|f_1|_{m,\mu} \|f_2|_{m,\delta}. \tag{31.13}$$

If $r > 1$ and the parameters θ, σ ν are chosen according to (31.10), then for all δ_1, $\delta_2 \ge 0$, $\delta_1 + \delta_2 \le m$

$$\rho(r, \delta_1, p, \tau)\, \rho(r, \delta_2, q, t) \le \rho(r, \delta_1 + \delta_2, \theta, \sigma).$$

Analogously to (31.13), inequality (31.12) then implies

$$\| f_1\, f_2 |_{m,\nu}^{(\theta,\sigma)} \le C_2 \, \| f_1 |_{m,\beta}^{(p,\tau)} \, \| f_2 |_{m,\alpha}^{(q,t)} . \tag{31.14}$$

Inequalities (31.13), (31.14) provide property 3).

Property 4) is now easy to derive from 3) by induction on k. □

Remark 31.1. Although the derived embedding properties are not sharp, they suffice for our further purposes.

Let ε be a formal small parameter. Given a function ψ, let us linearize the expression $F(\psi + \varphi\, h)$ with respect to ε. Adopt the notation

$$(a, b) = \sum_{i=1}^{N} a_i\, b_i ,$$

$$L(\psi, h) = |I + \nabla^* \psi| \operatorname{div}\left[(I + \nabla^* \psi)^{-1} h \right] + \left(\nabla |I + \nabla^* \psi| ,\ (I + \nabla^* \psi)^{-1} h \right).$$

Calculating $\lim_{\varepsilon \to 0} \varepsilon^{-1} [F(\psi + \varepsilon\, h) - F(\psi)]$, we obtain the following operator

$$G(\psi) h \equiv \{ G_1(\psi),\ G_2(\psi) \}\, h,$$

$$G_1(\psi) h \equiv (I + \nabla \psi)(h - r\, h_r) + (\psi - r\, \psi_r)(I + \nabla^* h) - 2\, L(\psi,\, h), \quad (r, \varphi) \in K,$$

$$G_2(\psi) h \equiv L(\psi, h)|_{\partial K} .$$

Lemma 31.3. *There exists a $\delta_* > 0$ depending only on m, N, and the properties of $\partial\omega$ such that in the open ball $P \subset X$ of radius δ_* centered at the point $x = 0$ the operator $F : X \mapsto Z \times S_\partial$ has Fréchet derivative coinciding with $G(\psi)$, $\psi \in P$.*

Proof. First, let us verify that the operator $G(\psi)$ represents the Gateaux (or weak) derivative of F. By definition, $G(\psi)$ is the principal linear part of F at the element ψ. Hence, it suffices to find a $\delta > 0$ such that the operators F, $G(\psi)$ map the ball $P_\delta = \{ h \in X : \ \| h \|_X < \delta \}$ into $Z \times S_\partial$.

We begin by checking that $F_1 : P_\delta \mapsto Z$. Consider the function $|I + \nabla^* f|$. By [147, Th. 1.3.18], given a matrix A of dimension $N \times N$, there holds the representation (Newton's formulas)

$$\det [\lambda\, I - A] = \sum_{k=0}^{N} (-1)^k \alpha_k\, \lambda^{N-k},$$

where

$$\alpha_0 = 1, \quad k\, \alpha_k = \sum_{i=1}^{k} \alpha_{k-i} \operatorname{trace}\left(A^i \right) \quad \forall k,\ 1 \le k \le N.$$

Adapting these formulas to the present case, we get

$$|I+\nabla^* f| = 1 + \left\{\operatorname{trace}(\nabla^* f) + \cdots + |\nabla^* f|\right\} \equiv 1 + \Phi(\nabla^* f).$$

As follows from Lemma 31.2, 4), $\Phi(\nabla^* f) \in S$, that is, $\nabla|I+\nabla^* f| \in Z$. Let us now discover what are the assumptions on f guaranteeing the validity of the representation (Neumann's formula)

$$\frac{1}{|I+\nabla^* f|} = 1 + \sum_{k=1}^{\infty}(-1)^k \left(\Phi(\nabla^* f)\right)^k. \tag{31.15}$$

Equality (31.15) makes sense if the series on its right-hand side is convergent. To estimate the members of the series, we apply part 4) of Lemma 31.2:

$$\begin{aligned}\|(\Phi(\nabla^* f))^k\|_S &\le Ck\,\|\Phi(\nabla^* f)\|_S^k \\ &\le kC\left(2N(N-1)^2\right)^k \left\{\sum_{i,j=1}^{N} \|(\nabla^* f)_{i,j}\|_S\right\}^k.\end{aligned} \tag{31.16}$$

By virtue of Lemma 31.2, part 1) and inequality (31.9), there holds the inequality

$$\sum_{i,j}^{N} \|(\nabla f)_{i,j}\|_S \le C_2\,\|f\|_X.$$

Fix an arbirtary $\varepsilon > 0$ and claim

$$\|f\|_X \le \left\{(1+C_1+C_2)\,2(1+\varepsilon)N(N-1)^2\right\}^{-1}. \tag{31.17}$$

The right-hand side of (31.17) we denote by δ_0. (31.16) implies the inequality

$$\|\Phi(\nabla^* f)\|_S^k \le C\frac{k}{(1+\varepsilon)^k}, \quad k = 1, 2, \ldots,$$

whence the convergence of the series $\sum_{k=1}^{\infty} \|\Phi(\nabla^* f)\|_S^k$ and the inequality $\||I+\nabla^* f|^{-1}\| < \infty$. Thus, we have proved the inclusion $\nabla \ln|I+\nabla^* f| \in Z$.

The inclusion $(I+\nabla f)(f - r f_r) \in Z$ is a direct consequence of parts 2)–3) of Lemma 31.2. Hence, $F_1(f) \in Z$, provided (31.17) holds with some $\varepsilon > 0$.

The above-derived inclusion $(1 - |I+\nabla^* f|) \in S$ implies $F_2(f) \in S_\partial$.

The analogous procedure proves the existence of some $\delta_1 > 0$ such that for $\|h\|_X < \delta_1$, $\|\psi\|_X < \delta_1$ the inclusion $G_1(\psi)h \in Z$ holds. The only difference between the latter and former assertions consists in the necessity to estimate the S-norms of the elements of the matrix $(I+\nabla^*\psi)^{-1}$. For this we use the representation

$$(I+\nabla^*\psi)^{-1}_{ij} = \frac{A_{ij}}{|I+\nabla^*\psi|}$$

where A_{ij} are the algebraic completions of the elements $(I + \nabla^* \psi)_{ij}$. For δ_* one may now take $\min\{\delta_0; \delta_1\}$.

The above arguments prove the existence of the Gateaux derivative of F. Note that an analogous consideration gives the following estimate: given $\psi_{1,2} \in P_{\delta_*}$

$$\|G(\psi_1) - G(\psi_2)\|_{Z \times S_\partial} \leq C' \, \|\psi_1 - \psi_2\|_X$$

with a constant C' depending only on m, N, δ_*, and the differentiability properties of $\partial\omega$. Thus, $G(\cdot)$ is Lipschitz continuous in P_δ implying [88, Ch. 10] that the Fréchet and Gateaux derivatives of F coincide. □

Lemma 31.4. *For any $p \geq 2 + [(N-1)/2]$ and arbitrary q, τ, γ, β, the functions $f \in V_{p,\gamma}^{(q,\tau)} \cap V_{p,\beta}$ are continuous in K.*

Proof. For any points (r', φ'), $(r'', \varphi'') \in K$, there hold the inequalities

$$\begin{aligned}|f(r'', \varphi'') - f(r', \varphi')| &\leq \int_{r'}^{r''} |f_s(s, \varphi')| \, ds + \int_L \left| \frac{df}{dl}(r'', \varphi) \right| \, dl \\ &\leq (r'' - r') \sup_{s \in (r', r'')} |f_s| + d(L) \left| \frac{df}{dl} \right|,\end{aligned}$$

where L is an arbitrary smooth curve from ω connecting the points φ', φ'' on the sphere S_{N-1}, and $d(L)$ is the arc length of L. The functions on the right-hand side of this inequality can be viewed as functions of the local variables on S_{N-1}. Applying to them Sobolev's embedding theorem, we get

$$|f(r', \varphi') - f(r'', \varphi'')| \leq C(r', r'') \left[|r' - r''| + d(L) \right] \|f\|_{V_{p,\gamma}^{(q,\tau)} \cap V_{p,\beta}}.$$

The last inequality means the continuity of f. The constant $C(r', r'')$ may grow unboundedly as $r \to 0$, $r \to \infty$, but remains finite for each pair of r', r'' bounded away from zero and infinity. □

31.4 Poisson's equation in weighted classes. This subsection is devoted to the study of the classical Dirichlet problem for the Poisson equation

$$\Delta U = f \quad \text{in } K, \tag{31.18}$$

$$U|_{\partial K} = 0. \tag{31.19}$$

By now, this problem has already been given a detailed study, including the situation when the problem domain is a cone [89, 117]. Nonetheless, we have to undertake our own investigation because of the special form of the function classes the solutions of problem (31.18)–(31.19) are assumed to belong to.

The usual way to deal with problems posed in the cone-like domains relies on the separation of the radial and angular variables. Assume that $f(r, \varphi) \in L_2(\omega)$ for each

$r > 0$. We say that a function $U(r, \varphi)$ is a weak solution of equation (31.18) if U possesses in K weak derivatives up to the second order, for each $r > 0$ U, $\Delta_S U$, U_r, $U_{rr} \in L_2(\omega)$, and for each function $\Phi(\varphi) \in C_0^\infty(\overline{\omega})$ and for all $r > 0$ there holds the identity

$$\int_\omega \Phi \left\{ U_{rr} + \frac{N-1}{r} U_r + \frac{1}{r^2} \Delta_S U - f \right\} d\omega = 0.$$

Theorem 31.2. *Assume that* (31.7) *is true, and let* $\partial\omega \in C^\infty$. *For each* $f \in S$ *problem* (31.18)–(31.19) *has a unique classical solution* $U \in T^0$. *This solution satisfies the estimate* $\|U\|_T \le C \|f\|_S$ *with a constant* C *depending only on* m, N, *and* $\partial\omega$.

Proof. Let us seek a solution in the form

$$U(r, \varphi) = \sum_{i=1}^{\infty} U_i(r)\psi_i(\varphi), \quad f(r, \varphi) = \sum_{i=1}^{\infty} f_i(r)\psi_i(\varphi),$$

where $\{\psi_i\}$ is the basis of $L_2(\omega)$ constituted by the eigenfunctions of problem (31.6). Substituting these series into equation (31.18), we get, still formally, the equations for the coefficients $U_i(r)$:

$$U_i'' + \frac{N-1}{r} U_i' - \frac{\lambda_i}{r^2} U_i = f_i, \quad r > 0. \tag{31.20}$$

The solutions of equation (31.20) bounded as $r \to 0$ and $r \to \infty$ admit the representations

$$U_i(r) = r^{\nu_{1,i}} \int_r^\infty \frac{s^{\nu_{2,i}} f_i(s)}{W_i(s)} ds + r^{\nu_{2,i}} \int_0^r \frac{s^{\nu_{1,i}} f_i(s)}{W_i(s)} ds, \tag{31.21}$$

where $r^{\nu_{1,i}}$, $r^{\nu_{2,i}}$ are the partial solutions of the corresponding homogeneous equation with

$$\nu_{j,i} = -\frac{N-2}{2} + (-1)^j \left\{ \lambda_i + \frac{(N-2)^2}{4} \right\}^{1/2}, \quad i, j = 1, 2, \ldots,$$

and

$$W_i(r) = \left\{ 4\lambda_i + (N-2)^2 \right\}^{1/2} r^{\nu_{1,i}+\nu_{2,i}-1}$$

is the Wronskian of the functions $r^{\nu_{1,i}}$, $r^{\nu_{2,i}}$. Recall that condition (31.7) implies the inequalities

$$\nu_{1,i} \le \nu_{1,1} \le -N, \quad \nu_{2,1} \ge \nu_{2,i} \ge 2. \tag{31.22}$$

This separation of variables needs justification. Introduce the functions

$$U^{(n)}(r, \varphi) = \sum_{i=1}^{n} U_i \psi_i, \quad f^{(n)}(r, \varphi) = \sum_{i=1}^{n} f_i \psi_i.$$

Assume that the sequence $\{U^{(n)}\}$ has been proven to be fundamental, i.e.,

$$\begin{aligned}
&\sup_{r>0}\left\{\left(1+\frac{1}{r^2}\right)\left\|U^{(n)}-U^{(k)}\right\|_2(r)\right\} \\
&+\sup_{r>0}\left\{\left(1+\frac{1}{r^2}\right)\left\|\Delta_S U^{(n)}-\Delta_S U^{(k)}\right\|_2(r)\right\} \\
&+\sup_{r>0}\left\{\left(r+\frac{1}{r}\right)\left\|D_r U^{(n)}-D_r U^{(k)}\right\|_2(r)\right\} \\
&+\sup_{r>0}\left\{\left(1+r^2\right)\left\|D_r^2 U^{(n)}-D_r^2 U^{(k)}\right\|_2(r)\right\} \\
&\to 0 \quad \text{as } n,k\to\infty.
\end{aligned} \tag{31.23}$$

Then from $\{U^{(n)}\}$ one can extract a subsequence converging to a function U such that

$$\begin{aligned}
&\sup_{r>0}\left\{\left(1+\frac{1}{r^2}\right)\|U\|_2(r)\right\}+\sup_{r>0}\left\{\left(1+\frac{1}{r^2}\right)\|\Delta_S U\|_2(r)\right\} \\
&+\sup_{r>0}\left\{\left(r+\frac{1}{r}\right)\|D_r U\|_2(r)\right\}+\sup_{r>0}\left\{\left(1+r^2\right)\left\|D_r^2 U\right\|_2(r)\right\}<\infty.
\end{aligned} \tag{31.24}$$

Let us show that U is a weak solution of equation (31.18). By (31.20), for each function $\Phi(\varphi)\in C_0^\infty(\overline{\omega})$ and every integer n

$$\int_\omega \Phi\left\{U_{rr}^{(n)}+\frac{N-1}{r}U_r^{(n)}+\frac{1}{r^2}\Delta_S U^{(n)}-f^{(n)}\right\}d\omega=0. \tag{31.25}$$

By virtue of Parseval's inequality, for every $r>0$

$$(1+\sigma^2)^2\|f^{(n)}\|_2^2(\sigma)\le(1+\sigma^2)^2\sum_{i=1}^{n}f_i^2(\sigma)\le(1+\sigma^2)^2\|f\|_2^2(\sigma)\le\|f\|_S^2,$$

whence

$$\sup_{r>0}\left\{(1+r^2)\|f^{(n)}-f^{(k)}\|_2(r)\right\}\to 0 \quad \text{as } n,k\to\infty,$$

implying the fundamental nature of $\{f^{(n)}\}$. Letting $n\to\infty$ in (31.25), we get the integral identity satisfied by the function $U(r,\varphi)$.

Suppose that we have established the estimate

$$\|U\|_T\le C\,\|f\|_S,\quad C=\text{const}. \tag{31.26}$$

Applying Lemma 31.4 we then get that in K the function U is twice continuously differentiable and hence satisfies equation (31.18) in the classical sense.

Another implication of (31.26) is that for each $r>0$ the function U has a trace on $\partial\omega$. If the series

$$\sum_{i=1}^{\infty}\lambda_i U_i^2(r) \tag{31.27}$$

is convergent, then for each $r > 0$ the sequence $\{U^{(n)}\}$ converges to U in the norm of $H^1(\omega)$. This means that for each $r > 0$, $U(r, \varphi) \in H_0^1(\omega)$, [110], whence the fulfillment of the boundary condition (31.19).

1) *Fundamentality of the sequence* $\{U^{(n)}\}$. It follows from Parseval's inequality and representations (31.21) that

$$\|U^{(n)}\|_2^2(r) = C_0 \sum_{i=1}^{n} \frac{1}{4\lambda_i + (N-2)^2} \times \left\{ \left(r^{\nu_{1,i}} \int_0^r s^{1-\nu_{1,i}} f_i(s)\, ds \right)^2 + \left(r^{\nu_{2,i}} \int_r^\infty s^{1-\nu_{2,i}} f_i(s)\, ds \right)^2 \right\}. \tag{31.28}$$

Taking into account (31.22), we get from this: for every $r > 0$

$$\left(1 + \frac{1}{r^2}\right) \|U^{(n)}\|_2(r) \le C_1 \sup_{\sigma > 0} \left\{ (1 + \sigma^2) \|f^{(n)}\|_2(\sigma) \right\}, \tag{31.29}$$

$$\left(1 + \frac{1}{r^2}\right) \|U^{(n)} - U^{(k)}\| \le C_2 \sup_{(\sigma > 0)} \left\{ (1 + \sigma^2)^2 \sum_{i=k}^{n} f_i^2(\sigma) \right\} = C_2 \sup_{\sigma > 0} \left\{ (1 + \sigma^2)^2 \|f^{(n)} - f^{(k)}\|_2^2(\sigma) \right\} \to 0 \quad \text{as } n, k \to \infty. \tag{31.30}$$

Differentiation in r of the representations (31.21) implies the estimate: for all $r > 0$

$$\left(r + \frac{1}{r}\right) \|D_r U^{(n)}\|_2(r) \le C_3 \sup_{\sigma > 0} \left\{ (1 + \sigma^2) \|f\|_2(\sigma) \right\}. \tag{31.31}$$

A second differentiation of (31.21) in r gives

$$U_i''(r) = \frac{\nu_{1,i}(\nu_{1,i} - 1)}{\nu_{2,i} - \nu_{1,i}} r^{\nu_{1,i} - 2} \int_0^r s^{1-\nu_{1,i}} f_i(s)\, ds + \frac{\nu_{2,i}(\nu_{2,i} - 1)}{\nu_{2,i} - \nu_{1,i}} r^{\nu_{2,i} - 2} \int_r^\infty s^{1-\nu_{2,i}} f_i(s)\, ds - f_i(r). \tag{31.32}$$

Observing that for all integers i the numbers $\nu_{1,i}(\nu_{1,i} - 1)$ and $\nu_{2,i}(\nu_{2,i} - 1)$ have the same sign, from (31.32) we derive the relations

$$|U_i''(r)| \le \frac{C_4}{\sqrt{\lambda_i}} \left(\frac{\lambda_i}{r^2} |U_i(r)| \right) + C_5 |f_i(r)|. \tag{31.33}$$

Since $\lambda_i \to \infty$ as $i \to \infty$, for all i beginning with some i_0 the inequality $2C_4 < \sqrt{\lambda_i}$ holds. Expressing $\lambda_i r^{-2} U_i(r)$ from (31.20) and substituting them into (31.33), for $i > i_0$ we get the estimate

$$(1 + r^2) \|D_r^2 U^{(n)}\|_2(r) \le C_6 \sup_{\sigma > 0} \left\{ (1 + \sigma^2) \|f^{(n)}\|_2(\sigma) \right\}. \tag{31.34}$$

For $i \le i_0$, relations (31.32) imply an inequality analogous to (31.34). (The value of i_0 depends on λ_i, i.e., it is implicitly defined by the properties of the domain ω.) With the aid of (31.31) we get: for all $r > 0$

$$(1+r^2)\|D_r^2 U^{(n)}\|_2(r) \le C_6 \sup_{r>0} \left\{(1+r^2)\|f^{(n)}\|_2(r)\right\}.$$

The norm $\|\Delta_S U^{(n)}\|_2$ can be estimated with the help of equation (31.18). Finally, we get: for all $r > 0$

$$\left(1+\frac{1}{r^2}\right)\left\{\|U^{(n)}\|_2(r) + \|\Delta_S U^{(n)}\|_2(r)\right\} + (1+r^2)\|D_r^2 U^{(n)}\|_2(r)$$
$$+\left(r+\frac{1}{r}\right)\|D_r U^{(n)}\|_2(r) \le C_7 \sup_{\sigma}\left\{(1+\sigma^2)\|f^{(n)}\|_2(\sigma)\right\}. \qquad (31.35)$$

Similarly to (31.30), inequality (31.35) implies the fundamentality of the sequence $\{U^{(n)}\}$ and the estimate for the limit function U.

2) By (31.21), the terms of series (31.27) satisfy inequalities (31.29), whence the convergence of the series for all $r > 0$.

3) Let us prove (31.26), the basic estimate of the above considerations. Differentiate equation (31.18) k times in r and then multiply by the smooth strictly positive function $\theta(r)$ of the form

$$\theta(r) = \begin{cases} r^\gamma & \text{for } r \in (0, 1/2), \\ 1 & \text{for } r \in (3/4, 5/4), \\ r^\delta & \text{for } r \in (3/2, \infty), \end{cases}$$

where γ, δ are real constants. As a result we get the equation for the function $V \equiv \theta(r) D_r^k U(r,\varphi)$: $\Delta V = F$ in K. The right-hand side has the form

$$F(r,\varphi) = \theta\, D_r^k f + \theta'' D_r^k U + 2\theta' D_r^{k+1} U + \frac{N-1}{r}\left[\theta' D_r^k U + \theta\, D_r^{k+1} U\right]$$
$$-\sum_{\substack{\alpha+\beta=k \\ \alpha\ge 0,\, 0\le\beta\le k-1}} \left\{(N-1) D_r^\alpha\left(\frac{1}{\alpha}\right) D_r^\beta U + D_r^\alpha\left(\frac{1}{r^2}\right) D_r^\beta(\Delta_S U)\right\}\theta.$$

As before, let us consider the functions

$$V^{(n)}(r,\varphi) = \sum_{i=1}^{n} V_i(r)\psi_i(\varphi), \quad F^{(n)}(r,\varphi) = \sum_{i=1}^{n} F_i(r)\psi_i(\varphi),$$

where $F_i(r)$ are the Fourier coefficients of F in the basis $\{\psi_i\}$. The Fourier coefficients V_i are expressed through F_i by formulas (31.21). In order to make equalities (31.21) consistent, the parameters γ and δ should be chosen in accordance with the value of k and the properties of $f, U, \ldots, D_r^{k+1}U$. For each $r > 0$ the functions $V^{(n)}$, $F^{(n)}$, $D_r V^{(n)}$, $D_r^2 V^{(n)}$, $\Delta_S V^{(n)}$ converge strongly in $L_2(\omega)$ to the corresponding derivatives of the limit function V. Then V satisfies inequality (31.35). Arguing by induction, we get thereby the estimates for the norms $\|D_r^{k+2}U\|_2(r)$, $1 \le k \le m+1$.

The functions $u = D_\varphi^\kappa U$, $2 \le |\kappa| \le m+1$, satisfy equation (31.18) with the right-hand sides $F = D_\varphi^\kappa f$, which can be represented in the form

$$F(r,\varphi) = \sum_{i=1}^{\infty} F_i(r)\psi_i(\varphi), \qquad F \in L_2(\omega) \text{ for each } r > 0.$$

The function u along with its derivatives can be represented by a Fourier series in the basis $\{\psi_i\}$. Furthermore, for the derivatives of u up to second order the Fourier coefficients are expressed with the help of (31.21). This provides the estimates through $\|D_\varphi^\kappa f\|_2(r)$ for the norms

$$\|D_r^2 D_\varphi^\kappa U\|_2(r), \quad \|D_r D_\varphi^\kappa U\|_2(r), \quad \|D_\varphi^\gamma U\|_2(r), \quad |\gamma| = |\kappa| + 2.$$

The higher-order mixed derivatives are estimated similarly.

For example, let us consider the simplest case $k = 1$. Let $r \to \infty$. Set $\gamma = 1$ and $\delta = -1$. Then in equation (31.18) for the function V the right-hand side takes the form

$$F = \theta\, D_r f + \left(\frac{N-1}{r}\theta + \theta''\right) D_r U + \left(2\theta' + \theta\right) D_r^2 U + 2\theta\,\frac{\Delta_S U}{r^3}.$$

Thus, $F \sim O(1)$ as $r \to 0$, $F \sim O(r^{-2})$ as $r \to \infty$. Having applied (31.35) to the solution of (31.18) with the right-hand side $\Delta_S f$, we get the estimates

$$\|\Delta_S U\|_2(r) + \|\Delta_S^2 U\|_2(r) + r^2\,\|D_r^2(\Delta_S U)\|_2(r) + r\,\|D_r(\Delta_S U)\|_2(r)$$
$$\le C \sup_{\sigma>0}\left\{(1+\sigma^2)\|\Delta_S f\|_2(\sigma)\right\}. \qquad (31.36)$$

4) *Uniqueness.* Let there exist two different solutions of problem (31.18)–(31.19) U_1, $U_2 \in T$. The function $W = U_1 - U_2$ satisfies the homogeneous equation (31.18), and its Fourier coefficients in the expansion in the basis $\{\psi_i\}$ are defined as the solutions of the homogeneous equation (31.20). Under condition (31.7), the only solution bounded as $r \to 0$ and $r \to \infty$ admitted by this equation is the trivial one. Hence $W \equiv 0$, which contradicts the assumption. □

Remark 31.2. It follows from the proof of Theorem 31.2 that for each $f \in Z$ problem (31.18)–(31.19) has a unique solution $U \in X^0$. This solution is classical and satisfies the inequality $\|U\|_X \le C\,\|f\|_Z$.

Given an arbitrary function $f \in V_{m-1,\alpha+2}^{(-4,-3)} \cap V_{m-1,\alpha}$, there exists a unique weak solution of problem (31.18)–(31.19) $U \in S^0$ such that $\|U\|_S \le C\,\|f\|$. (By the weak solution of the Dirichlet problem we mean the weak solution of equation (31.18) satisfying the boundary condition (31.19) in the sense of traces on ∂K).

32 Solvability of the nonlinear problem

32.1 Newton's method. We shall obtain a solution of problem (31.4)–(31.5) as the limit of the sequence $\{h_n\}$ defined by the recurrence formulas

$$h_{n+1} = h_n - \Gamma(h_n), \quad \Gamma(s) \equiv [G(0)]^{-1} F(s), \quad h_1 = 0. \tag{32.1}$$

Thus, the main attention will be focused on the study of the linear model. Having applied the operator $G(0)$ to both sides of (32.1), we write the problem in the following form:

$$G(0)\left[h_{n+1} - h_n\right] = F(h_n).$$

Given a vector h_n, we thus have to find a solution $z = h_{n+1} - h_n$ of the problem

$$2\,\nabla\,\mathrm{div}\, z + r\, z_r - z = (I + \nabla h_n)\,(h - r\, h_r) - 2\,\nabla \ln\left\{\frac{|I + \nabla^* h_n|}{1 + \theta_0(\varphi)}\right\} \quad \text{in } K, \tag{32.2}$$

$$\mathrm{div}\, z|_{\partial K} = \left\{|I + \nabla^* h_n| - 1\right\}\Big|_{\partial K}, \quad z \to 0 \text{ as } r \to \infty.$$

To verify the convergence of the approximation scheme we will rely on the classical Newton method [88], but in a form adjusted to our purposes.

Lemma 32.1. *Let $F(x)$ be defined in a ball $P_\delta \subset X$ of radius $\delta > 0$ centered at the point $x = 0$. Assume that the operator*

$$\Gamma \equiv [G(0)]^{-1}\, F : X \mapsto X,$$

has in P_δ the Fréchet derivative

$$\Gamma'(\psi) \equiv [G(0)]^{-1}\, G(\psi) : P_\delta \mapsto X, \quad \psi \in P_\delta.$$

and Γ' is Lipschitz continuous in P_δ:

$$\|\Gamma'(\psi_1) - \Gamma'(\psi_2)\|_X \le C_1\, \|\psi_1 - \psi_2\|_X.$$

Denote $\gamma \equiv \|\Gamma(0)\|_X$.

If $\gamma \le \min\{\delta/4;\ 1/4C_1\}$, the equation $F(x) = 0$ has in P_δ a unique solution h_*, obtained as the limit of the sequence $\{h_n\}$ given by (32.1).

Proof. Consider the auxiliary operator $A \equiv I - \Gamma$ in the ball $S = \{h \in X : \ \|h\|_X \le 2\gamma\}$.

1) A maps S into itself:

$$\begin{aligned}
\|Ax\|_X &= \|\,[G(0)]^{-1}\,\{G(0)h - F(h)\}\,\|_X \\
&\le \|[G(0)]^{-1}\,\{G(0)h + F(0) - F(h)\}\,\|_X + \|[G(0)]^{-1}\,F(0)\|_X \\
&\le \gamma + \sup_{\psi \in S} \|[G(0)]^{-1} G(0) - [G(0)]^{-1} G(\psi)\|_X\, \|h\|_X \\
&\le \gamma\{1 + 2C_1 \|\psi\|_X\} \le 2\gamma.
\end{aligned}$$

2) A is a contraction mapping of S into itself:

$$\begin{aligned}\|Ah_1 - Ah_2\|_X &\le \sup_{\theta\in(0,1)} \|A'(\theta h_1 + (1-\theta h_2))(h_1-h_2)\|_X \\ &\le \sup_{\theta\in(0,1)} \|[G(0)]^{-1}\{G(0) - G(\theta h_1 + (1-\theta)h_2)\}\|_X \|h_1 - h_2\|_X \\ &\le 2\,(\operatorname{diam} P_\delta)\,\|\Gamma'(0) - \Gamma'(\psi)\|_X \,\|h_1 - h_2\|_X \le \frac{1}{2}\|h_1 - h_2\|_X,\end{aligned}$$

whence the fundamentality of $\{h_n\}$, and the desired assertion follows. □

32.2 The linear model: potential and solenoidal parts. Like the nonlinear problem (31.4)–(31.5), the linear system (32.1) is not elliptic, and this feature forbids us from referring directly to any of the results of the classical theory about the solvability, qualitative properties of solutions, etc. To overcome this difficulty we break up problem (32.1) into two independent linear problems for the potential and curl components of the sought vector h_{n+1}.

Let us search for the solution in the form $z = \nabla v + \operatorname{rot} V$ with $v \in T^0$. This provides the orthogonality (and, correspondingly, independence) of the potential and curl parts of z in the sense of scalar product in $L^2(K)$. Indeed, we will have:

$$\int_K \nabla v(\operatorname{rot} V)\,dx = \int_{\partial K} v(\operatorname{rot} V\cdot\vec{n})\,dx - \int_K v\operatorname{div}(\operatorname{rot} V)\,dx = 0,$$

since the second term is identically zero, and $v|_{\partial K} = 0$.

Let us separate the potential and curl parts of $F_1(h_n)$. Since one of the terms of $F_1(h_n)$ is already potential, the needed decomposition has the form

$$H\,h_n \equiv (I + \nabla h_n)(h_n - r\,h_{nr}) = \nabla\Pi + \operatorname{rot}\Phi. \tag{32.3}$$

Applying to (32.3) the operations div and rot, we get the following necessary conditions the function Π has to satisfy:

$$\Delta\Pi = \operatorname{div} Hh_n \in Z \quad \text{in } K. \tag{32.4}$$

For the benefit of further use, let us claim

$$\Pi|_{\partial K} = \left\{-2\ln|I + \nabla^* h_n| - 2|I + \nabla^* h_n| + 2\right\}\Big|_{\partial K} \in S_\partial \tag{32.5}$$

Let $\sigma \in S$ be an arbitrary function satisfying the boundary condition (32.5). Such a function can be constructed with the aid of the standard procedure (see, for example, [110, Ch. 3]). For each $r > 0$ fixed, the function Π, thought of as a function of the variables $\varphi \in \partial\omega$ and depending parametrically on $r > 0$, can be continued from $\partial\omega$ into the interior of ω. The smoothness of the original and the continued functions coincide by the method of construction.

Now consider the function $\Theta(r,\varphi)$ satisfying the conditions

$$\Delta\Theta = \{\operatorname{div} Hh_n - \Delta\sigma\} \in V^{(-4,3)}_{m-1,\alpha+2} \cap V_{m-1,\alpha} \quad \text{in } K, \qquad \Theta|_{\partial K} = 0. \tag{32.6}$$

This problem has a unique weak solution $\Theta \in S^0$, therefore a solution of problem (32.4)–(32.5) can be defined by the equality $\Pi = \Theta + \sigma$. We now obtain the function Φ by just letting $\Phi = F_1 h_n - \Pi$ in $\overline{K}$.

The uniqueness of this decomposition is easy to verify. Given two different pairs of functions $\Pi_i \in S$, $\Phi_i \in X^0$, $i = 1, 2$, we get for the differences

$$\nabla(\Pi_1 - \Pi_2) + \operatorname{rot}(\Phi_1 - \Phi_2) = 0 \quad \text{in } K,$$

$$(\Pi_1 - \Pi_2)|_{\partial K} = 0, \quad (\Phi_1 - \Phi_2)|_{\partial K} = 0.$$

Applying to this equation the operations div and rot, we conclude that the differences must satisfy Laplace's equation in K. Under condition (31.7), the unique solution from T^0 (or S^0) admitted by this equation is trivial, whence the contradiction.

Substituting now $z = \nabla v + \operatorname{rot} V$ into (32.2) and separating the potential and the curl parts, we get the following problem for v and V:

$$\Delta v + \frac{r}{2} v_r - v = f(r,\varphi) \quad \text{in } K, \tag{32.7}$$

$$\Delta v|_{\partial K} = |I + \nabla^* h_n||_{\partial K}; \quad |\nabla v| \to 0 \text{ as } r \to \infty, \tag{32.8}$$

where

$$f(r,\varphi) = \left[-\frac{\Pi}{2} + \ln\left(\frac{|I + \nabla^* h_n|}{1 + \theta_0(\varphi)}\right)\right] \in S;$$

$$\operatorname{rot} V - r D_r(\operatorname{rot} V) = -\operatorname{rot}\Phi(r,\varphi) \quad \text{in } K, \tag{32.9}$$

$$|\operatorname{rot} V| \to 0 \quad \text{as } r \to \infty. \tag{32.10}$$

Problem (32.7)–(32.8) can be simplified with the use of (32.5). Let $v \in T$ be a solution to this problem. All functions entering equation (32.7) have traces on ∂K, whence

$$\left(\frac{r}{2} D_r v - v\right)\Big|_{\partial K} = f|_{\partial K} - (\Delta v)|_{\partial K} = 0.$$

Hence, $D_r\left(r^2 v\right)|_{\partial K}$ does not depend on r. Since $v \in T$, then necessarily $v|_{\partial K} = 0$. So, by our choice of the function $f(r,\varphi)$, the solution $v \in T$ of problem (32.7)–(32.8) is a solution of the problem

$$\Delta v + \frac{r}{2} v_r - v = f(r,\varphi) \quad \text{in } K, \tag{32.11}$$

$$v|_{\partial K} = 0, \quad |\nabla v| \to 0 \text{ as } r \to \infty.$$

Conversely, given the solution $v \in T^0$ of problem (32.11), it satisfies the boundary condition (32.8), whence the equivalence of these problems. Moreover, this statement guarantees that the potential and solenoidal parts of z are orthogonal (in $L^2(K)$).

32.3 Solenoidal component. Let us make use of the identity

$$\operatorname{rot}(r\, V_r - V) \equiv r\, D_r(\operatorname{rot} V)$$

and define the solution of equation (32.9) by the equality

$$V(r, \varphi) = \int_r^\infty \frac{\Phi(s, \varphi)}{s^3}\, ds \quad \text{in } K. \tag{32.12}$$

Obviously, the structure of the right-hand side of this equation shows that the smoothness of V in the angular variables φ is one order lower than in r, generally speaking. Thus, the smoothness of h_{n+1} will be one order lower than of h_n and, as a result, the iteration procedure fails: whatever the smoothness of h_n, it will be exhausted within a finite number of iterations. So, let us pay special attention to the right-hand side of (32.12). It will be convenient to pass to the Cartesian coordinate system in $\mathbb{R}^N$. Denote by $y_1, \ldots, y_N$ the projections of the vector y on the axis of the Cartesian coordinate system and adopt the notation

$$\vec{e} = \frac{\zeta}{|\zeta|}, \qquad (\vec{e}, \nabla_\zeta\, y_k) \equiv D_r\, y_k.$$

Let $N = 3$ and (i, j, k) be an arbitrary even permutation. There holds the equality

$$\begin{aligned} \operatorname{rot}_k \{(\nabla y)\, r\, D_r y\} &= \sum_{m=1}^{3} \left\{ \frac{\partial y_m}{\partial \zeta_j}\, e_i\, (\vec{e}, \nabla_\zeta y_m) - \frac{\partial y_m}{\partial \zeta_i}\, e_j\, (\vec{e}, \nabla_\zeta y_m) \right\} \\ &+ |\zeta| \sum_{m=1}^{3} \left\{ \frac{\partial}{\partial \zeta_j}\, (\vec{e}, \nabla_\zeta y_m)\, \frac{\partial y_m}{\partial \zeta_i} - \frac{\partial}{\partial \zeta_i} (\vec{e}, \nabla_\zeta y_m)\, \frac{\partial y_m}{\partial \zeta_j} \right\}. \end{aligned} \tag{32.13}$$

Substituting $y = h_n$ into (32.13), we get at once that the components of the vector $\operatorname{rot} H h_n$ do not contain the second derivatives of h_n in the angular variables. This observation along with the inclusion $\Delta\Phi \in Z$ imply

$$r^2 \int_r^\infty \frac{\Phi(s, \varphi)}{s^3}\, ds \in U.$$

This is even easier to verify in the case $N = 2$, where $\operatorname{rot} V \equiv (V_{\zeta_1}, -V_{\zeta_2})$.

The same arguments prove the validity of the inclusion

$$r^2 \int_r^\infty s^{-3}\, \Delta^{-1} \left[\operatorname{rot}\{G_1(\psi)\, h_n\}\right] ds \in U,$$

where $\Delta^{-1} : Z \mapsto X$ is the inverse to the Laplace operator.

By Theorem 31.31.2 equation (32.4) has a unique solution $\Phi \in X$ generating the solution of problem (32.9)–(32.10) given by (32.12). Moreover,

$$\|\operatorname{rot} V\|_X \le C \|V\|_U \le C_1\, \|h_n\|_X$$

with bounded constants C, C_1 depending only on m, N, and properties of $\partial\omega$.

32.4 Potential component. Now we turn to the study of problem (32.11). This part of our procedure is of crucial importance since it is the solution of problem (32.11) that defines the sophisticated function spaces to which the solution of the complete nonlinear problem is supposed to belong. In this subsection, we prove the existence and uniqueness theorems for problem (32.11). Moreover, for special choices of the right-hand side (32.11) the solutions of (32.11) can be constructed explicitly, thus explaining the necessity in the performed choice of the weighted function classes.

Let us search for a solution of problem (32.7)–(32.8) in the form of a series

$$v(r,\varphi)=\sum_{i=1}^{\infty} v_i(r)\psi_i(\varphi),$$

where $\{\psi_i\}$ is the basis of $L_2(\omega)$ constituted by the eigenfunctions of problem (31.6). The functions v_i satisfy the equations

$$L_{\lambda_i}\, v_i \equiv v_i''+\left(\frac{r}{2}+\frac{N-1}{2}\right)v_i'-\left(1+\frac{\lambda_i}{r^2}\right)v_i=f_i(r), \tag{32.14}$$

where f_i are the Fourier coefficients in the expansion of $f(r,\varphi)$ in the basis $\{\psi_i\}$. The solutions of the homogeneous equation admit the representation

$$z(r)=x^{-N/4}\exp\left(-\frac{x}{2}\right)y(x),\quad x=\frac{r^2}{4}.$$

Here $y(x)$ is the solution of the Whittaker equation [175]

$$4x^2\,y''=(x^2-4kx+4m_i^2-1)\,y$$

with the parameters

$$k=-1-1\frac{N}{4},\quad m_i=\frac{1}{4}\left[4\lambda_i+(N-2)^2\right]^{1/2}.$$

Choosing for $y(x)$ the first or the second Whittaker function, we obtain the solutions $z_{1,i}$ and $z_{2,i}$ of the homogeneous equation (32.14). If $2m_i$ is not an integer, the functions $z_{j,i}$ admit the following asymptotic representations:

$$\begin{aligned} z_{1,i}(r)&=\left(\frac{r}{4}\right)^{m_i+1/2-N/4}\left(1+O(r^2)\right)\quad \text{as } r\to 0,\\ z_{1,i}(r)&=r^2\frac{\Gamma(2m_i+1)}{4\Gamma(m_i+1/2-k)}\left(1+O(r^{-2})\right)\quad \text{as } r\to\infty, \end{aligned} \tag{32.15}$$

$$\begin{aligned} z_{2,i}(r)=&\ \frac{\Gamma(2m_i)}{4\Gamma(m_i+1/2-k)}\left(\frac{r^2}{4}\right)^{-m_i+1/2-N/4}\left(1+O(r^2)\right)\\ &+\frac{\Gamma(-2m)}{4\Gamma(m_i+1/2-k)}\left(\frac{r^2}{4}\right)^{m_i+1/2-N/4}\left(1+O(r^2)\right)\quad \text{as } r\to 0, \end{aligned}$$

$$z_{2,i}(r) = \left(\frac{r^2}{4}\right)^{-N/2-1} \exp\left(-\frac{r^2}{4}\right)\left(1 + O(r^{-2})\right) \quad \text{as } r \to \infty.$$

If $2m_i$ is an integer, the asymptotics of the function $z_{2,i}$ as $r \to 0$ is complicated and contains a logarithmic term.

The solution of equation (32.14) can be represented in the form

$$v_i(r) = z_{2,i}(r) \int_0^r \frac{z_{1,i}(s) f_i(s)}{W_i(s)}\, ds + z_{1,i}(r) \int_r^\infty \frac{z_{2,i}(s) f_i(s)}{W_i(s)}\, ds, \tag{32.16}$$

where $z_{j,i}$ are the solutions of the homogeneous equation, and

$$W_i(r) = -\frac{\Gamma(2m_i + 1)}{\Gamma(m_i + 1/2 - k)} \left(\frac{r^2}{4}\right)^{-N/2+1/2} \exp\left(-\frac{r^2}{4}\right)$$

is the Wronskian of the functions $z_{j,i}$.

Lemma 32.2. *Let the functions $v(r)$, $u(r) \in C^2(\mathbb{R}^+)$, $v, u \to 0$ as $r \to 0$ and $r \to \infty$, and*

$$L_{\overline{\lambda}} u = L_\lambda v \quad \text{for all } r > 0 \tag{32.17}$$

with some $\lambda > \overline{\lambda} > 2N$. Then $|v(r)| \le |u(r)|$ for all $r \ge 0$.

Proof. Let there exist points where $v \ge 0$ and $u - v < 0$. Then at some point r' the difference $u - v$ takes its minimal negative value and $v(r')$ still remains nonnegative. By this assumption, and due to (32.17), at that point

$$L_{\overline{\lambda}}(u - v)\big|_{r=r'} = v\,\frac{\overline{\lambda} - \lambda}{r'^2} \le 0.$$

On the other hand, since r' is a point of local negative minimum of $u - v$,

$$L_{\overline{\lambda}}(u - v)\big|_{r=r'} \ge \left(1 + \frac{\overline{\lambda}}{r'^2}\right)(u(r') - v(r')) > 0,$$

which contradicts the former inequality. Thus, the inequality $u - v \ge 0$ holds at each point where $v \ge 0$. In the same way we establish that $u - v \le 0$ at every point where $v \le 0$. □

The monotone character of dependence of the solutions to equation (32.14) on the parameter λ_i allows one to assume that none of the $2m_i$ is an integer. Thus, when deriving estimates for the functions v_i or their derivatives one can use formulas (32.15)–(32.16) without any restriction.

Theorem 32.1. *Let* (31.7) *hold, and $\partial\omega \in C^\infty$. For each function $f \in S$ problem* (32.7)–(32.8) *has a unique solution $v \in T^0$; also $\|v\|_T \le C\,\|f\|_S$ with a constant $C \equiv C(m, N, \partial\omega)$ not depending on v and f.*

Proof. As we said before, it would be enough to prove the unique solvability of problem (32.11). The proof follows the proof of Theorem 31.2 up to minor changes: now to derive the estimates for the function $v^{(n)} = \sum_{i=1}^{n} v_i \psi_i$ and its derivatives one has to employ the asymptotic representations (32.15)–(32.16) in place of the explicit formulas providing the solutions to the homogeneous equation. In particular, this implies the estimate

$$\left(\frac{1}{r} + r^3\right) \|D_r v^{(n)}\|_2(r) + \left(\frac{1}{r^2} + r^2\right) \|v^{(n)}\|_2(r) + \left(1 + \frac{1}{r^2}\right) \|\Delta_S v^{(n)}\|_2(r)$$
$$\leq C \sup_{\sigma > 0} \left\{(1 + \sigma^2) \|f^{(n)}\|_2(\sigma)\right\}.$$

□

Remark 32.1. Given a vector $h_n \in X$, in equation (32.7) $f \in S$ and, moreover, $\|f\|_S \leq C \|h_n\|_X$. That is, for each $h_n \in X$ problem (32.7)–(32.8) has a unique solution $v \in T^0$.

Remark 32.2. If in equation (32.7) the right-hand side does not depend on r, in the Fourier expansion of f the coefficients are constant. This simplifies the study of convergence of the series representing the solution of problem (32.11) and makes it possible to prove the following assertion. Given a function $F(\varphi) \in H_0^{m+1}(\omega)$, the problem

$$\Delta v + \frac{r}{2} v_r - v = F(\varphi) \quad \text{in } K, \quad (\Delta v)|_{\partial K} = 0; \quad |\nabla v| \to 0 \quad \text{as } r \to \infty, \tag{32.18}$$

has a unique solution $v \in T^0$. Also, $\|v\|_T \leq C \|F\|_{H_0^{m+1}(\omega)}$ with the constant C from the conditions of Theorem 32.1.

Remark 32.3. The above considerations prove the solvability of problem (32.2). The uniqueness of the solution $z = \nabla v + \operatorname{rot} V$ follows from standard considerations of the corresponding homogeneous problem.

32.5 The nonlinear problem. Now we have at our disposal everything needed for the proof of solvability of the nonlinear problem (31.4)–(31.5). We shall apply Lemma 32.1; therefore the radius δ of the ball P_δ, where all considerations are to be carried out, is immediately assumed as small as we please according to the conditions of the auxiliary propositions.

Consider the operator $\Gamma(h) \equiv [G(0)]^{-1} F(h)$. The properties of this operator are already known. The above-established unique solvability of problem (32.2) means that the operators

$$\Gamma : X \mapsto X, \quad \Gamma'(\psi) \equiv [G(0)]^{-1} G(\psi) : X \mapsto X, \quad \|\psi\|_X < \delta$$

are correctly defined.

The Lipschitz continuity of $G(\psi)$ in P_δ is established in Lemma 31.3. Due to the linearity, the Lipschitz continuity of $\Gamma'(\psi)$ in P_δ is then easy to derive.

The last thing needed to make use of Lemma 32.1 is the estimate on $\|\Gamma(0)\|_X$. It follows from Remark 32.2 to Theorem 32.1:

$$\|\Gamma(0)\|_X \equiv \gamma \le C_2 \,\| \ln(1+\theta_0(\varphi))\|_{H_0^{m+1}(\omega)} \to 0 \quad \text{as } \|\theta_0\|_{H_0^{m+1}(\omega)} \to 0.$$

Hence, under the conditions of Lemma 32.1 the parameter γ can be assumed as small as we need.

Referring to Lemma 32.1 we conclude that under the conditions of Theorem 31.1 the nonlinear problem (31.4)–(31.5) admits the unique solution h_* obtained via the approximation procedure of the modified Newton method.

33 Canonical Lagrangian coordinates

In this section we perform the passage to Lagrangian coordinates in the Cauchy problem (28.4)–(28.5). Equation (28.4) can be represented in the form

$$u_t + \operatorname{div}_x(u\, v) = 0,$$

where the vector v is defined by the equality

$$v = -m\, u^{m-2}\, \nabla_x u.$$

Viewing the last equality as the trajectory equation of some continuum with density u and velocity v, let us rewrite it in the form

$$X_t(t,p) = v(t, X(t,p)), \tag{33.1}$$

where $p = (p_1, \dots, p_n)$ is the set of geometrical Lagrangian coordinates

$$X(0,p) = p. \tag{33.2}$$

Introduce the notation

$$U(t,p) = u[t, X(t,p)], \quad M = (\nabla_p X)^*, \quad |M| = \det M.$$

Proceeding analogously to Section 29, we arrive at the Cauchy identity

$$(U\,|M|)_t = 0,$$

implying, by virtue of (28.5), (33.2),

$$U = u_0(p)\,|M|^{-1}.$$

This allows one to transform equation (33.1) into

$$M^* X_t = -\frac{m}{m-1} \nabla_p \left[u_0^{-1} |M|\right]^{1-m}. \tag{33.3}$$

Here we made use of the identity

$$\nabla_p U = M^* \nabla_x u.$$

Adjoining to (33.3) the initial conditions (33.2), we get the Cauchy problem for the system of n scalar quasilinear equations for the sought functions $X_i(t, p), i = 1, \ldots, n$. The transformation of problem (28.4)–(28.5) to the form (33.3), (33.2) was performed in paper [109].

As in the preceding consideration of the PME, it is assumed here

$$u_0(x) \geq 0, \quad x \in \mathbb{R}^n. \tag{33.4}$$

One should distinguish two cases. The first one corresponds to the situation when the inequality in (33.4) is strict. Then the transformation of equation (28.4) to the form (33.3) can be performed on each domain of the space $\mathbb{R}^n$. Moreover, the smoothness of the vector function X and condition (33.2) guarantee the bijectivity of the mapping $p \mapsto X(t, p)$, depending on the parameter t, on each compact set (for t small enough, at least).

In the second case, at some points of $\mathbb{R}^n$ in (33.4) the sign of equality is admitted. Let us denote by D the completion of this set in $\mathbb{R}^n$ so that $\overline{D} = \operatorname{supp} u_0(x)$, and assume that D is connected (the boundedness of D is not needed, generally speaking). Denote by $\Omega(t)$ and $\Gamma(t)$ the images of D and its boundary $B = \partial D$ under the mapping $p \mapsto X(t, p)$. Let $X(t, p)$ be a classical solution of problem (33.2), (33.3) such that $|M| > 0$ in the closure of the cylinder $Q_T = D \times (0, T)$. Assume that $u_0^{m-1} \in C^1\left(\overline{D}\right)$ and

$$\nabla\left(u_0^{m-1}\right) = 0 \quad \text{as } p \in B. \tag{33.5}$$

Then (33.3) implies the equality $X_t = 0$ for $p \in B, t \in [0, T]$. It then follows that for every t, $0 \leq t \leq T$, the origin of the free boundary (i.e. the null-level surface of the solution $u(x, t)$ of problem (28.4)–(28.5)), in the space of Lagrangian coordinates is motionless in the space of Lagrangian coordinates and in fact coincides with the surface B.

Condition (33.5) on the initial function $u_0(x)$ is quite restrictive. Besides, there is no reason to expect that in the general case solutions of system (33.3) would be regular up to the lateral boundary of the cylinder Q_T where this system becomes degenerate. Nonetheless, the assertion about the motionlessness of the origin $\Gamma(t)$ in the space (t, p) holds also for a wider class of functions $u_0(x)$, which need not satisfy (33.5). In particular, it is sufficient to require that $u_0 \in C\left(\overline{D}\right)$, and that u_0^{m-1} has in D the first-order derivatives in the variables $x_1, \ldots, x_n$. Then problem (28.4), (28.5) has a weak solution in the sense of Definition 17.3 [21]. In this solution "the mass flux"

$-m\,u^{m-1}\,\nabla_x u\,\cdot\nu$ across the boundary $\Gamma(t)$ of $\Omega(t)$ equals zero for all $t\in[0,T]$. At the boundary itself the kinematic condition holds:

$$V_\nu = -m\,u^{m-2}\,\nabla_x u\,\cdot\nu, \tag{33.6}$$

where ν is the unit outward normal vector to $\Gamma(t)$, and V_ν is the velocity of displacement of this surface in the direction ν. Equality (33.6) means that the surface $\Gamma(t)$ is transformed into the time-independent surface B under the mapping $x\mapsto X^{-1}(t,x)=p$.

Thus, we have passed from the Cauchy problem (28.4)–(28.5) for the PME to problem (33.2) for system (33.3). After this reformulation the initial condition takes the standard form while the system itself carries some information about the initial data of the original Cauchy problem. To attain a canonical form of this system, let us pass to the new independent space variables $\xi=(\xi_1,\dots,\xi_n)$ instead of $p=(p_1,\dots,p_n)$, although keeping the same notation for the sought function X.

Let us assume that $p_i(\xi)$ are smooth functions and denote by J the Jacobian matrix with the elements $\partial p_i/\partial\xi_j$, $i,j=1,\dots,n$. By N we denote the Jacobian matrix $(\nabla_\xi X)^*$. Having subjected the vector function p to the relation

$$|J|\,u_0(p)=1, \tag{33.7}$$

and taking into account that $N=M\,J$, $J^*\,\nabla_p=\nabla_\xi$, we get, instead of (33.3), the equivalent system

$$N^*\,X_t=-\frac{m}{m-1}\,\nabla_\xi\left(|N|^{1-m}\right) \tag{33.8}$$

which already does not contain "an arbitrary" element u_0. The initial condition for (33.8) is condition (33.2) rewritten in the new variables:

$$X(0,\xi)=p(\xi). \tag{33.9}$$

Thus, the Cauchy problem (28.4)–(28.5) is reduced to problem (33.8)–(33.9), where the vector function $p(\xi)$ satisfies relation (33.7) but is arbitrary in all other aspects.

Let us call the variables $\xi_1,\dots,\xi_n$ the *canonical variables* in the Cauchy problem (28.4)–(28.5). These canonical variables of Lagrange are a natural multi-dimensional generalization of the mass Lagrangian coordinate often used in the preceding chapters. Although, while in the one-dimensional problems the connection between the geometrical and the mass Lagrangian coordinates is usually rigid, the canonical Lagrangian coordinates are not uniquely defined in dimensions $n\ge 2$. Indeed, equality (33.7) can be viewed as a single differential equation connecting n unknown functions $p_i(\xi)$. We shall show later on how one can manage the appearing arbitrariness to gain a certain simplification of problem (33.8)–(33.9). Now, let us discuss some features of equation (33.7).

First of all, it is worth noting that the functions $p_i(\xi)$ connected by this equation have domain of dependence a priori unknown. Its boundary $S=\partial R$ is defined in the process of solving equation (33.7) as the surface where the equality $u_0\,[p(\xi)]=0$ is attained, while $u_0\,[p(\xi)]>0$ if $\xi\in R$. Moreover, at least one of the derivatives $\partial p_i/\partial\xi_k$, $i,k=1,\dots,n$, grows unboundedly as $\xi\to S$. The last peculiarity makes it difficult to solve equation (33.7) under the assumption that the function u_0 and the topological structure of its support $\overline{D}$ are arbitrary. Nonetheless, it is easy to present sufficient conditions which would guarantee the possibility of defining effectively the domain R and to solve equation (33.7) in this domain. Let us consider an example of such a situation.

Let $n = 2$. Let D be a bounded simply-connected domain star-like with respect to the origin on the plane p_1, p_2 so that its boundary B is given by the equation $\rho = f(\varphi)$, where $\rho = (p_1^2+p_2^2)^{1/2}$, $\varphi = \arctan(p_2/p_1)$ are the polar coordinates, $f(\varphi)$ is a smooth function. Let us denote

$$u_0(p) = q(\rho, \varphi), \quad r = \left(\xi_1^2 + \xi_2^2\right)^{1/2}, \quad \theta = \arctan\left(\xi_2/\xi_1\right).$$

With this notation, equation (33.7) takes the form

$$\rho\,(\rho_r\,\varphi_\theta - \rho_\theta\,\varphi_r)\,q(\rho, \varphi) = r.$$

It can be satisfied by letting $\varphi = \theta$ and defining ρ from the relation

$$2\int_0^\rho s\,q(s, \theta)\,ds = r. \tag{33.10}$$

Due to the condition $q > 0$ in D, the function $\rho = \rho(r, \theta)$ is uniquely defined by equality (33.10) in the domain

$$R = \left\{(r, \theta) : \ 0 \le r < 2\int_0^{\rho(\theta)} s\,q(s, \theta)\,ds, \ 0 \le \theta < 2\pi\right\}.$$

If $u_0(p)$ is a smooth function of the variables p_1, p_2 in D, so are the functions

$$p_1(\xi_1, \xi_2) = \rho(r, \theta)\cos\theta, \qquad p_2(\xi_1, \xi_2) = \rho(r, \theta)\sin\theta$$

in the domain R. Near the curve $S = \partial R$ the character of singularity of the derivatives $\partial p_i/\partial \xi_k$, $i, k = 1, \ldots, n$, is defined by the rate of vanishing of u_0 near the boundary B of the domain D where u_0 vanishes.

Just as simply, the discussed problem can be solved in the case when D is a planar sector (this situation is treated in Section 34). Let us point out a few conditions sufficient for the solvability of equation (33.7) for arbitrary $n \ge 2$: D is diffeomorphic to a planar or spherical layer; D is bounded and star-like with respect to one of its interior points; D is bounded, the function u_0 has in D a unique point of nondegenerate maximum and does not have other stationary points.

Let us now pay attention to the fact that equation (33.7) and system (33.8) are invariant with respect to every smooth invertible transformation in the space ξ with the Jacobian equal to unity. Sometimes this may be of help in finding the solution of equation (33.7) that possesses the simplest domain of definition R of the vector-valued function $p(\xi)$. An example of such a situation is presented below.

Let D be given by the inequality $p_n > h(p')$, where the smooth function h is defined for all $p' = (p_1, \dots, p_n) \in \mathbb{R}^{n-1}$. Set $p_k = \xi_k$ for $k = 1, \dots, n-1$ and define p_n from equation (33.7) and the condition $p_n = h(\xi')$ as $\xi_n = 0$; here $\xi' = (\xi_1, \dots, \xi_{n-1})$. Obviously,

$$\int_{h(\xi')}^{p_n} u_0(\xi', \eta)\, d\eta = \xi_n. \tag{33.11}$$

In this case R becomes the half-space $\xi_n > 0$. Problem (33.2), (33.3) is reduced to the following one: it is required to find in the domain $R \times (0, T)$ a solution of system (33.8) which satisfies (33.9). In the last condition $p' = \xi' \in \mathbb{R}^{n-1}$ and the function $p_n(\xi', \xi_n)$ is uniquely defined for $\xi_n > 0$ by relation (33.11).

In conclusion of this section, let us present an example of exact solution to problem (33.8). Let us seek for the components X_i of the vector X in the form

$$X_i = \alpha_i(t)\, Y_i(\xi), \quad i = 1, \dots, n. \tag{33.12}$$

The substitution of (33.12) into (33.8) leads to the following relations:

$$\sum_{i=1}^{n} \alpha_i \frac{d\alpha_i}{dt} Y_i \frac{\partial Y_i}{\partial \xi_k} = -\frac{m}{m-1} \left(\prod_{i=1}^{n} \alpha_i \right)^{1-m} \frac{\partial}{\partial \xi_k} \left(A^{1-m} \right),$$

where $k = 1, \dots, n$ and $A = \det(\partial Y_i / \partial \xi_k)$. Let us subject the unknown functions to the system of equations

$$\alpha_1 \frac{d\alpha_1}{dt} = \alpha_2 \frac{d\alpha_2}{dt} = \cdots = \alpha_n \frac{d\alpha_n}{dt} = \frac{2m}{m-1} \left(\prod_{i=1}^{n} \alpha_i \right)^{1-m}. \tag{33.13}$$

The relations connecting the functions $\alpha_i(t)$, $Y_i(\xi)$ will be identically satisfied if we set

$$A = \left(C - \sum_{i=1}^{n} Y_i^2 \right)^{-1/(m-1)}$$

with $C = \text{const.} > 0$. In this construction, the functions Y_i can be chosen with a large arbitrariness but their proper dependence on ξ is of little interest. This is because to find a solution $u(x, t)$ of equation (28.4) it suffices to know $|N|$ as a function of $x = X(\xi, t)$ and t. Since $\xi_1, \dots, \xi_n$ are the canonical Lagrangian coordinates, we have $u = |N(x, t)|^{-1}$. Using formulas (33.12) and the expression for A in terms of Y_i, we find

$$u = \left(\prod_{i=1}^{n} \alpha_i \right)^{-1} \left(C - \sum_{i=1}^{n} \frac{x_i^2}{\alpha_i^2} \right)^{1/(m-1)}, \tag{33.14}$$

where $\alpha_i(t)$ are defined from system (33.13). In what follows, we restrict ourselves to nonnegative solutions of this system, thus providing the nonnegativity of $u(x, t)$.

Equality (33.14) defines a solution of equation (28.4) inside the ellipsoid $E_n(t)$, centered at the origin with the semi-axis $C^{1/2}\alpha_i(t)$ $(i = 1, \dots, n)$. Continuing $u(x, t)$

by zero outside $E_n(t)$, we get a generalized solution of equation (28.4) defined in the whole space.

System (33.13) possesses the $n-1$ first integrals

$$\alpha_j^2 = \alpha_n^2 + \gamma_j, \quad j = 1, \dots, n-1, \tag{33.15}$$

where γ_j are constants. We shall assume all γ_j nonnegative, thereby providing the possibility to construct for our system a solution defined for all $t > 0$ and such that $\alpha_n(0) = 0$. Moreover, without loss of generality one may assume that

$$\gamma_{n-1} \le \gamma_{n-2} \le \dots \le \gamma_1. \tag{33.16}$$

Let us define the function $\alpha_n(t)$ for $t > 0$ as the inverse to the formula

$$\int_0^{\alpha_n} \beta^m \left[\prod_{j=1}^{n-1}(\beta^2 + \gamma_j)\right]^{(m-1)/2} d\beta = \frac{2mt}{m-1}, \tag{33.17}$$

and the values of α_j ($j = 1, \dots, n-1$) as the positive solutions of equation (33.15). Then the set of functions $\alpha_1(t), \dots, \alpha_n(t)$ forms the solution of the Cauchy problem for system (33.13):

$$\alpha_j(0) = \gamma_j^{1/2} \quad \text{as } j = 1, \dots, n-1, \quad \alpha_n(0) = 0.$$

Assume that all inequalities (33.16) are strict and that $\gamma_{n-1} > 0$. In this case, equality (33.14) defines some n-parameter family of solutions of equation (28.4). According to (33.15), here all semi-axes of the ellipsoid $E_n(t)$ are different and $C^{1/2}\alpha_n(t)$ is the minimal one. By (33.17), the asymptotics of α_n as $t \to 0$ has the form

$$\alpha_n = \left[\frac{2m(m+1)}{m-1}\left(\prod_{j=1}^{n-1}\gamma_j\right)^{(m-1)/2} t\right]^{1/(m+1)} + o\left(t^{2/(m+1)}\right).$$

Let us now pass in solution (33.14) to the limit as $t \to +0$. We get

$$u(x,t) \to K_{m,0}\left(\prod_{j=1}^{n-1}\gamma_j\right)^{-1/2}\left[\left(C - \sum_{j=1}^{n-1}\frac{x_j^2}{\gamma_j}\right)_+\right]^{(m+1)/2(m-1)} \delta(x_n), \tag{33.18}$$

as $t \to 0$. Here $\delta(x_n)$ is the Dirac measure, the symbol r_+ denotes $\max(r, 0)$, and

$$K_{m,0} = 2\int_0^1 (1-\eta^2)^{1/(m-1)}\, d\eta.$$

Thus, we have constructed for the Cauchy problem for the PME a solution having as initial data a measure concentrated inside the ellipsoid $E_n(0)$ of dimension $n-1$. The initial condition (33.18) is fulfilled in the sense of distributions. This solution of the Cauchy problem (28.4), (33.18) is unique in a properly defined class of generalized solutions. This follows from the general uniqueness theorem of M. Pierre [130].

Let now $\gamma_{n-1} = \cdots = \gamma_{n-l} = 0$ but $\gamma_{n-l-1} > 0$, where $1 \le l \le n-2$. Then the initial distribution (28.5) corresponding to solution (33.14) is the measure of the form

$$u_0 = K_{m,l}\left(\prod_{j=1}^{n-l-1}\gamma_j\right)^{-1/2}\left[\left(C-\sum_{j=1}^{n-l-1}\frac{x_j^2}{\gamma_j}\right)_+\right]^{(2+(l+1)(m-1))/2(m-1)} \times \delta(x_{n-l})\ldots\delta(x_n),$$

where

$$K_{m,l} = \Omega_{l+1}\int_0^1 (1-\eta^2)^{1/(m-1)}\eta^l\, d\eta$$

(Ω_n is the surface area of the unit sphere in $\mathbb{R}^n$). The singular support of this measure is of dimension $n-l-1$ (it is assumed that $n \ge 2$). Being invariant with respect to the group O_{l+1}, the solution (33.14) itself contains $n-l$ free parameters.

Finally, if all $\gamma_1, \ldots, \gamma_{n-1}$ are zero, solution (33.14) becomes invariant with respect to the complete group of rotations O_n and yields the well-known Barenblatt solution [19] which corresponds to the initial data

$$u_0 = K_{m,n-1}\, C^{(2+n(m-1))/2(m-1)}\delta(x),$$

with $\delta(x)$ the Dirac function in $\mathbb{R}^n$. In the latter case, this solution, as we point out in passing, acquires the property of self-similarity which the solutions (33.14) lack.

The solutions to equation (33.1) with ellipsoidal support corresponding to the integer values of m were constructed in [170]. In the general case, these solutions are independent and were found practically simultaneously by many authors [58, 84, 91, 138, 149].

34 Boussinesq's equation in filtration theory

34.1 Group analysis of Boussinesq's equation in Lagrangian coordinates. The object of our considerations will be the equation

$$u_t = \Delta\left(u^2\right), \tag{34.1}$$

where $u = u(x, y, t)$, $\Delta = \partial^2/\partial x^2 + \partial^2/\partial y^2$, and the Cauchy problem for this equation

$$u(x, y, 0) = u_0(x, y) \quad (x, y) \in \mathbb{R}^2, \tag{34.2}$$

where u_0 is a nonnegative function. The case when the support of u_0 does not coincide with $\mathbb{R}^2$ is the most interesting. Then the solutions of (33.1) display free boundaries, whose dynamics is the most natural to describe in terms of Lagrangian coordinates (see Section 33).

In application to equation (34.1), Lagrangian coordinates are also useful from another point of view. Since the passage from the original Euler coordinates to Lagrangian variables is a nonlocal transformation, the group properties of the system of equations equivalent to (33.1) may be different and, particularly, richer than those displayed by the original equation. (The same situation for one-dimensional evolution equations was already discussed in Section 10.)

For equation (28.4), whose special case is (34.1), the transformation to Lagrangian coordinates was undertaken in the previous section. Here we assume at once that the Lagrangian coordinates ξ, η are canonical in the sense of the definition of Section 33. This constraint does not diminish the generality of the study of the group properties of equation (34.1) in variables of Lagrange. Indeed, the property of a system of differential equations to possess a Lie group is local, while a local reduction of system (33.3) to the form (33.8) is always possible – in a neighborhood of every point $(x, y) \in \mathbb{R}^2$ where $u_0 \neq 0$, at least.

Letting in (33.8) $m = 2$, $n = 2$, we get

$$|N|^2 N^* z_t = 2\,\nabla |N|. \tag{34.3}$$

Here $z = (x(\xi, \eta, t), y(\xi, \eta, t))$ is the sought function, $N = \partial z/\partial\xi$ is the Jacobian matrix, ∇ is the gradient in the variables $\zeta = (\xi, \eta)$, and $|N| = x_\xi\, y_\eta - x_\eta\, y_\xi$. The calculations of the full Lie group admitted by equation (34.1) are performed by known methods [121] and are so cumbersome that we omit them here. Let us only formulate a few results. The maximal group of transformations admitted by equation (34.1) proves to be 6-parameter. The basis of its Lie algebra is formed by the following infinitesimal operators [43]:

$$X_0 = \frac{\partial}{\partial t}, \quad X_1 = \frac{\partial}{\partial x}, \quad X_2 = \frac{\partial}{\partial y}, \quad X_3 = y\,\frac{\partial}{\partial x} - x\,\frac{\partial}{\partial y},$$

$$X_4 = 2t\,\frac{\partial}{\partial t} + x\,\frac{\partial}{\partial x} + y\,\frac{\partial}{\partial y}, \quad X_5 = t\,\frac{\partial}{\partial t} - u\,\frac{\partial}{\partial u}. \tag{34.4}$$

As regards system (34.3), it admits the infinite-dimensional group G generated by the operators

$$\Sigma_0 = \frac{\partial}{\partial t}, \quad \Sigma_1 = \frac{\partial}{\partial x}, \quad \Sigma_2 = \frac{\partial}{\partial y}, \quad \Sigma_3 = y\,\frac{\partial}{\partial x} - x\,\frac{\partial}{\partial y},$$

$$\Sigma_4 = 4t\,\frac{\partial}{\partial t} + x\,\frac{\partial}{\partial x} + y,\frac{\partial}{\partial y}, \quad \Sigma_5 = 2t\,\frac{\partial}{\partial t} - \xi\,\frac{\partial}{\partial \xi} - \eta\,\frac{\partial}{\partial \eta},$$

$$\Sigma_\psi = \psi_\eta\,\frac{\partial}{\partial \xi} - \psi_\xi\,\frac{\partial}{\partial \eta}. \tag{34.5}$$

(Although the operators X_k in (34.4) and Σ_k in (34.5) with $k = 0, 1, 2, 3$ are of the same form, we prefer to denote them by different symbols since the groups they generate act in different spaces.) The "infinite" part of the group G arises due to the presence in

(34.5) of the operator Σ_ψ in which $\psi(\xi, \eta)$ is an arbitrary C^∞ function. Each of these operators corresponds to some transformation of the plane ξ, η, or its portion which preserves the area element. One might predict beforehand the fact itself that the Lie algebra would contain the group G of operators Σ_i, $i = 0, \ldots, 5$, and Σ_ψ, but it is far from trivial to assert that the group they form turns out to be the maximal among all those admitted by system (34.3).

Having calculated the group admitted by system (34.3), we get the possibility to construct its exact solutions which possess the property of full or partial invariance. We do not aim to describe here the entire class of such solutions and confine ourselves to a detailed consideration of the self-similar solutions. The three following subsections are devoted to their study. Now, let us pay attention to the feature that makes all the more interesting the study of the group properties of Boussinesq's equation in Lagrangian coordinates. Had we performed with the help of the group transformation generated by the operator Σ_ψ the passage from the variables ξ, η to the new variables, we would preserve system (34.3) but change, generally speaking, the domain in which its solution was defined. Suppose we are studying the initial-boundary value problem in the cylinder $Q_T = D \times (0, T)$, with $D \subset \mathbb{R}^2$ a bounded domain. At the lower face of Q_T the values of the functions x and y are given; the conditions on the lateral boundary of Q_T need not be specified. The question is: does there exist a possibility to make use of the arbitrariness of ψ in order to map by a group transformation the domain D onto a simpler domain $D' \subset \mathbb{R}^2$ of the same square and of the same topological structure?

In the general case, the answer is unknown. On the other hand, the question can be given an affirmative (and quite easy) answer in the case when D is a curvilinear circle. Introduce the polar coordinates $\rho = (\xi^2 + \eta^2)^{1/2}$, $\varphi = \arctan(\eta/\xi)$, and assume that D is given by the inequalities $f(\varphi) < \rho < g(\varphi)$, where f and g are smooth positive 2π-periodic functions. Denote by S the area of D and by (r, θ) the polar coordinates on the plane ξ', η'. We claim that under a mapping preserving the area element the image D' of D is a round circle $r_1 < r < r_2$, and also $\pi(r_2^2 - r_1^2) = S$. Since the Jacobian of this transformation is assumed to be equal to unity, we obtain that

$$\rho(\rho_r \varphi_\theta - \rho_\theta \varphi_r) = r. \tag{34.6}$$

Equation (34.6) that connects the two functions ρ and φ can be satisfied by letting

$$\varphi = \lambda(\theta), \qquad \rho = \left[\frac{r^2}{\lambda'(\theta)} + \omega(\theta)\right]^{1/2}. \tag{34.7}$$

The functions λ and ω are to be defined from the conditions $\rho = f(\varphi)$ as $r = r_1$, $\rho = g(\varphi)$ as $r = r_2$; $\lambda'(\theta)$ and $\omega(\theta)$ are 2π-periodic; $\lambda(0) = 0$, $\lambda(2\pi) = 2\pi$. Substituting (34.7) into (34.6) and using the above-listed conditions, we find:

$$\int_0^\lambda \left[g^2(\zeta) - f^2(\zeta)\right] d\zeta = \frac{S}{\pi}\theta, \qquad \omega = \frac{r_2^2 f[\lambda(\theta)] - r_1^2 g[\lambda(\theta)]}{r_1^2 - r_1^2}.$$

Let us now consider the case of a simply-connected bounded domain D. Assume that it is convex, and let its boundary ∂D be given by the equation $F(\xi, \eta) = 0$. Let F

be a smooth function and $F_\xi^2 + F_\eta^2 \neq 0$. It is natural to search for a mapping of D onto the circle

$$D' = \left\{\xi', \eta' : \xi'^2 + \eta'^2 < a^2 = \frac{S}{\pi}\right\},$$

where S is the area of D. The functions ξ, η of the variables ξ', η' are defined from the conditions

$$\frac{\partial \xi}{\partial \xi'} \frac{\partial \eta}{\partial \eta'} - \frac{\partial \xi}{\partial \eta'} \frac{\partial \eta}{\partial \xi'} = 1 \quad \text{as } \xi'^2 + \eta'^2 < a^2,$$

$$F(\xi, \eta) = 0 \quad \text{as } \xi'^2 + \eta'^2 = a^2.$$

These conditions are easy to satisfy by letting

$$\xi = \frac{\partial \sigma}{\partial \xi'}, \qquad \eta = \frac{\partial \sigma}{\partial \eta'},$$

where the function σ is a solution of the second boundary value problem for the Monge–Ampère equation

$$\sigma_{\xi'\xi'} \sigma_{\eta'\eta'} - \sigma_{\xi'\eta'}^2 = 1, \quad \xi'^2 + \eta'^2 < \alpha^2, \tag{34.8}$$

$$F(\sigma_{\xi'}, \sigma_{\eta'}) = 0 \quad \xi'^2 + \eta'^2 = \alpha^2. \tag{34.9}$$

It follows from results of [131] that problem (34.8)–(34.9) has a solution; moreover, the corresponding surface $\sigma' = \sigma(\xi', \eta')$ is convex in the direction $\sigma' < 0$. The convexity condition defines the function σ up to an additive constant. If $\partial D \in C^{m+1+\alpha}$ (m is an integer, $0 < \alpha < 1$), then $\sigma(\xi', \eta') \in C^{m+\alpha}\left(\overline{D'}\right)$. These properties of the solutions of problem (34.8)–(34.9) provide the bijectivity of the mapping of the circle D' onto the domain D. So, when solving the initial-boundary value problem for system (34.3) in a cylinder Q_T with the convex lower face D one does not lose generality by assuming that this domain is merely a disk.

34.2 Similarity solutions of first and second type. Equation (34.1) admits the two-parameter group of dilatations generated by the operators X_4, X_5 (34.4). This provides the possibility to search for its self-similar solutions. The general form of the dilatation operator, where t is not invariant, is

$$Y_\alpha = \alpha X_4 + (1 - 2\alpha) X_5,$$

α is an arbitrary parameter. The family of operators Y_α corresponds to the family of self-similar solutions of equation (34.1) of the form

$$u = t^{2\alpha-1} U\left(\frac{x}{t^\alpha}, \frac{y}{t^\alpha}\right). \tag{34.10}$$

Adopt the notation

$$p = x t^{-\alpha}, \quad q = y t^{-\alpha}.$$

The substitution of (34.10) into (34.1) leads to the equation for the function $U(p,q)$:

$$\Delta(U^2) + \alpha\, r\, U_r - (2\alpha - 1)\, U = 0, \tag{34.11}$$

where Δ is the Laplace operator in the variables p, q, and $r = (p^2 + q^2)^{1/2}$. By convention we shall call the solutions (34.10) *self-similar solutions of the first type.*

Letting $\alpha \to \infty$ in the expression $\alpha^{-1}\, Y_\alpha$, we obtain the operator $X_4 - 2X_5$. The solutions of equation (34.1) invariant with respect to this operator have the form

$$u = (x^2 + y^2)\, V(\theta, t), \quad \theta = \arctan(y/x), \tag{34.12}$$

and V satisfies the equation

$$V_t = (V^2)_{\theta\theta} + 16\, V^2. \tag{34.13}$$

Let us term the solutions (34.12) *second-type self-similar solutions* of equation (34.1). A substantial difference between the first and second type self-similar solutions is that the function U in representation (34.10) satisfies an elliptic equation while equation (34.13) for the function U remains parabolic. It is natural to pose for this equation the Cauchy problem,

$$V(\theta, 0) = V_0(\theta), \tag{34.14}$$

with a nonnegative function V_0.

We will assume $V_0 > 0$ for $|\theta| < \gamma < \pi$ and $V_0 = 0$ for $-\pi \le \theta \le -\gamma$, $\gamma \le \theta \le \pi$. According to (34.12), this induces the initial condition for the function u of the form $u(x, y, 0) = u_0 \equiv (x^2 + y^2)\, V_0(\theta)$. The quadratic growth of $u_0(x, y)$ at infinity corresponds to the so-called critical exponent in the theory of the Cauchy problem for the PME with $m = n = 2$ [21]. In this case the positive solution of problem (34.1), (34.2) decays within a finite time.

It is not as simple to pose an initial condition for the first-type self-similar solutions. A priori it is clear only that the support of u_0 must be a homogeneous domain on the plane x, y (we exclude the situation when $u_0 > 0$ for all $(x, y) \in \mathbb{R}^2$). In the general case, this is a sector with apex at the origin; a special case is when the singular support of u_0 is the origin $x = y = 0$ and

$$u_0 = M\, \delta(x)\, \delta(y), \tag{34.15}$$

where $M = \text{const.} > 0$. Necessarily, here $\alpha = 1/4$, and the function u has the form

$$u = \frac{1}{t^{1/2}} \left[\left(\frac{M}{8\pi} \right)^{1/2} - \frac{x^2 + y^2}{16\, t^{1/2}} \right]_+, \tag{34.16}$$

where $s_+ = \max\{s;\, 0\}$. Solution (34.16) represents a partial case of the well-known Barenblatt solution for equation (28.4). The uniqueness of a solution for problem (34.1), (34.15) follows from the general uniqueness theorem for the Cauchy problem (28.4), (28.5) with $u_0 = M\, \delta(x)$, where $\delta(x)$ is the Dirac function in $\mathbb{R}^n$ [130].

Let us now consider the case $\alpha \neq 0$, $\alpha \neq 1/4$, and assume that $\operatorname{supp} u_0$ is the sector $|\theta| < \gamma < \pi$. The condition of invariance of relation (34.2) with respect to the dilatation transformation induced by the operator Y_α, along with the natural requirement of existence of a finite nonzero limit as $t \to \infty$ of the function u of the form (34.10), lead to the equality

$$u_0 = (x^2 + y^2)^{1-1/2\alpha}\, \chi(\theta), \tag{34.17}$$

where χ is a nonnegative function vanishing as $-\pi \leq \theta \leq -\gamma$, $\gamma \leq \theta \leq \pi$. The function u_0 is bounded at the origin if $\alpha \geq 1/2$. Its behavior as $x^2 + y^2 \to \infty$ is compatible with the asymptotics of the solution to equation (34.11) for large r.

34.3 Existence of self-similar solutions. We start with consideration of the first-type solutions. Their construction is reduced to solving the Cauchy problem (34.1), (34.2) with u_0 of the form (34.17).

Proposition 34.1. *Let in condition* (34.17)

$$\alpha > 1/4 \quad \textit{and}\ \chi \in L^1(-\pi, \pi).$$

Then problem (34.1), (34.2), (34.17) *admits a unique solution* $u(x, y, t)$ *in the sense of distributions, existing for all* $t > 0$ *and admitting representation* (34.10).

The proof presented below is due to J. L. Vazquez.

Proof. Under the conditions of Propositions 34.1 $u_0 \in L^1_{\mathrm{loc}}(\mathbb{R})^2$ and, besides, it grows like $R^{2-1/\alpha}$ as $R = (x^2 + y^2)^{1/2} \to \infty$, i.e. the exponent of its growth is less than the critical one, which equals 2. By Theorem (E) of [21], for the Cauchy problem under discussion this guarantees the existence of a solution in the layer $\mathbb{R}^2 \times (0, T)$ for each $T > 0$. Theorem (U) of the work cited asserts that this solution is unique. It remains to show that it can be represented in the form (34.10). For this we note that due to the invariance with respect to the operator Y_α, equation (34.1) admits the solutions $u(x, y, t)$ and

$$\tilde{u}(x, y, t; \lambda) = \lambda^{-2+1/\alpha}\, u(\lambda x, \lambda y, \lambda^{1/\alpha} t) \quad \forall \lambda > 0.$$

Further, initial condition (34.2) with u_0 given by equality (34.17) is invariant with respect to the dilatation transformation induced by the operator Y_α in the subspace x, y, u. By virtue of the uniqueness of solution to the Cauchy problem, $\tilde{u}(x, y, t; \lambda) = u(x, y, t)$ for $\lambda \in \mathbb{R}_+$. Letting $\lambda = t^{-\alpha}$, we then have

$$u(x, y, t) = t^{2\alpha-1} u\left(\frac{x}{t^\alpha}, \frac{y}{t^\alpha}, 1\right) \equiv t^{2\alpha-1} U\left(\frac{x}{t^\alpha}, \frac{y}{t^\alpha}\right),$$

and the assertion of Proposition 34.1 follows. □

This result, along with the estimate for solutions of the Cauchy problem (34.1), (34.2), (34.17) following from Theorem (E) [43], means that the degenerate elliptic equation (34.11) has the nontrivial nonnegative solution $U(p,q)$ defined on the whole plane and such that

$$t^{2\alpha-1}U\left(\frac{x}{t^\alpha},\frac{y}{t^\alpha}\right)\to(x^2+y^2)^{1-1/2\alpha}\chi(\theta)\quad\text{as } t\to 0$$

in the norm of $C([0,T];\ L^1(\rho_\beta))$. Here $\beta>1$ and $L^1(\rho_\beta)$ is the completion of the set of functions $f(p,q)\in L^1_{\text{loc}}(\mathbb{R}^2)$ in the norm

$$\|f\|_{L^1(\rho_\beta)}=\int_{\mathbb{R}^2}|f|\left(1+p^2+q^2\right)^{-\beta}dp\,dq.$$

It is worth noting also that Proposition 34.1 remains valid in the case of the Cauchy problem (34.2), (34.17) for the equation

$$u_t=\Delta\left(|u|\,u\right)$$

without any assumption on the positivity of u_0.

Let us now assume that under condition (34.17) $\alpha\le 1/4$. If also $\alpha>0$, then u_0 has a strong singularity at the origin: it does not even belong to $L^1_{\text{loc}}(\mathbb{R}^2)$. On the other hand, if $\alpha<0$, this function grows too rapidly at infinity. Thus, a plausible conjecture is that in the case $\alpha\le 1/4$, $\alpha\ne 0$ the Cauchy problem (34.1), (34.2), (34.17) has no solutions.

Let us turn to the self-similar solutions of the second type which are described in terms of solutions of the Cauchy problem (34.13), (34.14).

Proposition 34.2. *Let $V_0(\theta)$ be a nonnegative 2π-periodic function not identically equal to zero and such that $V_0^2\in C^1[-\pi,\pi]$. Then:*

a) *there exists $\tau>0$ such that problem* (34.13), (34.14) *has a unique weak solution $V(\theta,t)$ in the strip $\mathbb{R}\times(0,\tau)$;*

b) *there exists $T>\tau$ such that*

$$\max_{|\theta|\le\pi}V(\theta,t)\to\infty\quad \textit{as } t\to T-0, \tag{34.18}$$

and the following estimate holds:

$$T\le\frac{1}{16\overline{V_0}}, \tag{34.19}$$

where $\overline{V_0}$ is the mean value of V_0 on the interval $(-\pi,\pi)$.

Proof. The unique local solvability of problem (34.13), (34.14) follows from known results of the theory of quasilinear degenerate parabolic equations (see, for instance, [97]). Therefore, we just have to verify the assertion b).

Given $t>0$ fixed, let us denote by $\overline{V}(t)$ the mean value of the function $V(\theta,t)$ on the interval $(-\pi,\pi)$. Integrating equation (34.13) in θ between the limits $-\pi$, π, we get

$$\frac{d\overline{V}}{dt}=\frac{8}{\pi}\int_{-\pi}^{\pi}V^2(\theta,t)\,d\theta.$$

Estimating from below the right-hand side of this relation with the help of the Cauchy inequality, we obtain:

$$\frac{d\overline{V}}{dt} \geq 16\,\overline{V}^2.$$

By the conditions of the Proposition, $\overline{V}(0) = \overline{V}_0 > 0$; it then follows from the previous inequality that

$$\overline{V}(t) \geq \frac{\overline{V}_0}{1 - 16\,\overline{V}_0 t}. \tag{34.20}$$

Due to (34.20), there exists $T \in (0, 1/16\,\overline{V}_0]$ such that $\overline{V} \to \infty$ as $t \to T - 0$. This implies relation (34.18) and simultaneously gives estimate (34.19) on the "life span" of the solution to problem (34.13), (34.14). □

So far, we have considered the self-similar solutions of problem (34.1)–(34.2) with the support of u_0 a planar sector or the whole plane (x, y). Meanwhile, for the first-type solutions another situation is admissible where u_0 is a measure concentrated at the origin and the function $U(p, q)$ in (34.10) is compactly supported. Let us show that this is possible only with $\alpha = 1/4$.

Let us term a function $U(p, q)$ a generalized positive solution of equation (34.11) if U is nonnegative and continuous in $\mathbb{R}^2$, $\nabla(U^2) \in L^2_{\text{loc}}(\mathbb{R}^2)^2$, and for each test function $f \in C_0^\infty(\mathbb{R}^2)$ the following integral identity holds:

$$\int_{\mathbb{R}^2} \Big\{\nabla f \cdot \Big[\nabla(U^2) + \alpha\, U\, \vec{r}\Big] + (4\alpha - 1)\, f\, U\Big\}\, dp\, dq = 0, \tag{34.21}$$

where $\vec{r} = (p, q)$.

If a generalized positive solution U is compactly supported, (34.21) immediately implies that $\alpha = 1/4$. As was already mentioned in 34.2, in this case U depends only on $|\vec{r}|$ and the corresponding self-similar solution of equation (34.1) is of the form (34.16).

34.4 Computation of self-similar solutions. An example. Let us consider the second-type self-similar solution of equation (34.1). In the Euler variables it is represented by (34.12), where the function $V(\theta, t)$ is a solution of the Cauchy problem (34.13), (34.14). Assume that

$$\begin{aligned} &V_0^2 \in C^1[-\pi, \pi], \quad V_0 > 0 \quad \text{for } |\theta| < \gamma < \pi, \\ &V_0 = 0 \quad \text{for } -\pi \leq \theta \leq -\gamma, \gamma \leq \theta \leq \pi. \end{aligned} \tag{34.22}$$

By virtue of Proposition 34.2, problem (34.13), (34.14), (34.22) admits on some interval $(0, \tau)$ a unique solution $V(\theta, t)$. For the sake of brevity, let us assume that $V_0(\theta)$ is an even function. It then follows from the uniqueness result that so is the solution $V(\theta, t)$ in the variable θ.

According to results of the theory of degenerate parabolic equations (see Chapter III), there exists a continuous function $\alpha(t)$ defined on an interval $[0, \tau_*)$, $\tau_* \leq \tau$, such that $\alpha(0) = \gamma$, $V(\theta, t) = 0$ for $-\pi \le \theta \le -\alpha(t)$ and $\alpha(t) \le \theta \le \pi$, $0 \le t \le \tau_*$; $V(\theta, t) > 0$ for $|\theta| < \alpha(t), 0 \le t \le \tau_*$; $\alpha(t)$ is monotone increasing for $t \in [0, \tau_*]$. On the plane θ, t the curves $\theta = \pm\alpha(t)$ are the levels of degeneracy of equation (34.13). Since the function $\alpha(t)$ is a priori unknown, it is natural to term these curves the free boundaries. Besides the condition $V(\alpha(t), t) = 0$, at the free boundary $\theta = \alpha(t)$ the following "zero flux" condition holds:

$$\lim_{\theta \to \alpha(t)-0} V\, V_\theta = 0, \quad t \in [0, \tau_*). \tag{34.23}$$

Moreover, for almost all t from the interval $[0, \tau_*)$ we have

$$\frac{d\alpha}{dt} = -2 \lim_{\theta \to \alpha(t)-0} V_\theta. \tag{34.24}$$

Relations similar to (34.23), (34.24) hold as well at the other free boundary $\theta = -\alpha(t)$.

The existence and uniqueness theorems for problem (34.13), (34.14), (34.22) were proven without having recourse to Lagrangian coordinates. However, in the original Euler variables its numerical treatment is made much more complicated by the presence of the free boundary. In what follows, we will reduce this problem to a problem in a fixed domain on the plane of Lagrangian coordinates. Note that here any direct introduction of Lagrangian coordinates is difficult because equation (34.13) does not have divergence form. Meanwhile, the original equation (34.1) is of divergence form, so for the equivalent system (34.3) we can construct a self-similar solution which corresponds to the second-type self-similar solution (34.12).

To obtain such solutions of system (34.3) one has to take the admissible dilatation operator that leaves t invariant. Note that the group of dilatations preserving system (34.3) is larger than the analogous group for equation (34.1): it is three-parameter. According to (34.5), its generators have the form Σ_4, Σ_5, and Σ_6, where $\psi = \xi\,\eta$. Therefore, the dilatation operators which leave the variable t invariant form a one-parameter family. We shall make use of the appearing arbitrariness to obtain the most simple representation for the self-similar solution we are interested in. For this purpose, let us consider the operator

$$Z = \Sigma_4 - 2\Sigma_5 - 2\Sigma_\psi \equiv x\,\frac{\partial}{\partial x} + y\,\frac{\partial}{\partial y} + 4\eta\,\frac{\partial}{\partial \eta}.$$

The solutions of system (34.3) invariant with respect to the operator Z are sought in the form

$$x = \eta^{1/4} f(\xi, t), \quad y = \eta^{1/4} g(\xi, t). \tag{34.25}$$

Substituting (34.25) into (34.3) we get a system of equations which can be reduced to the following:

$$\begin{aligned} f\, f_t + g\, g_t &= -16\, h^{-1}, \\ f_\xi\, f_t + g_\xi g_t &= 8\, h^{-2}\, h_\xi, \\ h &= g\, f_\xi - f\, g_\xi. \end{aligned} \tag{34.26}$$

Now note that due to (34.25)

$$\frac{g}{f} = \frac{y}{x} = \tan\theta,$$

and then introduce, instead of f and g, the new sought functions $v \geq 0$ and θ by the relations

$$f = 4\,v^{1/4}\cos\theta, \quad g = 4\,v^{1/4}\sin\theta. \tag{34.27}$$

It follows from the first and third equations of (34.26) that

$$h = -4\,\frac{v^{1/2}}{v_t},$$

$$\theta_\xi = \frac{1}{4\,v_t}. \tag{34.28}$$

Substituting these relations into the second equation of (34.26), with the substitution (34.27) already performed, we arrive at the following expression for θ_t:

$$\theta_t = \frac{v}{4}\left(\frac{v_t^2}{v^2}\right)_\xi. \tag{34.29}$$

The constraint of compatibility of (34.28) and (34.29) leads to the following equation for the function $v(\xi, t)$:

$$\left(\frac{1}{v_t}\right)_t = \left[v\left(\frac{v_t^2}{v^2}\right)_\xi\right]_\xi. \tag{34.30}$$

Our next objective will be to proceed from (34.14), (34.23) in order to pose an initial-boundary value problem for equation (34.30). For this one has to establish the connection between the functions $V(\theta, t)$ and $v(\xi, t)$. Since ξ, η are the canonical Lagrangian coordinates, the Cauchy identity has the form

$$u\,|N| = 1,$$

where $|N|$ is the determinant of the Jacobian matrix of the mapping $(\xi, \eta) \rightarrow (x, y)$. Substituting into this relation the expression for u (34.12) and calculating $|N|$ with the help of (34.25), (34.27), (34.28), we will have

$$V\,[\theta(\xi, t), t] = -\frac{v_t}{16\,v}. \tag{34.31}$$

As was already explained in Chapter III, the passage to Lagrangian coordinates "stops" the free boundary in the porous medium equation. Therefore, for the solutions of equation (34.30) the domain of dependence is a rectangle $|\xi| < A, 0 < t < \tau_*$, where $A > 0$ is some constant. Besides, the evenness of V in the variable θ makes it possible to restrict our consideration to the solutions v even with respect to ξ. Finally, equation (34.30) admits the rescaling $\xi = A\,\xi'$, $t = t'$, $v = A\,v'$. This provides the possibility to reduce the matter to the initial-boundary value problem for equation (34.30) in the rectangle

$$D_* = \{\xi,\, t:\ 0 < \xi < 1,\ 0 < t < \tau_*\},$$

having posed on its left lateral boundary the symmetry condition

$$v_\xi(0, t) = 0, \quad 0 \le t < \tau_*. \tag{34.32}$$

Equation (34.30) is of second order in t and, hence, two initial conditions are required:

$$v(\xi, 0) = v_0(\xi), \quad 0 \le \xi \le 1,$$
$$v_t(\xi, 0) = v_1(\xi), \quad 0 \le \xi \le 1. \tag{34.33}$$

Let us introduce the notation $\theta(\xi, 0) = \theta_0(\xi)$. Letting in (34.28), (34.31) $t = 0$, we get two relations connecting three functions v_0, v_1, θ_0 of the variable ξ (recall that $V(\theta, 0)$ is assumed given). This leaves the choice of $v_0(\xi)$ at our disposal; thereafter the dependence of v_1 and θ_0 on ξ is uniquely defined. (This is a reflection of the functional arbitrariness in the choice of the canonical Lagrangian coordinates, already used in another situation in 34.1.) We simply set

$$v(\xi, 0) = 1, \quad 0 \le \xi \le 1. \tag{34.34}$$

The last equality and (34.28), (34.31) taken at $t = 0$ imply that

$$\frac{d\theta_0}{d\xi} = \frac{1}{4v_1(\xi)}, \tag{34.35}$$

$$V_0\,[\theta_0(\xi)] = -\frac{v_1(\xi)}{16\,v_0(\xi)}, \tag{34.36}$$

where $V_0(\xi)$ is an even function subject to (34.32). It follows from (34.35), (34.36) and the natural condition $\theta_0(0) = 0$ that

$$\int_0^{\theta_0} V_0(\theta)\,d\theta = -\frac{\xi}{64}, \tag{34.37}$$

which allows one to define $\theta_0(\xi)$ as an odd function on the interval $[-1, 1]$. Thereby, relation (34.36) defines also the function $v_1(\xi)$. Due to (34.22), this function satisfies the conditions

$$v_1(\xi) < 0 \quad \text{if } 0 \le \xi < 1; \quad v_1(1) = 0. \tag{34.38}$$

Letting in (34.37) $\xi = 1$ and taking into account the evenness of $V_0(\theta)$, we get

$$\int_0^{\gamma} V_0(\theta)\,d\theta = \frac{1}{64}. \tag{34.39}$$

The last relation seems to pose a restriction on the choice of the function V_0, but this restriction turns out to be illusory. Indeed, if the value of the integral in (34.39) is equal to some $B \neq 1/64$, in problem (34.13), (34.14) we can perform the dilatation transformation $\theta = \theta'$, $t = t'/64\,B$, $V = 64\,B\,V'$, $V_0 = V_0'$. Although the height of the rectangle D_* changes, equation (34.13) transforms into itself and the new sought function will satisfy condition (34.39). We will always assume that such a transformation has already been performed.

To complete the statement of the initial-boundary value problem one has to pose conditions on the right lateral boundary of the rectangle D_*, which corresponds to the line of degeneracy $\theta = -\alpha(t)$ of equation (34.13). One of them follows from (34.31):

$$v_t(1, t) = 0, \quad 0 \le t < \tau_*. \tag{34.40}$$

Thus, as should have been anticipated, equation (34.30) degenerates at the segment $\xi = 1, 0 \le t < \tau_*$. Another condition follows from (34.23) and the equality

$$V_\theta = -\frac{v_t}{4}\left(\frac{v_t}{v}\right)_\xi, \tag{34.41}$$

holding by virtue of (34.28), (34.31). Here the left-hand side is considered as a function of the variables ξ and t. Like "the zero flux" condition (34.23), condition (34.41) controls the growth of the space derivative of v_t near the line of degeneracy:

$$\lim_{\xi \to 1-0} v_t \left(\frac{v_t^2}{v^2}\right)_\xi = 0, \quad t \in [0, \tau_*).$$

This relation among with (34.34), (34.40) yields that $v_\xi\, v_t^2 \to 0$ as $\xi \to 1-0, 0 \le t < \tau_*$, which allows one to rewrite the required condition in the simpler form

$$\lim_{\xi \to 1-0} v_t^2\, v_{t\xi} = 0, \quad t \in [0, \tau_*). \tag{34.42}$$

Finally, let us introduce instead of v the new sought function $w(\xi, t) = -v_t$. According to (34.34), we get

$$v(\xi, t) = 1 - \int_0^t w(\xi, \tau)\, d\tau. \tag{34.43}$$

Due to (34.30), (34.43), w satisfies in the rectangle D_* the equation

$$\left(-\frac{1}{w}\right)_t = \left(\left[1 - \int_0^t w(\xi, \tau) d\tau\right]\left\{\left(\frac{w}{1 - \int_0^t w(\xi, \tau)\, d\tau}\right)^2\right\}_\xi\right)_\xi. \tag{34.44}$$

Thus, we have passed from the third-order mixed-type equation (34.30) to equation (34.44). Although it is an integro-differential equation, it remains structurally quite close to a parabolic second-order equation.

The initial condition for equation (34.44) follows from (34.33), (34.43):

$$w(\xi, 0) = -v_1(\xi), \quad 0 \le \xi \le 1. \tag{34.45}$$

The function $v_1(\xi)$ satisfies condition (34.38), the compatibility condition (34.32) $v_1'(0) = 0$, and some smoothness restriction which follow from (34.22), (34.36), (34.37) and are not made concrete here. Relations (34.32), (34.40) imply the boundary condition for w,

$$w_\xi(0, t) = 0, \quad 0 \le t < \tau_*, \tag{34.46}$$

$$w(1, t) = 0, \quad 0 \le t < \tau_*. \tag{34.47}$$

An additional condition at the line of degeneracy follows from (34.42):

$$\lim_{\xi \to 1-0} w^2 w_\xi = 0, \quad 0 \le t < \tau_*. \tag{34.48}$$

The problem of defining the function w is formulated. Given a solution of this problem, the formula

$$\frac{d\alpha}{dt} = -\frac{1}{2} \lim_{\xi \to 1-0} w\, w_\xi, \quad \alpha(0) = \gamma \tag{34.49}$$

then recovers the function $\alpha(t)$ which defines the trajectory of the free boundary on the plane θ, t. (To derive this relation from condition (34.24), we used equalities (34.34), (34.41), (34.42), and (34.43).) The solution $V(\theta, t)$ of problem (34.13, (34.14) is given parametrically by relations (34.31) and

$$-\theta(\xi, t) = \frac{1}{4} \int_0^\xi \frac{d\eta}{w(\eta, t)}. \tag{34.50}$$

The last equality is provided by (34.28), (34.43), and by the requirement of the oddness of θ in the variable ξ.

It was problem (34.44)–(34.48) that was treated numerically. However, before passing to the description of the methods and results of the computations, let us present the example of its exact solution used to test the numerical method. Let the initial function V_0 be of the form

$$V_0(\theta) = \left[\frac{\cos 2\theta - \cos 2\gamma}{32(\sin 2\gamma - 2\gamma \cos 2\gamma)}\right]_+ \tag{34.51}$$

and $0 < \gamma < \pi/2$. (Note that this function satisfies conditions (34.39).) Then the solution of the Cauchy problem (34.13), (34.14) is given by the formula

$$V(\theta, t) = \frac{\kappa\,[\cos 2\theta - \cos 2\alpha(t)]_+}{\sin^3 2\alpha(t)}, \quad 0 \le t < \tau_*, \tag{34.52}$$

where

$$\kappa = \frac{\sin^3 2\gamma}{32(\sin 2\gamma - 2\gamma \cos 2\gamma)},$$

$$\tau_* = \frac{(2\pi - 4\gamma + \sin 4\gamma)(\sin 2\gamma - 2\gamma \cos 2\gamma)}{\sin^3 2\gamma}, \tag{34.53}$$

and the function $\alpha(t)$ defining the position of the free boundary is given implicitly by the equality

$$4\alpha - \sin 4\alpha - 4\gamma + \sin 4\gamma = 32\kappa t.$$

This function increases monotonically on the interval $[0, \tau_*]$ from the value γ up to $\pi/2$. When $t \to \tau_* - 0$, solution (34.52) becomes infinite simultaneously at every point of the interval $\theta \in (-\pi/2, \pi/2)$. It is curious to observe that for this solution the dependence of the "life span" on the parameter γ happens to be very close to the linear one: $\tau_* \to 2\pi/3$ as $\gamma \to 0$, $\tau_* = \pi$ for $\gamma = \pi/4$, $\tau_* \to 4\pi/3$ if $\gamma \to \pi/2$.

Proceeding from (34.36), (34.37), (34.51), it is easy to find the following representation for the function $v_1(\xi)$ which enters the initial condition (34.45) of problem (34.44)–(34.48):

$$\xi = \frac{\sin 2\theta - 2\theta\cos 2\gamma}{\sin 2\gamma - 2\gamma\cos 2\gamma}, \quad v_1 = \frac{\cos 2\gamma - \cos 2\theta}{2(\sin 2\gamma - 2\gamma\cos 2\gamma)}, \tag{34.54}$$

where $0 \le \theta \le \gamma$. Not as simple is the representation for the solution itself $w(\xi, t)$ corresponding to solution (34.52) of problem (34.13), (34.14). However, it need not be obtained since the required test consists in the numerical solution of problem (34.44)–(34.48) with the function $v_1(\xi)$ given by formulas (34.54), the recalculation of the found solution from the variables ξ, t, w to the variables θ, t, V with the help of formulas (34.31), (34.50), and the comparison of the result with the exact solution (34.52).

One other possibility of testing is based on the following observation. For small values of the parameter t the integro-differential equation (34.44) is well approximated by the differential equation

$$\left(-\frac{1}{w}\right)_t = \left(w^2\right)_{\xi\xi}. \tag{34.55}$$

Equation (34.55) possesses the family of self-similar solutions

$$\tilde{w} = \left[\frac{3}{16(\beta + t)}\right]^{1/3} \zeta(\xi), \tag{34.56}$$

where $\beta = \text{const.} > 0$, and the function $\zeta(\xi)$ is uniquely defined by the relation

$$(2 + \zeta)\sqrt{1 - \zeta} = 2\xi$$

and the condition $\zeta > 0$ as $\xi \in [0, 1)$. The function $\tilde{w}(\xi, t)$ satisfies the boundary conditions (34.46)–(34.48) and the initial condition (34.45), where

$$v_1 = -\left(\frac{3}{16\beta}\right)^{1/3} \zeta(\xi). \tag{34.57}$$

Our anticipation is that for small t the solution w of problem (34.44)–(34.48), (34.57) will be close to $\tilde{w}$. Having had recourse to formulas (34.36), (34.37), one can establish the correspondence between the function v_1 of the form (34.57) and the function $V_0(\theta)$ from the initial condition for equation (34.13)

$$V_0 = \frac{3}{128\gamma}\left[1 - \left(\frac{\theta}{\gamma}\right)^2\right]_+ , \tag{34.58}$$

where $\gamma = (9\beta/32)^{1/3}$; due to (34.22), γ cannot attain the value π, which poses the restriction on the parameter β: $\beta < 32\pi^3/9$.

Problem (34.44)–(34.49), (34.57) with $\beta = 32\gamma^3/9$ was solved numerically by the implicit finite-difference scheme with nonlinear iterations. The first derivatives were approximated to the first order of exactness, and the approximation of the second-order derivatives was of second order. The finite-difference equations were solved by the double sweep method. Within the main domain of calculations, the steps in time and space were chosen equal to 0.01. These steps decreased near the blow-up instant. The calculations were performed by V. V. Pukhnachov in collaboration with G. B. Volkova.

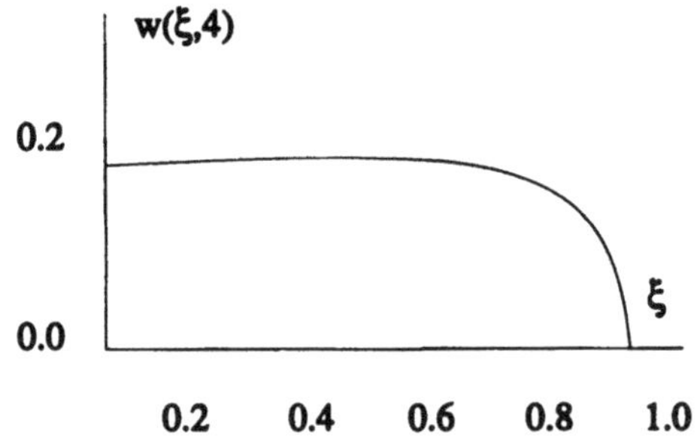

Figure 1

Figure 1 demonstrates the graph of the function $w(\xi, t)$ corresponding to the value $\gamma = \pi/2$ at the instant $t = 4$, and Figure 2 presents the graph of the function $\tilde{w}(\xi, t)$ (34.56) at the same instant.

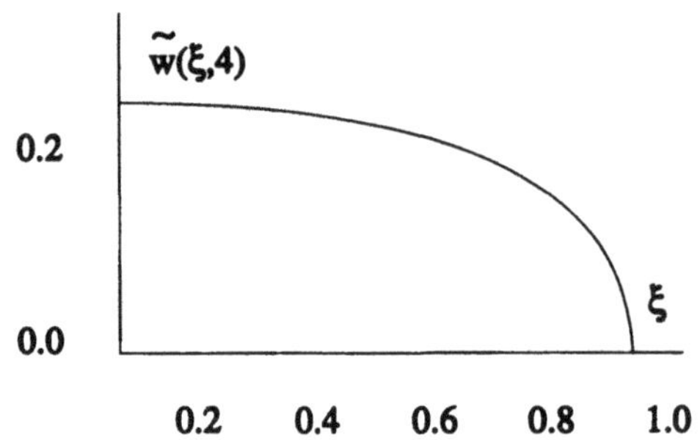

Figure 2

The profile of function $w(\xi, t)$ at the instant $t = 5$ close to the blow-up is shown in Figure 3.

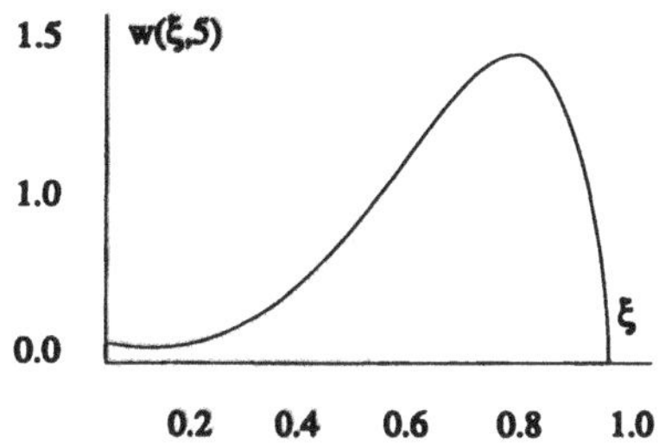

Figure 3

Figures 4 and 5 exhibit the behavior of $V(\theta, t)$ at the instants $t = 4$, $t = 5$, correspondingly.

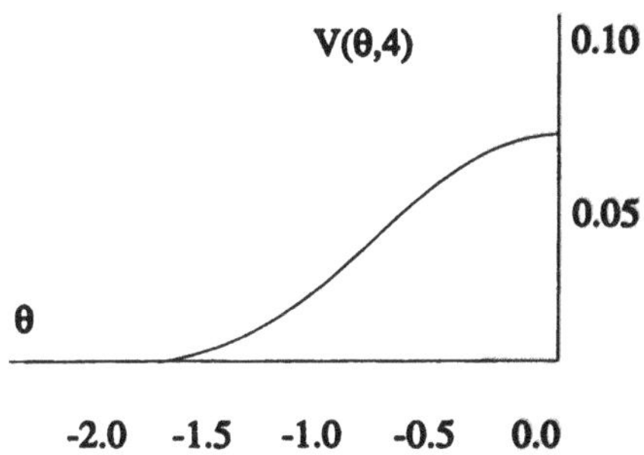

Figure 4

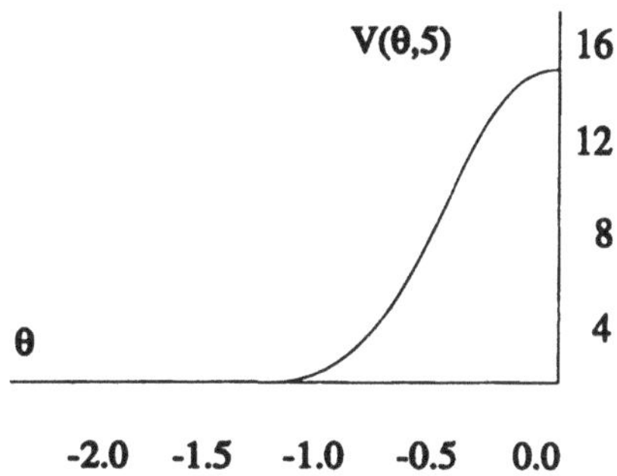

Figure 5

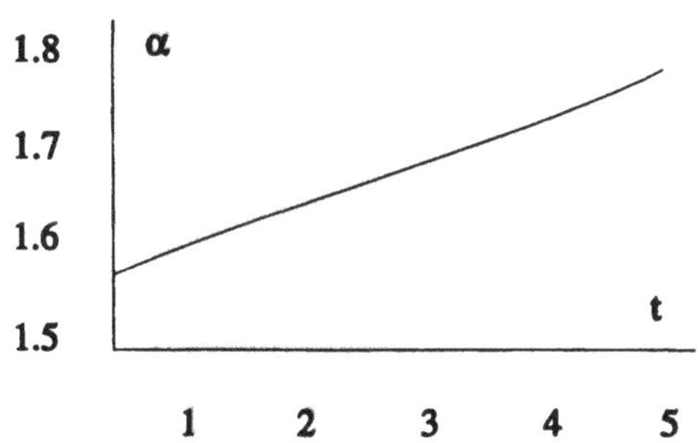

Figure 6

The numerical solution of problem (34.44)–(34.49), (34.57) with $\gamma = \pi/2$ decays near the instant $t = 5.0112$.

34.5 How could one linearize the Boussinesq equation? Here we study the linear model of problem (34.1), (34.2) under the assumption that the function u_0 is compactly supported. The analysis of the linear approximation in Euler variables is made complicated by the degeneracy of equation (34.1) and by the presence of the free boundary. One can avoid the latter difficulty by passing to Lagrangian coordinates where Boussinesq's equation corresponds to system (34.3) with the initial condition (34.2) as the Cauchy data for this system:

$$x = x_0(\xi, \eta), \quad y = y_0(\xi, \eta), \quad (\xi, \eta) \in R. \tag{34.59}$$

The functions x_0, y_0 are connected by the relation

$$\frac{\partial(x_0, y_0)}{\partial(\xi, \tau)} u_0(x_0, y_0) = 1, \tag{34.60}$$

and the domain R is defined from the conditions

$$u_0\left[x_0(\xi, \eta), y_0(\xi, \eta)\right] > 0 \text{ if } (\xi, \eta) \in R,$$

$$u_0\left[x_0(\xi, \eta), y_0(\xi, \eta)\right] = 0 \text{ if } (\xi, \eta) \in S = \partial R$$

(see Section 33 for details). The solution of system (34.3) is to be found in the cylinder $Q_T = R \times (0, T)$. Its lateral boundary corresponds to the free boundary Γ_t in the solution of problem (34.1), (34.2). It follows from the Cauchy identity and the equality $u = 0$ on Γ_t that

$$|N| \to \infty \quad (\xi, \eta) \in S, \ t \in (0, T), \tag{34.61}$$

where $|N| = x_\xi y_\eta - x_\eta y_\xi$ is the determinant of the Jacobian matrix $N = \partial(z)/\partial(\zeta)$.

The procedure of linearization of system (34.3) is very similar to, and about as cumbersome, as that performed in Section 30 in the study of the linear model of the one-phase Stefan problem. However, it much simplifies if the linearization is carried out on an exact solution of the system. This is the case that is considered below.

Let us take a solution $z = (x, y)$ of system (34.3) and denote $\tilde{z} = z + \varepsilon Z$, where ε is a small parameter. Let us substitute the function $\tilde{z}$ into system (34.3) and keep in the resulting relations the terms of zero and first order with respect to ε. Suppose that $|N| \neq 0$ in the domain Q_T and define the vector function W by the equality

$$W = N^{-1} Z. \tag{34.62}$$

Repeating almost literally the calculations of Section 30, and taking into account that z is an exact solution of (34.3), we arrive at the following equation for the function W:

$$A\, W_t = 2 \nabla \left(|N|^{-1} \operatorname{div} W \right), \tag{34.63}$$

where $A = N^* N$. To equation (34.63) we adjoin the initial condition

$$W = W_0(\xi, \eta) \quad \text{for } (\xi, \eta) \in \overline{R}, \quad t = 0, \tag{34.64}$$

where W_0 is a given vector-valued function. As regards the boundary condition, the degeneracy of equation (34.63) at the surface $S \times [0, T]$ (see (34.61)) does not allow one to assign the sought function any fixed value on this surface.

If a solution of problem (34.63), (34.64) is found, the perturbation εZ of the principal solution z of system (34.3) is defined with the help of (34.62), and on the Euler plane the perturbation of the free boundary $\tilde{\Gamma}_t$ is given in the parametric form $\tilde{z}(\xi, \eta, t) = z + \varepsilon Z$, $(\xi, \eta) \in S$, $t \in [0, T]$. Finally, for the perturbed solution of problem (34.1), (34.2) we get the expression $\tilde{u} = |N|^{-1}(1 - \varepsilon \operatorname{div} W)$.

Let us adopt the notation $q = \operatorname{div} W$. Multiplying both sides of equation (34.63) by the matrix A^{-1} and applying to the resulting equality the operator div, we get the equation for the function q:

$$q_t = 2 \operatorname{div}\left[A^{-1} \nabla \left(|N|^{-1}\right) q\right], \quad (\xi, \eta) \in R,\ t \in (0, T). \tag{34.65}$$

According to (34.64), the initial condition for this equation has the form

$$q = q_0(\xi, \eta), \quad (\xi, \eta) \in R,\ t = 0, \tag{34.66}$$

where $q_0 = \operatorname{div} W_0$.

Equation (34.65) degenerates at the lateral boundary of the cylinder Q_T. It belongs to the class of second-order equations with nonnegative characteristic form. The theory of such equations is developed in [119]. As a rule, the boundary conditions for such equations are formulated in terms of inclusion of the sought function into some weighted classes. Having solved problem (34.65), (34.66), the vector function W is recovered by integration from the equalities $W_t = 2A^{-1}\nabla\left(|N|^{-1} q\right)$ and (34.64).

Let us consider an important special case of problem (34.63), (34.64) which occurs in the linearization of system (34.3) on the solution $z = (x, y)$ corresponding to the Barenblatt solution of equation (34.1). Since this solution has a singularity at $t = 0$, we impose the initial condition at the instant $t = 1$. (We omit rewriting (34.65), (34.66) with $t = 0$ replaced by $t = 1$.) Relation (34.60) here takes the form

$$\frac{\partial(x_0, y_0)}{\partial(\xi, \eta)} \left[\left(\frac{M}{8\pi}\right)^{1/2} - \frac{x_0^2 + y_0^2}{16}\right] = 1.$$

The initial conditions (34.59) adjusted to these relations can be chosen in the form

$$x_0 = r_0(\rho) \cos \varphi, \quad y_0 = r_0(\rho) \sin \varphi,$$

where ρ, φ are the polar coordinates on the plane ξ, η, and the function r_0 on the interval $0 < \rho < \rho_0 \equiv \sqrt{M/\pi}$ satisfies the equation

$$r_0 \left[\left(\frac{M}{8\pi}\right)^{1/2} - \frac{r_0^2}{16}\right] \frac{d r_0}{d \rho} = \rho. \tag{34.67}$$

In is convenient to pass in equation (34.65) to the polar coordinates which make the matrix A diagonal. Observing that at the solution (34.16) $r \equiv (x^2 + y^2)^{1/2} = t^{1/4} r_0(\rho)$, we get for this matrix the following representation

$$A = \sqrt{t}\begin{pmatrix} (dr_0/d\rho)^2 & 0 \\ 0 & (r_0/\rho)^2 \end{pmatrix}.$$

In polar coordinates, equation (34.65) becomes

$$q_t = 2t^{-1}\left\{\frac{1}{\rho}\frac{\partial}{\partial\rho}\left[\frac{\rho}{r_0'^2}\frac{\partial}{\partial\rho}\left(\frac{\rho}{r_0 r'_0}q\right)\right] + \frac{\rho}{r_0^3 r'_0}\frac{\partial^2 q}{\partial\varphi^2}\right\}.$$

One can simplify it by passing to the new independent variables

$$s = \left(\frac{M}{8\pi}\right)^{-1/2}\frac{r_0(\rho)}{4}, \qquad \tau = \frac{1}{8}\ln t,$$

and replacing $q(\rho, \varphi, t)$ by the new sought function $v(s, \varphi, \tau) = (1 - s^2)q$. Here the dependence $r_0(\rho)$, $0 \leq r_0 \leq 4(M/8\pi)^{1/4}$, is uniquely defined by equation (34.67) and the condition $r_0 = 0$ as $\rho = 0$. Also, s can vary within the segment $[0, 1]$. In the new variables, for the function v we get the equation

$$\frac{\partial v}{\partial \tau} = \frac{1}{s}\frac{\partial}{\partial s}\left[s(1 - s^2)\frac{\partial v}{\partial s}\right] + \frac{1 - s^2}{s^2}\frac{\partial^2 v}{\partial\varphi^2}. \tag{34.68}$$

Equation (34.68) is to be solved in the cylinder $0 < s < 1$, $0 < \tau < (\ln T/8)$ with the initial condition following from (34.66) and under the condition of boundedness of v on the lateral boundary of the cylinder.

Equation (34.68) admits a total separation of variables, making it possible to represent its solution in the form of the Fourier series

$$v = \sum_{i=1}^{\infty} \exp(-\lambda_{0k}\tau)\, a_{0k} v_{0k}(s) + \sum_{n,k=1}^{\infty} \exp(-\lambda_{nk}\tau)\, (a_{nk}\cos n\varphi + b_{nk}\sin\varphi)\, v_{nk}(s). \tag{34.69}$$

Here v_{nk} are the normalized eigenfunctions of the Sturm–Liouville operator

$$L_n = -\frac{1}{s}\frac{d}{ds}\left[s(1 - s^2)\frac{d}{ds}\right] + \frac{n^2(1 - s^2)}{s^2}$$

on the interval $(0, 1)$, λ_{nk} are the corresponding eigenvalues; a_{nk}, b_{nk} are constants. For each nonnegative integer n the system of eigenfunctions $\{v_{nk}\}$ of the operator L_n is complete and orthogonal in $L^2(0, 1)$ with the weight s. All its eigenvalues λ_{nk} are positive, except for $\lambda_{01} = 0$. It is the normalized eigenfunction $v_{01} = \sqrt{2}$ that corresponds to the last eigenvalue.

Further, assume that in condition (34.66) the function q_0 satisfies the equality

$$\int_R q_0 \, d\xi \, d\eta = 0. \tag{34.70}$$

This relation expresses the fact that the perturbed and the principal solutions of problem (34.1), (34.2) have the same "mass". Then in expansion (34.69) we get $a_{01} = 0$ and all its terms becomes exponentially decreasing in τ. Let us additionally claim that

$$\int_R \xi \, q_0 \, d\xi \, d\eta = \int_R \eta \, q_0 \, d\xi \, d\eta = 0. \tag{34.71}$$

The latter relations can be interpreted as conserving "the mass center" in the perturbed solution of problem (34.1), (34.2). They imply the equalities $a_{11} = b_{11} = 0$ and thereby permit one to exclude from consideration the perturbations obtained by dilatations of solution (34.16) along the axis x and y.

Let the sought solution v of equation (34.68) have as the initial value a function $v_0(s, \varphi)$, square integrable in the circle $0 \le s \le 1$ with weight s. The series (34.69) will be absolutely convergent in the closed disk for every $\tau \ge \delta > 0$. Under conditions (34.70), (34.71), the rate of decrease of v as τ grows is defined by the minimum of the eigenvalues $\{\lambda_{nk}\}$, where $k \ge 2$ if $n = 0, 1$ and $k \ge 1$ if $n \ge 2$. The variational principle for the Sturm–Liouville problem imply that the eigenvalues λ_{nk} are monotone increasing functions of the integer arguments n, k. Thus, the minimum is attained at the two-element set $\{\lambda_{02}, \lambda_{21}\}$.

The eigenvalues of the operator L are computed in explicit form and its eigenfunctions are expressed through the Jacobi polynomials. In particular, $\lambda_{02} = 8$, $\lambda_{21} = 4$, and the corresponding eigenfunctions are $v_{02} = \sqrt{6}\,(2s^2 - 1)$, $v_{21} = \sqrt{6}\, s^2$. This implies the estimate for the solution of equation (34.68) having in the Fourier expansion (34.69) $a_{01} = a_{11} = b_{11} = 0$:

$$v = O\left(e^{-4\tau}\right) \quad \text{as } \tau \to \infty$$

uniformly in s, φ.

Coming back to the original variables we get the estimate of the difference $u - \tilde{u}$ between the principal and perturbed solutions of problem (34.1)–(34.2):

$$\varepsilon |N|^{-1} q = O\left(t^{-1}\right), \qquad t \to \infty, \tag{34.72}$$

uniformly in ρ, φ in the circle R: $0 \le \rho \le \rho_0 = \sqrt{M/\pi}$. (We use here that the product $|N|^{-1} q$ remains bounded despite the unboundedness of q near the boundary of R.) Since the function u defined by equality (34.16) decreases proportionally to $t^{-1/2}$ as t grows, estimate (34.72) means the stability of the source type solutions of the Boussinesq equation with respect to linear approximation.

In the light of papers [23, 178, 8], this fact itself is not unexpected. In the cited works, for the one-dimensional porous medium equation the stability of an analog of solution (34.16) was established. Also, in the first two works the stability problem was studied in the linearized statement, while in [8] the complete nonlinear problem was treated. Estimate (34.72) is apparently new. It allows one to estimate the closeness of the boundaries of supports of the functions u and $\tilde{u}$ for large t. The first one of these curves on the plane x, y is the circumference $(x^2+y^2)^{1/2} \equiv r = 4(Mt/(8\pi))^{1/4}$. Let the second curve in polar coordinates r, θ be given by the equation $r = 4(Mt/8\pi)^{1/4} + \omega(\theta, t)$. Then $\omega = O(t^{-1/4})$ as $t \to \infty$.

35 Local regularity of interfaces

Here we study the regularity properties of weak nonnegative solutions of the equation

$$u_t = \Delta u^m, \quad (x,t) \in \mathbb{R}^N \times (0,\infty), \tag{35.1}$$

with N positive integer and $m > 1$. This equation was already discussed above. In Section 17 we presented a physical model leading to this equation and described the phenomenon of finite speed of propagation of disturbances from the data, the key feature of the nonlinear equations of this type. Due to this effect, solutions of equation (35.1) display interfaces, i.e. hypersurfaces separating the regions where the solution is positive from those where it is equal to zero. Let us refer to [21]: given a nonnegative function $u_0 \in L^1(\mathbb{R}^N) \cap L^\infty(\mathbb{R}^N)$ compactly supported in $\mathbb{R}^N$, the Cauchy problem for equation (35.1) always has a unique weak solution $u(x,t)$, which is non-negative and has compact support for each $t > 0$, increasing with time.

The regularity of the occurring interfaces was already studied for the one-dimensional version of (35.1) and for several related equations (also one-dimensional). Apart from the one-dimensional case, the multi-dimensional case has still not been given an adequate analysis. It was proven in paper [33] that the moving part of the interface was a Lipschitz hypersurface and then shown in [34] that it was $C^{1+\alpha}$. To the best of our knowledge, these results are the strongest so far obtained.

To simplify matters, as in the one-dimensional case we discuss here only the Cauchy problem for (35.1). As follows from the forthcoming considerations, all our conclusions are applicable to any of the solutions to (35.1) which exhibit interfaces.

The following result is proven: the moving part of the interface admits a local in time parametrization which inherits the regularity of the initial interface in spatial variables and is C^∞ in time.

35.1 Lagrangian coordinates. We begin by reformulating the Cauchy problem for equation (35.1) viewed as the mathematical description of the process of propagation of a polytropic gas in a porous medium. The density of the gas is equal to $u(x,t)$ and the velocity is given by

$$\vec{v} = -\frac{m}{m-1} \nabla u^{m-1}. \tag{35.2}$$

With these definitions, equation (35.1) takes the form

$$u_t + \operatorname{div}(u\,\vec{v}) = 0 \tag{35.3}$$

and can be considered as the mass balance law of the fluid motion with density $u(x,t)$ and velocity given by Darcy's law (35.2).

Let us translate the mass balance law (35.3) into Lagrange coordinates. Let $\Omega(0)$, $\Omega(t) \subset \mathbb{R}^N$ be the domains occupied by a polytropic gas at the instants $t = 0$ and $t > 0$. This correspondence defines the mapping

$$x = X(\xi, t), \quad \xi \in \Omega(0),$$

which assigns the position $X(\xi,t)$ to the particle initially located at a point $\xi \in \Omega(0)$. Given the velocity field $\vec{v}(x,t)$, the motion of this particle is controlled by the *trajectory equation*

$$\begin{cases} X_t(\xi,t) = \vec{v}\,[X(\xi,t),t], & t > 0, \\ X(\xi,0) = \xi, \quad \xi \in \Omega(0). \end{cases} \tag{35.4}$$

Another ingredient of the description is the mass balance law. Let $\omega(t)$ be an arbitrary gas volume which contains the same particles at each instant $t \geq 0$. Then the mass of $\omega(t)$ is conserved in time, e.g.

$$\frac{d}{dt}\left\{\int_{\omega(t)} u(x,t)\,dx\right\} = 0.$$

By convention we shall use capital letters to denote magnitudes considered as functions of Lagrangian coordinates ξ. Thus, $U(\xi,t) = u[X(\xi,t),t]$ will be the pressure, $\vec{V}(\xi,t) = \vec{v}[X(\xi,t),t]$ the velocity, etc. Let

$$J = [J_{ij}], \quad J_{ij} = \frac{\partial X_i}{\partial \xi_j}, \quad i,j = 1,\dots,N$$

be the Jacobian matrix of the mapping $\xi \mapsto X$ and $|J| \equiv \det(\partial X/\partial \xi)$. Using the trajectory equation (2.1) and applying the rule of differentiation of determinants, it is easy to verify the validity of the relation

$$\frac{d}{dt}(|J|) = |J|\,\operatorname{div}\vec{v},$$

called the *Cauchy identity*. Passing formally to coordinates ξ, we have now:

$$\begin{aligned} 0 &= \frac{d}{dt}\left\{\int_{\omega(t)} u(x,t)\,dx\right\} = \int_{\omega(0)} \frac{d}{dt}\{U\,|J|\}\,d\xi \\ &= \int_{\omega(0)} \left[U_t + \nabla_X u[X(\xi,t),t]\,\vec{v}[X(\xi,t),t] + U \operatorname{div}_X \vec{v}\right]|J|\,d\xi \\ &= \int_{\omega(0)} \left[U_t + \operatorname{div}_X(u\vec{v})\right]|J|\,d\xi = \int_{\omega(t)} \left[u_t + \operatorname{div}_x(u\vec{v})\right]dx. \end{aligned}$$

Since the volume $\omega(t)$ is assumed arbitrary, it follows that in the Euler coordinates the mass balance law is expressed by (35.3) and its counterpart in Lagrangian coordinates is given by the equation

$$\frac{d}{dt}(U\,|J|) = 0, \quad t \geq 0.$$

After integration we get from this:

$$U(\xi, t)\,|J| = U(\xi, 0) \quad \text{in } \Omega(0),\ t \geq 0. \tag{35.5}$$

We are able to state in this framework the Lagrangian analog of (35.1). Suppose that the gas velocity follows the law (35.2). Recalculating the derivatives in x,

$$\nabla_x = \left(J^{-1}\right)^* \nabla_\xi, \quad (J^{-1})^* \text{ is the matrix transpose to } J^{-1},$$

we obtain the following system of equations.

$$X_t(\xi, t) = -\frac{m}{m-1}\nabla_x u^{m-1} \equiv -\frac{m}{m-1}\left(J^{-1}\right)^* \nabla_\xi U^{m-1}(\xi, t) \tag{35.6}$$

$$\text{in } Q = \Omega(0) \times (0, T),\ T > 0,$$

$$U(\xi, t)\,|J| = U(\xi, 0) \quad \text{in } Q, \tag{35.7}$$

$$X(\xi, 0) = \xi; \quad U(\xi, 0) = u(\xi, 0) \quad \text{in } \Omega(0).$$

System (35.6)–(35.7) contains $N+1$ equations for defining N components of the particle trajectory $X = (X_1, \ldots, X_N)$ and the density U along the trajectory.

Given a solution $(X(\xi, t), U(\xi, t))$ of problem (35.6)–(35.7), it is easy to recover a weak solution to (35.1). Assume that

1) $U(\xi, t)$ is non-negative, bounded and continuous in $\overline{Q}$;
2) $|\nabla_\xi U^{m-1}| \in L^\infty(Q)$;
3) for every $t \in (0, T]$ the mapping $\overline{\Omega(0)} \ni \xi \mapsto X(\xi, t) \in \overline{\Omega(t)}$ is a bijection.

Then we can define

$$x = X(\xi, t), \quad (\xi, t) \in Q; \quad X(\xi, 0) = \xi$$

and set

$$u(x, t) = \begin{cases} U(\xi, t) & \text{if } x = X(\xi, t), \\ 0 & \text{otherwise.} \end{cases}$$

This definition implies that the function $u(x, t)$ is non-negative, bounded and continuous in $\mathbb{R}^N \times [0, T]$, takes on the initial values by continuity and satisfies equation (35.1) in a weak sense. Indeed, given an arbitrary function $\varphi(x,t) \in C^\infty(\mathbb{R}^N \times [0,T])$, vanishing for large $|x|$ and as $t = T$, the following relations hold:

$$
\begin{aligned}
-\int_{\mathbb{R}^N} & u(x,0)\varphi(x,0)dx \\
&= \int_{\Omega(0)} \int_0^T \frac{d}{dt}\left(U|J|\varphi[X(\xi,t),t]\right) dt\, d\xi \\
&= \int_0^T \int_{\Omega(0)} \frac{d}{dt}(U|J|)\,\varphi\, d\xi\, dt + \int_0^T \int_{\Omega(0)} U|J|\frac{d\varphi}{dt} d\xi\, dt \\
&= \int_0^T \int_{\Omega(0)} \left[\varphi_t + \nabla_X \varphi \cdot \frac{dX}{dt}\right] U|J|\, d\xi\, dt \\
&= \int_0^T \int_{\Omega(0)} \left[\varphi_t - \frac{m}{m-1}\nabla_X u^{m-1} \cdot \nabla_X \varphi\right] U|J|\, d\xi\, dt \\
&= \int_{\mathbb{R}^N \times (0,T)} \left(\varphi_t u - \nabla_x \varphi \cdot \nabla_x u^m\right) dx\, dt,
\end{aligned}
$$

where we made use of (35.4) and (35.5). Since the weak solution to the Cauchy problem for (35.1) is already known to be unique, to study its regularity amounts to studying the regularity properties of a solution to (35.6)–(35.7).

The fulfillment of conditions 1)–2) will be provided just by the way of construction of a solution to problem (35.6)–(35.7). To satisfy condition 3) we proceed as follows. First, we shall search for such solutions of problem (35.6)–(35.7) that

$$0 < |J| < \infty \quad \text{in } \overline{Q},$$

which would guarantee local invertibility of the transformation $\xi \mapsto X(\xi,t)$ in a neighborhood of each inner point of Q. Second, we directly check that none of the hypersurfaces $\partial\Omega(t) \cap \{t = \text{const.}\}$ has a point of self-intersection, meaning that the mapping $\partial\Omega(0) \mapsto \partial\Omega(t)$ is a bijection. The verification of this fact relies on certain properties of the constructed solution $X(\xi,t)$ and is postponed until the final stage of the study.

Let us indicate the precise assumptions on the initial data. We use the notation

$$\Omega(t) := \operatorname{supp} u(x,t) \cap \{t = \text{const.}\}, \quad \Omega(0) := \Omega,$$

$$u^{m-1}(\xi,0) := P_0(\xi),$$

$$Q = \Omega \times (0,T], \quad \Gamma \text{ is the parabolic boundary of } Q.$$

The following is assumed:

$$
\text{A.1} \quad \begin{cases} \Omega \subset \mathbb{R}^N \text{ is simply-connected and convex}, \\ \partial\Omega \in C^{p+\alpha}, \quad p \geq 2 \text{ is a given integer, } \alpha \in (0,1]. \end{cases}
$$

$$\text{A.2}\quad \begin{cases} P_0(\xi) \in C^{1+\alpha}(\overline{\Omega}),\ \alpha \in (0,1], \\ P_0 \text{ is strictly positive inside } \Omega \text{ and } P_0 = 0 \text{ on } \partial\Omega, \\ \varepsilon^* \geq \|P_0\|_{C^{1+\alpha}(\Omega)} \geq |\nabla P_0| + |P_0| \geq \varepsilon_* > 0 \quad \text{on } \overline{\Omega}, \\ \varepsilon^*,\ \varepsilon_* \text{ are positive constants.} \end{cases}$$

Note that under conditions A.2 the interface starts moving immediately at each point of $\partial\Omega$ ([36, 33]). It will also be convenient for further purposes to characterize Ω and P_0 in the following terms. Let $\sigma(\xi)$ be the solution of the problem

$$\Delta\sigma = -1 \quad \text{in } \Omega; \quad \sigma = 0 \quad \text{on } \partial\Omega. \tag{35.8}$$

It is strictly positive in Ω and by Hopf's principle possesses the property

$$\left.\frac{\partial\sigma}{\partial n}\right|_{\partial\Omega} < -\gamma, \quad \gamma = \text{const.} > 0.$$

Thus, there exist positive constants K_1, K_2 such that

$$K_1 P_0(\xi) \leq \sigma(\xi) \leq K_2 P_0(\xi) \quad \text{in } \overline{\Omega}. \tag{35.9}$$

Under assumptions A.1, at each point of $\xi_0 \in \partial\Omega$ the inner normal vector $\vec{n}(\xi_0)$ is uniquely defined and there exists the set Ω_* such that if

$$\rho(\xi) := \min_{\xi_0 \in \partial\Omega} |\xi_0 - \xi| \equiv \operatorname{dist}(\xi, \partial\Omega), \quad \xi \in \Omega_*,$$

is assumed at a point $\eta_0 \in \partial\Omega$, then

$$\xi = \eta_0 + \rho\,\vec{n}(\eta_0).$$

Let us subject Ω_* to the restriction

$$|\nabla\sigma| \geq \frac{\gamma}{2} \quad \text{in } \overline{\Omega_*}$$

and denote

$$d_* = \min\{|x - y| :\ x \in \partial\Omega,\ y \in \partial\Omega_* \setminus \partial\Omega\} > 0.$$

Assumption A.2 implies the existence of positive constants C_1, C_2 such that

$$C_1 \leq \max_{\xi \in \Omega} \left(\frac{P_0(\xi)}{\rho(\xi)}\right) \leq C_2. \tag{35.10}$$

Our main result is the following.

Theorem 35.1. *Let* Ω *satisfy* A.1 *and* P_0 *satisfy* A.2. *Then*

a) *there exists* $T > 0$ *depending on* P_0 *and* Ω *such that in the cylinder* $Q = \Omega \times (0, T]$ *problem* (35.6)–(35.7) *has a unique solution* $(X(\xi, t), U(\xi, t))$, *generating the weak solution of* (35.1);

b) *given a point* $x_0 \in \partial\Omega(t_0)$, *there exist a neighborhood*

$$U(x_0, t_0) \subset \partial\Omega(t) \times \{(t_0 - \delta, t_0 + \delta)\},$$

$(\delta > 0,\ t_0 - \delta > 0)$, *a simply-connected open domain* $W \subset \mathbb{R}^{N-1}$, *and functions*

$$\varphi_i(t, y_1, \ldots, y_{N-1}) : W \mapsto U(x_0, t_0), \quad i = 1, \ldots, N,$$

such that

$$x_i = \varphi_i(t, y), \quad y \in W, \quad x \in U(x_0, t_0),$$

$$\varphi_i(t, y) \in C^{\infty}\left((t_0 - \delta, t_0 + \delta);\ C^{p+\alpha}(W)\right).$$

The idea of proof consists in the following. Having constructed a solution to (35.6)–(35.7), we obtain a natural parametrization for the hypersurface $\partial\Omega(t) \times (0, T)$,

$$\overline{\Omega} \ni \xi \mapsto X(\xi, t) = x \in \overline{\Omega(t)}, \quad t > 0,$$

since in the Lagrangian description it is constituted by those particles which were initially located on $\partial\Omega$. Next, since $\partial\Omega$ is assumed to admit parametrization from $C^{p+\alpha}(\mathbb{R}^{N-1})$, we get the desired parametrization for $\partial\Omega(t) \times (0, T)$ as their superposition. So, the proof will be reduced to checking that the vector function $X(\xi, t)$ possesses the required smoothness on $\partial\Omega$.

35.2 Newton's method. The principal difficulties of the further study are similar to those of Sections 31–32. First of all, we have to deal with a system of nonlinear equations. To circumvent this difficulty, we employ Newton's method, thus reducing the nonlinear problem to a sequence of linear ones. Next, the linear problems do not belong to a definite type. Therefore we divide each of them into two independent problems for the potential and solenoidal parts, which can be effectively studied on the basis of the existing theory.

Here the Newton approximations are different from those of Section 32. Let U, V, and $W \subset V$ be Banach spaces, and

$$F(x) : W \mapsto U$$

be a given nonlinear operator. Suppose that $\{0\} \in W \subset V$, and that $F(x)$ has in a ball $B(0) \subset V$ the Fréchet derivative

$$F'(x) : V \mapsto U.$$

Let there exist a sequence of *linear* operators

$$M_n : V \mapsto W$$

such that $F[M_n(x)]$ is defined for $x \in B(0)$.

Lemma 35.1. *Let under the above conditions the operator $F'(x)$ be Lipschitz continuous in the ball $B(0)$,*

$$\|F'(x_1) - F'(x_2)\|_U \le \lambda \, \|x_1 - x_2\|_V, \quad \lambda = \text{const}.$$

Set

$$\mu = \| \left[F'(0)\right]^{-1} \|_V, \quad \kappa = \| \left[F'(0)\right]^{-1} F(0) \|_V,$$

$$\sup_n \|M_n\|_V = \nu, \quad \gamma = \nu \, \mu \, \kappa \, \lambda.$$

If $\gamma < 1/4$ and τ_0 is the minimal root of the equation $\gamma \tau^2 - \tau + 1 = 0$, in the ball $\|x\|_V \le \kappa \tau_0$ the equation $F(x) = 0$ has a unique solution x^*. This solution is the limit of the sequence $\{x_n\}$ defined by the formula

$$x_{n+1} = x_n - \left[F'(x_0)\right]^{-1} (F(M_n[x_n])).$$

Proof. To prove this assertion we adapt the arguments used in [88, Ch. 10, Sec. 3, Th. 1] in the case $M \equiv I$. The proof will consist in checking that the operator

$$Ax = x - \left[F'(0)\right]^{-1} (F(M_n[x]))$$

maps $B(0)$ into itself and is a contraction operator in $B(0)$. Evidently,

$$\begin{aligned}\|Ax\|_V \le \; & \| \left[F'(0)\right]^{-1} \|_V \cdot \|F'(0)(x) - F(M_n[x]) + F(0)\|_U \\ & + \| \left[F'(0)\right]^{-1} F(0)\|_V,\end{aligned}$$

whence

$$\|Ax\|_V \le \mu \, \|F'(0)(x) - F(M_n[x]) + F(0)\|_U + \kappa.$$

Consider the mapping

$$\Phi(x) \equiv F'(0)(x) - F(M_n[x]) + F(0) \; : \; V \mapsto U.$$

It is differentiable and

$$\Phi'(x) = F'(0) - F'(M_n[x])$$

since M is assumed to be linear. Then

$$\begin{aligned}\|\Phi'(x)\|_U & \le \|F'(M_n[x]) - F'(0)\|_U \\ & \le \lambda \|M_n[x]\|_W \le \lambda \, \nu \, \|x\|_V \le \lambda \, \nu \, \kappa \, \tau_0,\end{aligned}$$

whence

$$\|\Phi(x)\|_U = \|\Phi(x) - \Phi(0)\|_U \le \|x\|_V \cdot \|\Phi'(x)\|_U \le \nu\lambda\tau_0^2\kappa^2.$$

If now $\|x\|_V \le \kappa\tau_0$,

$$\|Ax\|_V \le \mu\nu\lambda\tau_0^2\kappa^2 + \kappa = \kappa\left(1 + \mu\nu\lambda\tau_0^2\kappa\right) = \kappa\,\tau_0,$$

which means that A maps the ball $B(0) = \{x \in V : \|x\|_V \le \kappa\tau_0\}$ into itself. Let us check that A is a contraction mapping. We have

$$\begin{aligned} A'(x) &= I - \left[F'(0)\right]^{-1} F'(M_n[x]) \\ &\equiv \left[F'(0)\right]^{-1} \left(F'(0) - F'(M_n[x])\right), \end{aligned}$$

whence

$$\|A'(x)\|_V \le \mu\lambda\kappa\nu\tau_0 \quad \text{in } B(0).$$

The minimal root τ_0 of the equation $\gamma\tau^2 - \tau + 1 = 0$ can be directly calculated through γ. Substituting it into the last inequality, we have

$$\|A'(x)\|_V \le \frac{1}{2}$$

and conclude that A is a contraction mapping:

$$\|Ax\|_V \le \frac{1}{2}\|x\|_V.$$

Hence, A has in $B(0)$ a unique fixed point, the limit of the sequence $\{x_n\}$. □

We shall search for a solution of (35.6)–(35.7) of the form

$$X(\xi, t) = \xi + Y(\xi, t).$$

Substituting (35.7) into (35.6) we get the following system for defining $Y = (Y_1, \dots, Y_N)$:

$$F(Y) \equiv Y_t(I + \nabla^* Y) + \frac{m}{m-1}\nabla\left[\frac{P_0(\xi)}{|I + \nabla^* Y|^{m-1}}\right] = 0 \quad \text{in } Q, \tag{35.11}$$

where I denotes the identity matrix, $\nabla^* Y$ is the matrix with components $\partial Y_i/\partial\xi_j$. (Unless specially indicated, from now on all differential operators are assumed acting in the variables (ξ, t).) The initial conditions for Y have the form

$$Y(\xi, 0) = 0 \quad \text{in } \Omega. \tag{35.12}$$

Instead of the boundary conditions on $\partial\Omega \times [0, T]$ we pose the condition which will guarantee the local invertibility of the coordinate transformation inside Q:

$$0 < |I + \nabla^* Y| < \infty \quad \text{in } \overline{Q}. \tag{35.13}$$

Let us calculate the derivative $F'(x)$. First of all, note that the condition of Lipschitz continuity of the operator $F'(x)$ in $B(0)$ makes the Gateaux and Fréchet derivatives equal [88, Ch. X, Sec. 1,Th. 1]. Thus, we have to calculate

$$F'(0)(Y) = \lim_{\varepsilon\to 0} \frac{F(\varepsilon Y) - F(0)}{\varepsilon}.$$

With the use of Newton's formulas (see Section 32) we get: if $\lambda = 1$ and $(Y_i)_{\xi_j}$ are bounded,

$$|I + \nabla^*(\varepsilon Y)| = 1 + \varepsilon \operatorname{div} Y + o(\varepsilon),$$

$$|I + \nabla^*(\varepsilon Y)|^{1-m} = 1 - \varepsilon(m-1) \operatorname{div} Y + o(\varepsilon).$$

Hence,

$$\begin{cases} F'(0)(Y) \equiv Y_t - m\nabla\left[P_0(\xi)\operatorname{div} Y\right], & (\xi, t) \in Q, \\ Y(\xi, 0) = 0 \quad \text{in } \Omega. \end{cases} \tag{35.14}$$

To make use of Lemma 35.1 and to get a solution of problem (35.6)–(35.7) as the limit of the sequence

$$Y_{n+1} = Y_n - \left[F'(0)\right]^{-1}\left(F\left(M_n\left[Y_n\right]\right)\right) \tag{35.15}$$

amounts to

1) estimating $\|[F'(0)]^{-1}\|_V$, that is, to solve the problem

$$F'(0)\left(Y_{n+1} - Y_n\right) = \varphi \in U;$$

and then to estimate

$$\sup_{\varphi,\ \|\varphi\|_U \neq 0} \frac{\|Y_{n+1} - Y_n\|_V}{\|\varphi\|_U};$$

2) estimating $\|s\|_V$, where s is the solution of the problem

$$F'(0)(s) = F(0);$$

3) verifying the Lipschitz continuity of the operator $F'(x)$.

Once the function spaces U, V, W are introduced, the verification of condition 3) becomes merely a technical question. As for conditions 1)–2), their verification constitutes the major part of all further considerations.

One can pose at once a condition which would provide the fulfillment of restriction (35.13). Once again using Newton's formulas, we get: to fulfill (35.13) it suffices to claim that $|D_{\xi_i}(M_n[Y_n])_j|$, $i, j = 1, \dots, N$, are small in $\overline{Q}$. More precisely, we shall choose M_n so as to provide

$$\frac{1}{2} < |I + \nabla^* M_n[Y_n]| < \frac{3}{2}. \tag{35.16}$$

35.3 The linear model. Here we divide problem (35.14) into two independent linear problems for the potential and curl components of the sought vector Y_{n+1}.

1. *Decomposition of vector space* $L^2(\Omega)$. Let $\Omega \subset \mathbb{R}^N$ be an open bounded domain. Denote by $\boldsymbol{L}^2(\Omega)$ the Hilbert space of vector-valued functions

$$\vec{v} = (v_1(x), \dots, v_N(x)), \quad v_i(x) \in L^2(\Omega),$$

with the scalar product

$$(\vec{v}, \vec{u}) = \sum_i \int_\Omega v_i u_i \, dx.$$

As is well known (we refer to [96, Ch. 1, Sec. 2]), $\boldsymbol{L}^2(\Omega)$ admits the representation

$$\boldsymbol{L}^2(\Omega) = G(\Omega) \oplus J^0(\Omega)$$

as the direct sum of two orthogonal (in the sense of scalar product in $\boldsymbol{L}^2(\Omega)$) subspaces $G(\Omega)$ and $J^0(\Omega)$. $G(\Omega)$ is constituted by the vectors $\nabla\varphi$, with φ being such that

$$\varphi \in L^2_{\text{loc}}(\Omega), \quad \frac{\partial \varphi}{\partial x_i} \in L^2(\Omega), \quad i = 1, \dots, N.$$

$J^0(\Omega)$ is the closure in the norm of $\boldsymbol{L}^2(\Omega)$ of the set of solenoidal vectors, finitely supported in Ω. Given a sufficiently smooth vector field $\vec{v}$, the functions φ, $\vec{\psi}$ realizing the desired decomposition

$$\vec{v} = \nabla\varphi + \vec{\psi}, \quad \nabla\varphi \in G(\Omega), \quad \vec{\psi} \in J^0(\Omega),$$

can be found as follows: φ is defined as the solution of the problem

$$\Delta\varphi = \operatorname{div}\vec{v} \quad \text{in } \Omega, \quad \left.\frac{\partial \varphi}{\partial n}\right|_{\partial\Omega} = \vec{v}\cdot\vec{n}|_{\partial\Omega}, \tag{35.17}$$

and

$$\vec{\psi} = \vec{v} - \nabla\varphi, \tag{35.18}$$

where $\vec{n}$ denotes the unit outer normal vector to $\partial\Omega$. As is known ([98, Ch. 3, Sec. 3]), problem (35.17) is uniquely solvable in $C^{2+\gamma}(\overline{\Omega})$, $\gamma \in (0,1)$, if $\operatorname{div}\vec{v} \in C^\gamma(\overline{\Omega})$ and $\vec{v}\cdot\vec{n}|_{\partial\Omega} \in C^{1+\gamma}(\partial\Omega)$. Let φ be defined in this way. Given an arbitrary function $\nabla g \in \boldsymbol{L}^2(\Omega)$, it is easy to verify

$$(\vec{\psi}, \nabla g) = \int_\Omega \vec{\psi}\,\nabla g\,dx = \int_{\partial\Omega} g\,(\vec{\psi}\cdot\vec{n})\,d\Gamma - \int_\Omega g \operatorname{div}\vec{\psi}\,dx, \tag{35.19}$$

where Γ is the square element of $\partial\Omega$. By (35.17), $\operatorname{div}\vec{\psi} = 0$ providing that the second term of the right-hand side of (35.19) is zero. It follows from (35.18) and the boundary conditions in (35.17) that

$$\left.\vec{\psi}\cdot\vec{n}\right|_{\partial\Omega} = \vec{v}\cdot\vec{n}|_{\partial\Omega} - \left.\frac{\partial\varphi}{\partial n}\right|_{\partial\Omega} = 0,$$

whence the first term in the left-hand side of (35.19) is also zero. Hence, $\psi \perp \nabla g$ and the desired decomposition is completed.

If $v(\xi, t)$ also depends on t, we perform this decomposition considering v as depending on t like a parameter.

2. *Representation of the principal operator.* Let us consider problem (35.15) of finding the 'error' vector $s = Y_{n+1} - Y_n$. Rewrite this problem in the form

$$F'(0)(s) = F(M_n[Y_n]) \tag{35.20}$$

and then divide $F(M_n[Y_n])$ into potential and curl parts: $F(M_n[Y_n]) = \nabla\varphi + \vec{\psi}$. Making use of the structure of F, one can rewrite this representation in the form

$$\begin{aligned}(M_n[Y_n])_t\,&\left(I + \nabla^*(M_n[Y_n])\right)\\ &= \vec{\psi} + \nabla\left[\varphi - \frac{m}{m-1}\,\frac{P_0(\xi)}{|I + \nabla^*(M_n[Y_n])|^{m-1}}\right].\end{aligned} \tag{35.21}$$

Thus, it is required to decompose only the left-hand side of (35.21). One has to find functions φ and $\vec{\psi}$ such that

$$\Phi = \varphi - \frac{m}{m-1}\,\frac{P_0(\xi)}{|I + \nabla^*(M_n[Y_n])|^{m-1}},$$

$$\Delta\,\Phi = \operatorname{div}\left((M_n[Y_n])_t\left(I + \nabla^*(M_n[Y_n])\right)\right) \quad \text{in } \Omega, \quad t > 0, \tag{35.22}$$

$$\left.\frac{\partial\Phi}{\partial n}\right|_{\partial\Omega} = \vec{n}\cdot\left[(M_n[Y_n])_t\left(I + \nabla^*(M_n[Y_n])\right)\right]\Big|_{\partial\Omega},$$

$$\vec{\psi} = (M_n[Y_n])_t\left(I + \nabla^*(M_n[Y_n])\right) - \nabla\Phi. \tag{35.23}$$

Let us seek the function s of the form

$$s = \nabla v + \vec{z}, \quad \nabla v \in G(\Omega), \quad \vec{z} \in J^0(\Omega).$$

Substituting this function into (35.14) and separating the potential part, we get the following *parabolic* problem for defining $v(\xi, t)$:

$$Lv \equiv v_t - mP_0(\xi)\Delta\, v = f(\xi, t) \text{ in } Q\,, \quad v = 0 \text{ on } \Gamma \tag{35.24}$$

with the right-hand side

$$f(\xi, t) \equiv \Phi + \frac{m}{m-1}\,\frac{P_0}{|I + \nabla^*(M_n[Y_n])|^{m-1}}.$$

The boundary condition is still our constraint. It does not follow from (35.14) and we impose it to simplify the further proceedings. Meanwhile, this condition is motivated by condition (35.13): in view of (35.13), it is natural to expect that the second-order space derivatives of v are bounded in $\overline{Q}$; the boundary condition in (35.24) then follows by virtue of the equation. For the curl part $\vec{z}$ we get the Cauchy problem

$$\vec{z}_t = \vec{\psi} \quad \text{in } (0, T), \quad \vec{z}|_{t=0} = 0, \tag{35.25}$$

where $\xi \in \Omega$ enters as a parameter.

3. *Function spaces. The choice of operators M_n.* Resolution of problem (35.24), (35.25) constitutes only one step in the iteration process. Since the right-hand sides of these problems are generated, in their turn, by solutions of the same problems, one has to be sure that the operators $F(\cdot)$ and $\left[F'(0)\right]^{-1}(\cdot)$ are compatible. Under ideal circumstances, these operators should be defined on the same pair of function spaces. Unfortunately, this is not the case here. For instance, let us consider problem (35.25). Its solution is given explicitly by the formula

$$\vec{z}(\xi, t) = \int_0^t \vec{\psi}(\xi, \tau)\, d\tau \quad \text{in } (0, T) \tag{35.26}$$

and $\vec{z}$ is understood as depending on ξ as a parameter. According to (35.23), (35.25) the smoothness of $\vec{z}$ is the same as that of the function $\vec{\psi}$ and we have to conclude that the smoothness of $\vec{z}$ in ξ is, generally speaking, two orders lower than that of the function $\vec{z}$ found at the previous step. Thus, any initial smoothness will be exhausted in a finite number of iterations. To compensate for this eventual loss of regularity we consider the approximation scheme (35.15) where M_n are linear operators increasing the regularity of Y_n.

Let us specify the choice of the spaces U, V, W. Set

$$\begin{aligned} W \equiv & \left\{Y = \nabla\varphi : \varphi, \varphi_t \in C^{3+\alpha,(3+\alpha)/2}\left(\overline{Q}\right), \ \varphi = 0 \text{ on } \Gamma\right\} \\ & \oplus \left\{z \in J^0(\Omega) \cap \left(C^{2+\alpha,(2+\alpha)/2}(\overline{Q})\right)^N\right\}, \end{aligned}$$

with α from condition A.2,

$$\begin{aligned} V \equiv & \left\{Y = \nabla\varphi : \varphi \in C^{3+\beta,(3+\beta)/2}\left(\overline{Q}\right), \ \beta \in (0, 1], \ \varphi = 0 \text{ on } \Gamma\right\} \\ & \oplus \left\{z \in J^0(\Omega) \cap \left(C^{\alpha,\alpha/2}(\overline{Q})\right)^N\right\}, \\ U \equiv & \left\{Y = (Y_1, \dots, Y_N) : Y_i, (Y_i)_t \in C^{\alpha,\alpha/2}(\overline{Q})\right\}. \end{aligned}$$

Of course, the choice of these spaces in not entirely arbitrary. At the very first step of the iteration process one has to solve problem (35.24) with $f \equiv \frac{m}{m-1} P_0$ and the assumptions made on P_0 indicate what the space V should be. A further analysis of this problem shows that the exponent β in the definition of V depends on α, but they need not coincide and any estimation of β through α fails.

Let us proceed to choose the operators M_n. Given a vector field $\vec{y} \in V$,

$$\vec{y} = \nabla q + \vec{y_2}, \quad q = 0, \; \vec{y_2} = 0 \text{ on } \Gamma,$$

set

$$M[\vec{y}(\xi, t)] = P_0(\xi) \left\{ \nabla \int_0^t \left[\mathcal{L}^{-1} q(\xi, \tau) \right] d\tau + \int_0^t \left[\mathcal{L}^{-1} \vec{y_2}(\xi, \tau) \right] d\tau \right\},$$

and

$$M_n[\vec{y}(\xi, t)] = \underbrace{\int_0^t \int_0^{t_1} \cdots \int_0^{t_{n-1}}}_{n} M[\vec{y}(\xi, t_{n-2})] \, dt_{n-1} \ldots d\tau,$$

where $\mathcal{L}^{-1}$ is the linear operator assigning to a given function $\vec{\psi}$ the solution of the problem

$$\mathcal{L}\vec{\sigma} \equiv \vec{\sigma}_t - \Delta \vec{\sigma} = \vec{\psi} \quad \text{in } Q; \quad \vec{\sigma} = 0 \text{ on } \Gamma. \tag{35.27}$$

Lemma 35.2 ([98, Ch. 4, Sec. 5, Th. 2]). *Given a vector*

$$\vec{\psi}, \quad \psi_i \in C^{l+\delta, (l+\delta/2)} \left(\overline{Q} \right), \quad \delta \in (0, 1],$$

l an integer, $\psi_i(\xi, 0) = 0$, satisfying the compatibility conditions of l-th order, there exists a unique solution $\vec{\sigma}$ of problem (35.27). *Moreover,*

$$\|\sigma_i\|_{C^{l+2+\delta, (l+2+\delta)/2}\left(\overline{Q}\right)} \leq c \|\psi_i\|_{C^{l+\delta, (l+\delta)/2}\left(\overline{Q}\right)}, \quad i = 1, \ldots, N, \tag{35.28}$$

with a finite constant $c > 0$ depending only on l, δ, N, and the geometrical properties of $\partial\Omega \in C^{l+2}$.

Collecting the definitions of M, M_n, and Lemma 35.2, we have

$$\|M_n[y]\|_W \leq \frac{T^{n-1}}{n!} \|M[y]\|_W, \quad \|M[y]\|_W \leq c\, T \max_{\xi \in \Omega} P_0(\xi) \, \|y\|_V. \tag{35.29}$$

Thus, if under the conditions of Lemma 35.1 the constants λ, μ, κ are finite, the required condition on γ is provided by the choice of T. Also, (35.29) certainly implies the fulfillment of (35.16). It is essential that if the simplest approximation procedure with $M_n \equiv M$ gives a solution to problem (35.6)–(35.7), so do all approximation procedures with the above-defined M_n.

Now we may come back to problem (35.22). Suppose that $Y_n \in B(0) \subset V$. Denote by $\Phi(\xi, t)$ the right-hand side of (35.22). Recall that in condition A.1 $p \geq 2$. Then Lemma 35.2 is applicable and the regularity of the curl part $\vec{z}$ of Y_n can be improved as follows:

$$z_i, \; (z_t)_i \in C^{2+\alpha, (2+\alpha)/2} \left(\overline{Q} \right), \quad i = 1, \ldots, N.$$

Due to the definition of V, the potential part ∇v of Y_n is such that

$$\nabla v, \; \nabla v_t \in C^{2+\beta, (2+\beta)/2} \left(\overline{Q} \right).$$

This implies

$$\Phi,\ \Phi_t\ \in\ C^{1+\gamma,(2+\gamma)/2}\left(\overline{Q}\right),\quad \gamma=\min\{\alpha;\beta\},$$

and the norms of Φ, Φ_t depend on the norm $\|Y_n\|_V$ and N. Hence, regardless of any connection between α, β, the functions Φ, Φ_t are bounded in the norm of $C^{\alpha,(1+\alpha)/2}\left(\overline{Q}\right)$, and by [98, Ch. 3, Sec. 1, Th. 1] problem (35.22) has a unique classical solution φ such that

$$\nabla\varphi,\nabla\varphi_t\ \in\ C^{\alpha,(1+\alpha)/2}\left(\overline{Q}\right).$$

Since condition (35.16) is fulfilled by the choice of $B(0)$, due to condition A.2 we have $f,\ f_t\in C^{1+\alpha,(1+\alpha)/2}\left(\overline{Q}\right)$. By (35.23) the corresponding regularity is assigned to the curl part of $F(M_n[Y_n])$. Note that although the choice of M is at our disposal, the regularity of f in (35.24) is also controlled by P_0.

35.4 The potential part. Here we construct the solution to problem (35.24). It is assumed that $f(\xi,t)$ possesses the properties:

$$f,\ f_t\ \in\ C^{1+\alpha,(1+\alpha)/2}\left(\overline{Q}\right);\quad f(\xi,0)=\frac{m}{m-1}P_0\ \text{in}\ \overline{\Omega};\tag{35.30}$$

$$\max_{\xi\in\overline{Q}}\left\{\frac{|D_t^s f|}{P_0(\xi)};\ |D_t^s(\nabla f)|\right\}\le M_0,\quad s=0,1;\quad M_0=\text{const.}>0.\tag{35.31}$$

We use the standard regularization procedure. Consider the sequence $\{v_\varepsilon\}$ of classical solutions to the problems

$$\begin{cases} Lv_\varepsilon\equiv v_{\varepsilon t}-m(P_0+\varepsilon)\Delta\, v_\varepsilon=f+\varepsilon & \text{in } Q,\\ v_\varepsilon=0 \quad \text{on } \Gamma. \end{cases}\tag{35.32}$$

For each $\varepsilon>0$ problem (35.32) is nondegenerate parabolic and always has a unique classical solution $v_\varepsilon\in C^{3+\alpha,(3+\alpha)/2}\left(\overline{Q}\right)$ ([97, Ch. 4, Sec. 5, Th. 2]).

Lemma 35.3. *Under assumption* (35.31) *with* $s=0$

$$|v_\varepsilon|\le L_0\,t\,(P_0+\varepsilon)\quad in\quad \overline{Q}\tag{35.33}$$

with a finite constant $L_0>0$ *not depending on* ε.

Proof. To simplify the notation we omit here the subindex ε. Due to the linearity of equation (35.32), to get (35.33) it suffices to establish only the upper estimate. Consider the function $\omega(\xi,t)=\delta t\sigma(\xi)$, where $\delta=\text{const.}>0$ is to be defined, and $\sigma(\xi)$ is the solution of problem (35.8). Set $\omega_\varepsilon:=\delta t(\sigma(\xi)+\varepsilon)$. Evidently, $\omega_\varepsilon\ge v$ on Γ . Next,

$$L\omega_\varepsilon\equiv\delta(\varepsilon+\sigma)+m(P_0+\varepsilon)\delta t\ge\delta K_1P_0+\delta\varepsilon\ >\ M_0P_0\quad \text{in } Q,$$

provided δ is chosen large enough. The assertion now follows by the maximum principle. □

Lemma 35.4. *Each of v_ε possesses the estimate*

$$\max_{\overline{Q}} |\nabla v_\varepsilon| \le l_0 \left(L_0 t \max(C_2, 1) + \max_{\overline{Q}} |f| \right) \tag{35.34}$$

with an absolute constant $l_0 > 0$ not depending on ε; the constant L_0 is taken from (35.33) *and C_2 is from condition* (35.10).

Proof. Fix an arbitrary point $(\xi_0, t_0) \in Q$ and denote

$$\rho = \operatorname{dist}(\xi_0, \partial\Omega).$$

Consider the cylinder

$$G_\rho = \{(\xi, t) \in Q : |\xi - \xi_0| < \rho + \varepsilon, \ t < t_0\}.$$

Introduce the new independent variables (ζ, τ) and the new function $V(\zeta, \tau)$ by the formulas

$$\xi = \xi_0 + (\rho + \varepsilon)\zeta, \quad t = t_0 + (\rho + \varepsilon)\tau, \quad V(\zeta, \tau) = v_\varepsilon(\xi, t).$$

This change of variables renders G_ρ the cylinder

$$S = \left\{ (\zeta, \tau) : |\zeta| < 1, \ 0 > \tau > -\frac{t_0}{\rho + \varepsilon}, \ \xi_0 + (\rho + \varepsilon)\zeta \in \Omega \right\}.$$

The definition of ζ and τ gives

$$\frac{\partial}{\partial t} = \frac{1}{\rho + \varepsilon} \frac{\partial}{\partial \tau}, \quad \frac{\partial}{\partial \xi_i} = \frac{1}{\rho + \varepsilon} \frac{\partial}{\partial \zeta_i}, \quad i = 1, \ldots, N,$$

and the conditions for $V(\zeta, \tau)$ take the form:

$$V_\tau - m \frac{\varepsilon + P_0(\xi)}{\rho + \varepsilon} \Delta_\zeta V = F(\zeta, \tau) \quad (\equiv (\rho + \varepsilon) f(\xi, t)) \quad \text{in } S,$$

$$|V| \le L_0 t P_0(\xi_0 + (\rho + \varepsilon)\zeta) \quad \text{on the lateral boundary of } S,$$

$$V(\zeta, -\frac{t_0}{\rho + \varepsilon}) = 0.$$

It will be convenient to introduce the new functions

$$W(\zeta, \tau) := q \frac{V(\zeta, \tau)}{\rho + \varepsilon}, \quad \Phi(\zeta, \tau) := q \frac{F(\zeta, \tau)}{\rho + \varepsilon} \equiv qf$$

with the factor

$$\frac{1}{q} = \max_Q \left| \frac{v_\varepsilon}{\rho + \varepsilon} \right| + \max_Q |f|.$$

This definition yields

$$\begin{cases} W_\tau - a(\zeta)\Delta_\zeta W = \Phi(\zeta, \tau) & \text{in } S, \\ |W| \le 1 & \text{on the parabolic boundary of } S, \end{cases} \tag{35.35}$$

with

$$a(\zeta) = m\frac{P_0(\xi) + \varepsilon}{\rho + \varepsilon}, \qquad |\Phi| \le 1 \text{ in } \overline{S}.$$

By (35.10) this is a uniformly parabolic equation. Since W is a classical solution of problem (35.35), we can apply to it the result of nondegenerate parabolic theory, [97, Ch. 5, Sec. 3, Th. 1], to conclude:

$$|\nabla W(0, 0)| \le K, \quad K = \text{const.} > 0,$$

where K already depends only on $\max a$ and $\min a$ in S, that is, on the constants C_1, C_2 in (35.10). The desired estimate

$$|\nabla v_\varepsilon(\xi_0, t_0)| \le K\left(\max_Q\left(\frac{v_\varepsilon}{\rho + \varepsilon}\right) + \max_Q |f|\right)$$

now follows by returning to the original function v_ε. □

As follows from [93, 62], since v_ε are Lipschitz equicontinuos in ξ, they are Hölder equicontinuous in t. Hence, one can extract from the sequence $\{v_\varepsilon\}$ a subsequence (which will be assumed to coincide with the whole sequence), converging in $C\left(\overline{Q}\right)$ to a function $v(\xi, t)$. We are going to prove now that this function is the sought classical solution of problem (35.24).

Lemma 35.5. *$v(\xi, t)$ is the unique classical solution of problem* (35.24).

Proof. The uniqueness is an immediate byproduct of the linearity of (35.24). Let us verify that $v(\xi, t)$ is a weak solution of (35.24). Since $v_\varepsilon \to v$ and $|\nabla v_\varepsilon|$ are bounded in Ω uniformly in ε and t, $\nabla v_\varepsilon \to \nabla v$ weakly in $L^2(Q)$. Let us multiply equation (35.32) by an arbitrary function $h \in C_0^\infty(Q)$, $h(\xi, T) = 0$, and integrate the result by parts:

$$\int_Q \{h_t v_\varepsilon - mh\nabla P_0 \nabla v_\varepsilon - m(P_0 + \varepsilon)\nabla h \nabla v_\varepsilon + h(f + \varepsilon)\}\, d\xi dt = 0. \tag{35.36}$$

The passage to the limit as $\varepsilon \to 0$ in (35.36) gives the integral identity satisfied by the limit function v. To verify that the weak solution of (35.24) $v(\xi, t)$ is also a classical solution amounts to observing that on each subdomain $Q' \subset Q$, separated from Γ on a positive distance, equation (35.24) is nondegenerate parabolic and the smoothness of its solution v can be improved in accordance with results of the general theory [97]. □

Theorem 35.2. *Assume that* Ω *and* P_0 *satisfy conditions* A.1–A.2 *and let conditions* (35.30)–(35.31) *be fulfilled. There exists a unique classical solution* $v(\xi, t)$ *of problem* (35.24) *such that everywhere in* Q

$$\begin{cases} \text{a)}\ \ |v| \le k_1 t\sigma(\xi), & \text{b)}\ \ |\nabla v| \le k_2(1+t), \\ \text{c)}\ \ |v_t| \le k_3(1+t)\sigma(\xi), & \text{d)}\ \ |\Delta\, v| \le k_4(1+t), \\ \text{e)}\ \ \|\nabla(v_t)\|_{C^{\beta,\beta/2}(Q)} \le k_5, & \text{f)}\ \ P_0 \sum_{\delta=3} \|D_\xi^\delta v\|_{C^{\beta,\beta/2}(Q)} \le k_6, \end{cases} \tag{35.37}$$

with $k_1, \dots, k_6$, $\beta \in (0, 1]$, *finite constants depending only on* P_0, α, *and linearly on the constant* M_0 *in* (35.31); $\sigma(\xi)$ *is the solution of* (35.8).

Proof. Estimates (35.37) a)–b) follow from Lemmas 35.3, 35.4. Consider the problem for the function $s := v_t$:

$$\begin{cases} s_t - mP_0\Delta\, s = f_t \quad \text{in } Q, \\ s|_{\partial Q} = 0; \\ s|_{t=0} = v_t|_{t=0} = mP_0\Delta\ v|_{t=0} + f|_{t=0} = f(\xi, 0) \equiv \frac{m}{m-1}P_0(\xi) \quad \text{in } \Omega. \end{cases}$$

The pointwise estimate for v_t holds

$$|v_t| \le C_4(1+t)\sigma(\xi) \quad \text{in } \overline{Q},$$

which readily follows by comparison with the function $C_4(1+t)\sigma(\xi)$, where $\sigma(\xi)$ is the solution to (35.8), C_4 is an appropriately chosen constant (cf. the proof of Lemma 35.3). This estimate provides (35.37) d) by virtue of equation (35.24). To get (35.37) e) we can repeat the rescaling procedure of Lemma 35.4. By [97, Ch. 5, Sec. 3, Th. 1] we then get

$$\|s\|_{C^{1+\beta,(1+\beta)/2}(Q)} \le k_5, \quad t \ge 0,$$

with constant k_5, $\beta \in (0, 1]$ depending on $\|P_0\|_{C^{1+\alpha}(\Omega)}$, $\max_Q |f|$, Ω, and N. Moreover, the same trick that was used in the proof of Lemma 35.4 shows that the dependence of k_5 on $\|P_0\|$ and $\max\, |f|$ is linear. □

35.5 Solution to the nonlinear problem. One can check now that the conditions of Lemma 35.1 are fulfilled and thus establish existence of a solution to problem (35.14). First of all, let us recall that having chosen T appropriately small, estimate (35.29) guarantees local invertibility of the coordinate transformation. Let us verify that operator $F'(0)$ is Lipschitz continuous in a ball $B(0) \subset V$. Given a vector $x \in W$, one has to calculate

$$F'(x)(Y) = \lim_{\varepsilon \to 0} \frac{F(x + \varepsilon\, Y) - F(x)}{\varepsilon}.$$

With the use of the formula

$$|A + \varepsilon\, B| = |A| \cdot |I + \varepsilon\, A^{-1} \cdot B| = |A| \cdot \left(1 + \varepsilon \operatorname{trace}\left(A^{-1} \cdot B\right) + o(\varepsilon)\right),$$

A, B are matrixes $N \times N$, it is easy to calculate

$$F'(x)(Y) = Y_t\left(I + \nabla^* x\right) + x_t \nabla^* Y - m\nabla\left[\frac{P_0(\xi)}{|I + \nabla^* x|^{m-1}} \operatorname{trace}\left((I + \nabla^* x)^{-1} Y\right)\right].$$

Then we derive, applying the mean value theorem,

$$\begin{aligned} F'(x_1)(Y) - F'(x_2)(Y) &= Y_t \nabla^*(x_1 - x_2) + (x_1 - x_2)_t \nabla^* Y \\ &+ m(m-1)\nabla\Big[P_0(\xi) \operatorname{trace}\left((I + \nabla^* x_1)^{-1} Y\right) \\ &\times \int_0^1 \{\theta|I + \nabla^* x_1| + (1-\theta)|I + \nabla^* x_2|\}^{-m}\, d\theta\, \nabla^*(x_1 - x_2)\Big] \\ &- m\nabla\left[\frac{P_0(\xi)}{|I + \nabla^* x_2|^{m-1}} \operatorname{trace}\left((I + \nabla^* x_1)^{-1} Y - (I + \nabla^* x_2)^{-1} Y\right)\right]. \end{aligned}$$

If $x_1, x_2, Y \in B(0)$, then the first three terms of the right-hand side of this equality need no further estimating. To estimate the last term amounts to evaluating

$$\Delta_{ij} \equiv (I + \nabla^* x_1)^{-1}{}_{ij} - (I + \nabla^* x_2)^{-1}{}_{ij}, \quad i, j = 1, \ldots, N.$$

The components of the inverse matrices admit the representations

$$(I + \nabla^* x_k)^{-1}{}_{ij} = \frac{B_{ij}^{(k)}}{|B^{(k)}|}, \quad k = 1, 2,$$

where $B^{(k)} = (I + \nabla^* x_k)^{-1}$, $B_{ij}^{(k)}$ is the algebraic completion of the element of $(B_{ij}^k)^{-1}$. Consider the matrix $C(\theta)$ with elements

$$c_{ij}(\theta) = \theta(I + \nabla^* x_1) + (1-\theta)(I + \nabla^* x_2)$$

and denote by $C_{ij}(\theta)$ the algebraic completion of $c_{ij}(\theta)$. By the mean value theorem

$$\begin{aligned} \Delta_{ij} &= \int_0^1 \frac{d}{d\theta}\left[\frac{C_{ij}(\theta)}{|C(\theta)|}\right] d\theta \\ &= \sum_{kl} \int_0^1 \frac{\partial}{\partial c_{kl}(\theta)}\left(\frac{C_{ij}(\theta)}{|C(\theta)|}\right) d\theta\, \left(\nabla^*(x_1 - x_2)\right)_{kl} \end{aligned}$$

and the required estimate follows:

$$\|F'(x_1) - F'(x_2)\|_U = \sup_{\|y\|_V \neq 0} \frac{\|F'(x_1)(y) - F'(x_2)(y)\|_U}{\|y\|_V} \leq \lambda \|x_1 - x_2\|_V$$

with λ depending on N and the radius of $B(0)$.

Let us estimate $\|s\|_V$, where s is the solution of the problem $F'(0)(s) = F(0)$. Since $F(0) = \frac{m}{m-1} \nabla P_0$, we get at once that $\vec{z} \equiv 0$ in the representation $s = \nabla v + \vec{z}$, ($\nabla v \perp \vec{z}$ in $L^2(\Omega)$). The potential part ∇v is estimated by Theorem 35.2, whence

$$\|s\|_V = \|\nabla v\|_V \le K\,(1+t), \quad K = \text{const.} > 0.$$

Given a function $g \in U$, the solution s of the problem $F'(0)s = g$ has the form $s = \nabla v + z$, $\nabla v \in G(\Omega)$, $z \in J^0(\Omega)$. The estimates for the potential part of s follow from Theorem 35.2 and for the curl part z from representation (35.26), whence

$$\| \left[F'(0)\right]^{-1} g\|_V \le \|\nabla v\|_V + \|z\|_V \le L\,\|g\|_V.$$

with $L > 0$ a constant depending on the constants in (35.37) and T.

35.6 Regularity of the solution. We are now in a position to make certain conclusions about the regularity of the constructed solution to problem (35.6) $X(\xi, t) \equiv \xi + Y(\xi, t)$, where

$$Y = \lim_{n\to\infty} Y_n, \quad Y_{n+1} = Y_n - [F'(0)]^{-1}\left(F\left(M_n[Y_n]\right)\right).$$

At each step n, $Y_n - Y_{n-1} = \nabla v_n + \vec{z}_n$, where v_n, $\vec{z}_n$, are the solutions of problems (35.24), (35.25), respectively. It is easy to see that the derivatives $D_t(Y_n - Y_{n-1})$ have finite limits as $\xi \to \partial\Omega$ for each $t > 0$. By the definitions of V and its subspace $J^0(\Omega)$, $\vec{z}_n \equiv 0$ on $\partial\Omega$, and $\nabla(v_n)_t$ have finite limits at each point $\xi \in \partial\Omega$, $t > 0$ by virtue of (35.37) e).

By assumption A.1, given a point $\xi_0 \in \partial\Omega$, there exist a neighborhood $V(\xi_0) \subset \partial\Omega$, an open bounded domain $W \subset \mathbb{R}^{N-1}$, and functions

$$\psi_i\,(x_1, \dots, x_{N-1}) : W \mapsto V(\xi_0), \quad \psi_i(x) \in C^p(W), \quad i = 1, \dots, N,$$

defining a bijective mapping $W \mapsto V(\xi_0)$. This allows one to perform the local change of variables

$$V(\xi_0) \ni \xi \mapsto x \in W.$$

Given a function $v(\xi, t)$, $u = 0$ on $\partial\Omega$, we introduce the function

$$v(\xi + y, t) \equiv U(x, y, t), \quad \xi \in \partial\Omega, \quad y = \rho \cdot \vec{n}(\xi).$$

The following assertion is an immediate result of application to $v(\xi, t)$ of the rule of differentiation of a superposition.

Lemma 35.6. *At each point $\xi_0 \in \partial\Omega$, $t_0 > 0$, the first p derivatives of $v_n(\xi, t)$ in the tangential directions $D_\tau^p v_n$ exist and equal zero. Moreover, for each $k = 0, 1, 2, \dots$, $D_t^k\left(D_\tau^p v_n\right)(\xi_0, t_0) = 0$,*

$$\left|D_t^k\left(D_\tau^p v_n\right)(\xi_0, t_0)\right|^{(\alpha)}_{V(\xi_0)}$$

$$\equiv \sup_{\substack{\xi \ne \xi_0 \\ \xi \in V(\xi_0)}} \frac{\left|D_t^k\left(D_\tau^p v_n\right)(\xi_0, t_0) - D_t^k\left(D_\tau^p v_n\right)(\xi, t_0)\right|}{|\xi - \xi_0|^\alpha} = 0. \quad (35.38)$$

The same is true for the limiting hypersurface $\xi + Y(\xi, t)$, $\xi \in \partial\Omega$. Moreover, for each natural number k and $0 \le l \le p$,

$$D_t^k\left(D_\tau^l v_n\right) = 0, \quad D_t^k\left(D_\tau^l v\right) = 0 \quad \text{on } \partial\Omega, \quad \left|D_t^k\left(D_\tau^l v\right)\right|_{\partial\Omega}^{(\alpha)} = 0.$$

An immediate byproduct of Lemma 35.6 is: all hypersurfaces $Y_k(\xi, t)$, $\xi \in \partial\Omega$, obey the equations

$$(Y_k(\xi, t) - Y_{k-1}(\xi, t)) = \nabla(v_{kt}), \quad (\xi, t) \in \partial\Omega \times (0, T), \tag{35.39}$$

where $\nabla v_{kt} \cdot \vec{\tau} = 0$, and $\vec{\tau}$ is an arbitrary vector belonging to the plane tangential to $\partial\Omega$ at the point ξ.

To complete the proof of Theorem 35.1 one has to verify, first, that the constructed solution of problem (35.6)–(35.7) indeed generates the solution of the original Cauchy problem for equation (35.1). As follows from considerations in 35.2, this amounts to ensuring that none of $\partial\Omega(t)$, $t \in [0, T]$, has a point of self-intersection. This property is implied by (35.39) and the assumption about the convexity of Ω. Indeed, by (35.39) each of the particles located on $\partial\Omega(t)$ advances in the direction $\vec{n}$, therefore their trajectories never intersect.

The proof of Theorem 35.1 is thus reduced to checking the differentiability of $D_t Y_k$ in t on $\partial\Omega$. As follows from (35.39), this amounts to verifying the differentiability of $\vec{n} \cdot \nabla(v_k)$ on $\partial\Omega$.

The definition of operators M_n implies the following properties of the right-hand sides f_n of problems (35.24).

$$\max_{\xi \in \overline{Q}} \left\{ \frac{1}{P_0(\xi)} \left|D_t^k f_n\right| ; \left|D_t^k (\nabla f_n)\right| \right\} \le M_1, \quad k = 1, \dots, s. \tag{35.40}$$

Moreover,

$$\forall k = 1, \dots, s, \quad s \le n, \quad |D_t^k f_n| \le M_2 P_0(\xi) \frac{k!\, T^{n-k}}{n!} \|Y_n\|_V \tag{35.41}$$

where M_2 is an absolute constant depending only on s, α, and N.

Fix some $0 < t_0 < 1$ and set

$$R = \{\Omega_* \times (0, T]\} \cap \{(\xi, t) : \rho = \operatorname{dist}(\xi, \partial\Omega) < t_0/2\}.$$

Theorem 35.3. *Let $v(\xi, t)$ be a solution of problem* (35.24) *with the right-hand side f satisfying* (35.40). *Given a point $(\xi_0, t_0) \in R$, for each $t \in (2t_0, t_0 + 1]$*

$$\left|D_t^{k+1} v(\xi_0, t)\right| \le \mu_{k+1} \frac{\rho}{t_0} \tag{35.42}$$

$$\left|\nabla\left(D_t^k v(\xi_0, t)\right)\right| \le \mu_k \frac{1}{t_0}, \tag{35.43}$$

$$\left|\Delta\left(D_t^k v(\xi_0, t)\right)\right| \le \mu_k \frac{1}{t_0}, \quad k = 0, 1, \dots, s, \tag{35.44}$$

where μ_k are finite constants depending on the constants M_0, M_1 from (35.31), (35.40), *the constant k_1 from* (35.37)*:*

$$\mu_{k+1} \le l' \left(\frac{\mu_k}{t_0} + M_1 \right), \quad k = 0, \ldots, s,$$

l' is an absolute constant not depending on v and f.

The main ingredient of the proof is the following assertion.

Lemma 35.7. *Let $v(\xi, t)$ be a classical solution of equation* (35.24) *with the right-hand side f satisfying conditions* (35.30), (35.40). *If $v(\xi, t)$ is bounded in $\Omega_* \times \{t \ge t_0\}$ by a finite constant $k_0 > 0$, then*

$$|v| \le L\,(k_0 + M_0) \frac{\rho}{t - t_0} \quad \text{in } \Omega_* \times \{t_0 \le t \le t_0 + 1\} \tag{35.45}$$

with a finite constant L depending only on m, the constants in (35.10), *and the geometrical properties of $\partial\Omega$.*

Proof. Let us consider the solution $v(\xi, t)$ of problem (35.24) in the domain $\Omega_* \times \{t \ge t_0\}$. Let $\{v_\delta\}$ be a sequence of solutions of the regularized problems (35.32), converging to v in the norm of $C\left(\overline{Q}\right)$. Each v_δ satisfies the equation

$$L v_\delta \equiv v_{\delta t} - m(P_0 + \delta)\Delta\, v_\delta = f + \delta \quad \text{in } H = \Omega_* \times (t_0, t_0 + 1], \tag{35.46}$$

$|v_\delta|$ is bounded in $\overline{H}$ by $k_0 + \delta$. Since equation (35.46) is linear, it will be convenient to consider the normalized functions

$$v_\delta := \frac{1}{q}\, v_\delta, \quad f + \delta := \frac{1}{q}\,(f + \delta)$$

with

$$q = \max_{\Omega_* \times (0,T]} |v_\delta| + \max_{\Omega_* \times (0,T]} |f + \delta| \equiv k_0 + M_0.$$

This trick provides the exact dependence on data in the final estimate (35.45). So, we consider the solution $|v_\delta| \le 1$ of problem (35.24) with $|f + \delta| \le 1$. Introduce the function

$$\varphi(z) = K_0\left(1 - e^{-z}\right), \quad K_0 = \text{const.} > 0.$$

Obviously,

$$\varphi'(z) = K_0\, e^{-z} > 0, \quad \varphi''(z) = -K_0\, e^{-z} < 0, \quad \varphi(0) = 0,$$

and

$$\frac{\varphi''}{\varphi'} = -1, \quad \frac{1}{\varphi'(z)} \equiv \frac{1}{K_0}\, e^{z} = \frac{1}{K_0 - \varphi(z)}. \tag{35.47}$$

Since $\varphi(z)$ is monotone, one can introduce the function $u(\xi, t)$ by

$$v_\delta = \varphi(u(\xi, t)), \quad K_0 = 2.$$

$u(\xi, t)$ has to satisfy the conditions

$$u_t - m(P_0 + \delta)\Delta u + \left(-m\frac{\varphi''}{\varphi'}\right)(P_0 + \delta)|\nabla u|^2 = \frac{f + \delta}{\varphi'(u)} \quad \text{in } H,$$

$$u = \varphi^{-1}(v_\delta) \le \varphi^{-1}(\delta + 1) \text{ as } t = t_0 \text{ and on the lateral boundary of } H.$$

Due to (35.47) and since

$$\varphi^{-1}(s) = \ln\left(\frac{K_0}{K_0 - s}\right),$$

the above conditions can be rewritten as follows:

$$\mathcal{M}u \equiv u_t - m(P_0 + \delta)\Delta u + m(P_0 + \delta)|\nabla u|^2 \le |f + \delta|, \tag{35.48}$$

$$u \le \ln 2 \text{ as } t = t_0 \text{ and on the inner lateral boundary of } H, u = 0 \text{ as } \xi \in \partial\Omega. \tag{35.49}$$

Let us construct an upper barrier for $u(\xi, t)$ in H. Let $\sigma(\xi)$ be the solution of (35.8). Recall that in our choice of Ω_*

$$\sigma(\xi) \ge C_3 \operatorname{dist}(\xi, \partial\Omega), \quad |\nabla\sigma| \ge \gamma/2, \quad \text{in } \overline{\Omega_*}, \quad C_3 = \text{const.} > 0.$$

Now consider the function

$$\omega(\xi, t) = L_0\frac{\sigma(\xi)}{t - t_0}, \quad L_0 = \text{const.} > 0.$$

Then fix L_0 so as to satisfy the condition

$$\omega \ge \ln 2 \quad \text{on the inner lateral boundary of } H,$$

that is, $C_3 d_* L_0 \ge \ln 2$. Since $v_\delta(\xi, t_0)$ is finite, $\omega > v_\delta$ as $t = t_0$. Next,

$$\begin{aligned} \mathcal{N}\omega &\equiv -L_0\frac{\sigma}{(t - t_0)^2} + mL_0\frac{P_0 + \delta}{t - t_0} + mL_0^2(P_0 + \delta)\frac{|\nabla\sigma|^2}{(t - t_0)^2} \\ &\ge L_0\left[L_0 m\frac{\gamma^2}{4} - \frac{\sigma}{P_0 + \delta}\right]\frac{P_0 + \delta}{(t - t_0)^2}. \end{aligned} \tag{35.50}$$

By (35.31) we get from (35.50) that

$$\mathcal{N}\omega \ge |f + \delta| \quad \text{in } \overline{Q},$$

provided L_0 taken large enough. Then by the maximum principle

$$u(\xi, t) \le L_0\frac{\sigma(\xi)}{t - t_0} \quad \text{in } \overline{H}.$$

With the use of the mean value theorem we have

$$v_\delta = \varphi(u) \equiv \varphi(u) - \varphi(0) = \int_0^1 \frac{d\,\varphi(\theta u)}{d\,\theta}\,d\theta$$
$$\le \max|\varphi'|\,u \le K_0\,u$$
$$\le 2L_0C_3\,\frac{\rho}{t-t_0} \quad \text{in } \overline{H}. \tag{35.51}$$

Letting in this inequality $\delta \to 0$, we get (35.45). □

Proof of Theorem 35.2. We shall proceed by induction. For $k = 0$ the desired estimate was already established in Theorem 35.3. Take the sequence of points

$$t_i = \sum_{j=1}^{i} \delta^j, \quad \delta = \frac{t_0}{1+t_0} \quad \text{so that } t_i \to t_0,$$

and introduce the sets $H_i = \Omega_* \cap \{t > t_i\}$. Assume that (35.42)–(35.44) are already established for some natural k. Consider the function $u(\xi, t) \equiv D_t^{k+1} v(\xi, t)$. By differentiation we deduce the equation for $u(\xi, t)$

$$u_t - mP_0\Delta\,u = D_t^{k+1} f \quad \text{in } H_{k+1}.$$

The initial and boundary values of u are estimated at the previous step: by the induction conjuncture

$$|u| \le \mu_{k+1}\,\frac{\rho}{t_0} \quad \text{in } \overline{H_k}. \tag{35.52}$$

Let us fix a point (ξ', t') and consider the rectangle

$$G_\rho = \left\{(\xi, t) \in H_{k+1} :\ |\xi - \xi'| < \rho,\ \frac{t'}{2} < t < t'\right\}.$$

Set

$$\xi = \xi' + \rho\,\zeta, \quad t = t' + \rho\,\tau, \quad W(\zeta, \tau) = u(\xi, t).$$

Then

$$W_\tau - m\,\frac{P_0}{\rho}\,\Delta\,W = \rho\,D_t^{k+1}\,f$$

in the cylinder

$$C_1 = \left\{(\zeta, \tau) :\ |\zeta| < 1,\ \tau \in \left(-\frac{t'}{2\rho}, 0\right)\right\},$$

and $|W| \le q\,\rho$ on the parabolic boundary of C_1 with a constant $q \le \max|u/\rho| \le \mu_{k+1}/t_0$. By [97, Ch. 5, Sec. 3, Th. 1],

$$|\nabla_\zeta W| \le \lambda_1 \tag{35.53}$$

$$\text{in} \quad C_{1/2} = \left\{(\zeta, \tau) :\ |\zeta| < \frac{1}{2},\ \tau \in \left(\frac{1}{2} - \frac{t'}{2\rho}, 0\right)\right\}.$$

The constant λ_1 depends *linearly* on data of the problem for W (cf. the proof of Lemma 35.4 and Lemma 35.7) and

$$|\nabla u| \le l_1 \left(M_1 + \frac{\mu_{k+1}}{t_0} \right) \quad \text{in } G_{\rho/2} = \left\{ (\xi, t) : \ |\xi| < \frac{\rho}{2}, \ \frac{t'}{2} + \frac{\rho}{2} < t < t' \right\}.$$

Now we estimate the derivatives of $w = \nabla_\zeta W$. The components of this vector are bounded on the parabolic boundary of $C_{1/2}$ and constitute a solution of the following parabolic system:

$$w_t - m \frac{P_0}{\rho} \Delta w - m \nabla P_0 \operatorname{div} w = \nabla_\zeta \left(D_t^{k+1} f \right) \quad \text{in } C_{1/2}.$$

By [97, Ch. 7, Sec. 4, Th. 1]

$$\left| (w_i)_{\zeta_j} \right| \le \lambda_2 \quad \text{in} \quad C_{1/4} = \left\{ (\zeta, \tau) : \ |\zeta| < \frac{1}{4}, \ \tau \in \left(\frac{1}{2} + \frac{1}{4} - \frac{t'}{2\rho}, 0 \right) \right\},$$

which gives the estimate

$$\rho \, |\Delta u| \le l_2 \, (M_1 + \max |\nabla u|) \quad \text{in } G_{\rho/4}$$

with a constant l_2 not depending on u and f. Hence,

$$\max \left\{ \rho \, |\Delta u(\xi', t')|; \ |\nabla u(\xi', t')| \right\} \le l_3 \left(M_1 + \frac{\mu_{k+1}}{t_0} \right).$$

By virtue of the equation for $u(\xi, t)$

$$\left| D_t^{k+2} v \right| \le l_4 \left(M_1 + \frac{\mu_{k+1}}{t_0} \right)$$

and it remains to apply Lemma 35.7. □

We are now in the position to complete the proof of Theorem 35.1. Let $(\xi_0, t_0) \in \partial\Omega \times (0, T)$. Given a natural number s and a sequence Y_k converging to the solution of problem (35.6)–(35.7), we can rely on the estimates of Theorem 35.3 to check that $|D_t^s Y_k(\xi_0, t)|$ are equibounded in k for t close to t_0. It follows then that $Y_k(\xi_0, t)$ contains a subsequence converging in the norm of C^{s-1} to a function Y_*. But since the whole sequence converges in V, this partial limit and the limit of the whole sequence necessarily coincide.

Let us introduce the function space V_s as the closure of V in the norm

$$\|u\|_{V_s} \equiv \|u\|_V + \sum_{q=0}^{s} \sup_{(t_0, T)} \left| \nabla D_t^q u(\xi_0, t) \right|, \quad \xi_0 \in \partial\Omega,$$

and consider the operator A introduced in the proof of Lemma 35.1 as assigning to a point $x \in B(0) \subset V$ an element of V_s:

$$\|Ax\|_{V_s} \le \|A'(y)\|_{V_s} \|x\|_V = \|[F'(0)]^{-1} \{F'(0) - F'(M_n[y])\}\|_{V_s} \|x\|_V.$$

The second factor of the right-hand side of this inequality is bounded by the radius of $B(0)$. The norm of the first factor is given by the norm in V, already estimated in the proof of convergence of the sequence $\{Y_k\}$, and by the norm of the trace on $\partial\Omega$. As we said above, to estimate the trace amounts to estimating the corresponding derivatives of the solution to problem (35.24) with the right-hand side $f \equiv F'(0) - F'(M_n[y])$. In our choice of M_n such f satisfies all conditions of Theorem 35.3 and the required estimates are given by (35.43):

$$\|Y_{n+1}\|_{V_s} \le \|A\, Y_n\|_{V_s} \le L(s, t_0), \quad n = s+1, \ldots,$$

uniformly in n. Since s was taken arbitrary, the proof of Theorem 35.1 is completed.

Bibliography

[1] I. Sh. Akhatov, R. K.Gazizov, N. H. Ibragimov, Nonlocal symmetries. A heuristic approach. In: Itogi Nauki i Tekhniki. Sovrem. Probl. Mat. Nov. Dostizhen. (Current problems of mathematics. Newest advances) 34, VINITI, Moscow 1989, pp. 3–83 (in Russian).

[2] I. Sh. Akhatov, R. K. Gazizov, N. H. Ibragimov, Group classification of the equations of nonlinear filtration. Soviet Math. Dokl. 35 (1987), no. 2, 384–386; translated from: Dokl. Akad. Nauk. SSSR 293 (1987), no. 5, 1033–1035.

[3] I. Sh. Akhatov, R. K. Gazizov, N. H. Ibragimov, Quasi-local symmetry of equations of nonlinear heat conduction. Soviet Phys. Dokl. 32 (1987), no. 7, 554–556; translated from: Dokl. Akad. Nauk SSSR 295 (1987), no. 1, 75–78.

[4] L. Alvarez, J. I. Diaz, R. Kersner, On the initial growth of the interfaces in nonlinear diffusion-convection processes. In: Nonlinear diffusion equations and their equilibrium states I, W.-M. Ni, L. A. Peletier and J. Serrin, eds., Springer-Verlag, New York 1988, Math. Sci. Res. Inst. Publ. 12, pp. 1–20.

[5] W. F. Ames, C. Rogers, Nonlinear boundary value problems in science and engineering. Academic Press, New York 1989.

[6] B. M. Anisiutin, A. M. Meirmanov, Two-dimensional self-similar solutions of the one phase Stefan problem. Dinamika Sploshn. Sredy 78 (1986), 10–20 (in Russian).

[7] S. B. Angenent, Analyticity of the interface of the porous media equation after the waiting time. Proc. Amer. Math. Soc. 102 (1988), 329–336.

[8] S. B. Angenent, Asymptotics for the porous medium equations. In: Nonlinear diffusion equations and their equilibrium states I, W.-M. Ni, L. A. Peletier, J. Serrin, eds., Springer-Verlag, New York 1988, Math. Sci. Res. Inst. Publ. 12, pp. 21–34.

[9] S. N. Antontsev, A. V. Kazhikhov, V. N. Monakhov, Boundary value problems in mechanics of nonhomogeneous fluids. Nauka, Novosibirsk 1983. English translation: North-Holland, 1990.

[10] D. G. Aronson, The porous medium equation. In: Nonlinear diffusion problems, A. Fasano, M. Primicerio, eds., Lecture Notes in Math. 1224, Springer-Verlag, Berlin 1986, pp. 1–46.

[11] D. G. Aronson, Regularity properties of flows through porous media: the interface. Arch. Rat. Mech. Anal. 37 (1970), 1–10.

[12] D. G. Aronson, Ph. Benilan, Régularité des solutions de l'équation des milieux poreux dans $\mathbb{R}^n$. C. R. Acad. Sci. Paris. Ser. I 288 (1979), 103–105.

[13] D. G. Aronson, L. A. Caffarelli, S. Kamin. How an initially stationary interface begins to move in porous medium flow. SIAM J. Math. Anal. 14 (1983), 639–658.

[14] D. G. Aronson, L. A. Caffarelli, J. Vazquez, Interfaces with a corner point in one-dimensional porous medium flow. Comm. Pure Appl. Math. 38 (1985), 375–404.

[15] D. G. Aronson, L. A. Peletier, Large time behavior of solutions of the porous medium equation in bounded domains. J. Differential Equations 39 (1981), 378–412.

[16] D. G. Aronson, L. A. Peletier, The asymptotic profile of solutions of degenerate diffusion equations. Arch. Rational Mech. Anal. 91 (1986), 207–229.

[17] D. G. Aronson, J. Vazquez, Eventual C^∞-regularity and concavity for flows in one-dimensional porous media. Arch. Rational Mech. Anal. 99 (1987), 329–348.

[18] C. Bardos, F. Golse, B. Perthame, R. Sentis, The nonaccretive radiative transfer equations. Existence of solutions and Rosseland approximation. J. Funct. Anal. 77 (1988), 434–460.

[19] G. I. Barenblatt, On some unsteady motions of a liquid and gas in a porous medium. Prikl. Mat. Mekh. 16 (1952), 67–78 (in Russian).

[20] J. Bear, Dynamics of fluids in porous media. Elsevier, New York 1972.

[21] P. Benilan, M. G. Crandall, M. Pierre, Solution of the porous medium equations in $\mathbb{R}^N$ under optimal conditions on initial values. Indiana Univ. Math. J. 33 (1984), 51–87.

[22] Ph. Benilan, J. L. Vazquez, Concavity of solutions of the porous medium equations. Trans. Amer. Math. Soc. 299 (1987), 81–93.

[23] J. G. Berryman, Evolution of a stable profile for a class of nonlinear diffusion equations. III. Slow diffusion on the line. J. Math. Phys. 21 (1980), 1326–1331.

[24] M. Bertsch, R. Dal Passo, A numerical treatment of a superdegenerate equation with applications to the porous media equation. Quart. Appl. Math. 48 (1990), 133–152.

[25] M.Bertsch, C. J. van Duijn, J. R. Esteban, Zhang Hongfei, Regularity of the free boundary in a doubly degenerate parabolic equation. Comm. Partial Differential Equations 14 (1989), 391–412.

[26] M. Bertsch, D. Hilhorst, The interfaces between regions where $u < 0$ and $u > 0$ in the porous medium equation. Appl. Anal. 41 (1991), 111–130.

[27] M. Bertsch, L. A. Peletier, Porous media equations: an overview. Report no. 7, Univ. of Leiden, Math. Institute, June 1983.

[28] G. Bluman, S. Kumei, On the remarkable nonlinear diffusion equation $(\partial/\partial x)[a(u+b)^{-2}\partial u/\partial x] - \partial u/\partial t = 0$. J. Math. Phys. 21 (1980), 1019–1023.

[29] J. Buckmaster, Viscous sheets advancing over dry beds. J. Fluid Mech. 81 (1977), 735–756.

[30] V. M. Budak, A. A. Samarskii, A. N. Tikhonov, A collection of problems of mathematical physics, second edition, Nauka, Moscow 1972. English translation: Pergamon Press, Oxford 1964.

[31] J. R. Burgan, A. Munier, M. R. Feix, E. Fialkov, Homology and the nonlinear heat diffusion equation. SIAM J. Appl. Math. 44 (1984), 11–18.

[32] L. A. Caffarelli, A. Friedman, Regularity of the free boundary for the one-dimensional flow of gas in a porous medium. Amer. J. Math. 101 (1979), 1193–1218.

[33] L. Caffarelli, J. L. Vazquez, N. Wolansky, Lipschitz continuity of solutions and interfaces of the N-dimensional porous medium equation. Indiana Univ. Math. J. 36 (1987), 373–401.

[34] L. Caffarelli, N. Wolansky, $C^{1,\alpha}$ regularity of the free boundary for the N-dimensional porous media equation. Comm. Pure Appl. Math. 43 (1990), 885–902.

[35] F. Calogero, Why are certain nonlinear PDE's both widely applicable and integrable? In: What is integrability, Springer Ser. Nonlinear Dynam., Springer-Verlag, Berlin 1991, 1–62.

[36] M. Chipot, T. Sideris, An upper bound for the waiting time for nonlinear degenerate parabolic equations. Trans. Amer. Math. Soc. 288 (1985), 423–427.

[37] J. D. Cole, On a quasilinear parabolic equation used in aerodynamics. Quart. Appl. Math. 3 (1950), 201–230.

[38] R. Courant, K. O. Friedrichs, Supersonic flow and shock waves. Interscience, 1948, Springer-Verlag, New York 1976.

[39] L. M. Degtyaryov, V. V. Krylov, A method for the numerical solution of problems of the dynamics of wave fields with singularities. U.S.S.R. Comput. Math. and Math. Phys. 17 (1977), no. 6, 172–179; translated from: Zh. Vychisl. Mat. i Mat. Fiz. 17, no. 6, (1977), 1523–1530.

[40] J. I. Diaz, R. Kersner, Non-existence of one of the free boundaries in a degenerate equation in filtration theory. C. R. Acad. Sci. Paris Sér I Math. 296 (1983), 505–508.

[41] J. I. Diaz, R. Kersner, On a nonlinear degenerate parabolic equation in infiltration or evaporation through a porous medium. J. Differential Equations 69 (1987), 368–403.

[42] J. I. Diaz, S. I. Shmarev, On the behavior of the interfaces in nonlinear processes with convection dominating diffusion via Lagrangian coordinates. Adv. Math. Sci. Appl. 1 (1992), 19–45.

[43] V. A. Dorodnitsyn, I. V. Knyaseva, S. R. Svirshchevskii, The group properties of nonlinear heat conduction equations with a source. Keldysh Institute of Appl. Math., Peprint no. 79, 1982, 24 p. (in Russian).

[44] V. A. Dorodnitsyn, S. R. Svirshchevskii, On the Lie-Bäcklund groups admitted by the heat equation with source. Institute of Appl. Math. AN SSSR, Preprint no. 101, 1983 (in Russian).

[45] L. C. Evans, A free boundary proble: a flow of two immissible fluids in a one-dimensional porous medium. I. Indiana Univ. Math J. 26 (1977), 915–932.

[46] L. C. Evans, A free boundary problem: a flow of two immiscible fluids in a one-dimensional porous medium. II. Indiana Univ. Math J. 27 (1978), 93–111.

[47] V. A. Florin, Some simplest nonlinear problems of consolidation of water-saturated ground medium. Izv. Akad. Nauk SSSR Otdel. Tekhn. Nauk 1948, no. 9, 1389–1402 (in Russian).

[48] A. S. Fokas, B. Fuchssteiner, Bäcklund transformations for hereditary symmetries. Nonlinear Anal. 5 (1981), 423–432.

[49] A. S. Fokas, Y. C. Yortsos, On the exactly solvable equation $S_t = [(\beta S + \gamma)^{-2} S_x]_x + \alpha(\beta S + \gamma)^{-2} S_x$ occuring in two-phase flow in porous media. SIAM J. Appl. Math. 42 (1982), 318–332.

[50] V. A. Fok, The hydrogen atom and non-Euclidean geometry. Izv. Akad. Nauk SSSR, Ser. VII, Otdel. Mat. Estestv. Nauk, 1935, no. 2, 169–179 (in Russian).

[51] A. Friedman, Variational inequalities and free boundary problems. Wiley Interscience, New York 1982.

[52] A. Friedman, Partial differential equations of parabolic type. Prentice-Hall, Englewood Cliffs, NJ, 1964.

[53] A. Friedman, S. Kamin, The asymptotic behavior of gas in an n-dimensional porous medium. Trans. Amer. Math. Soc. 262 (1980), 551–563.

[54] W. Fulks, R. B.Guenter, A free boundary problem and extention of Muskat's model. Acta Math. 122 (1969), 273–300.

[55] V. I. Fushchich, A. G. Nikitin, Symmetries of Maxwell's equations. Naukova Dumka, Kiev 1983. English translation: D. Reidel Publ., Dordrecht 1987.

[56] V. I. Fushchich, A new method of investigating the group properties of the equations of mathematical physics. Soviet Phys. Dokl. 24, no. 6, (1979), 437–439; translated from: Dokl. Akad. Nauk SSSR 246, no. 4, (1979), 846–850.

[57] V. I. Fushchich, Additional invariance of relativistic equations. Theoret. and Math. Phys. 7, no. 1, (1971), 323–330; translated from: Teoret. Mat. Fiz. 7, no. 1, (1971), 3–12.

[58] V. A. Galaktionov, Invariant subspaces and new explicit solutions to evolution equations with gradient nonlinearities. Proc. Roy. Soc. Edinburgh Sect. A 125 (1995), 225–246.

[59] B. H. Gilding, A nonlinear degenerate parabolic equation. Ann. Scuola Norm. Sup. Pisa Cl. Sci. (4) 4 (1977), 393–432.

[60] B. H. Gilding, The occurrence of interfaces in nonlinear diffusion- advection processes. Arch. Rational Mech. Anal. 100 (1988), 243–263.

[61] B. H. Gilding, Improved theory for a nonlinear degenerate parabolic equation. Ann. Scuola Norm. Sup. Pisa Cl. Sci. (4) 16 (1989), 165–224; 17 (1990), 479.

[62] B. H. Gilding, Hölder continuity of solutions of parabolic equations. J. London Math. Soc. 13 (1976), 103–106.

[63] B. H. Gilding, R. Kersner, The characterization of reaction-convection-diffusion processes by travelling waves. J. Differential Equations 124 (1996), 27–79.

[64] M. Guedda, D. Hilhorst, C. Picard, The one-dimensional porous medium equation with convection: continuous differentiability of interfaces after the waiting time. Appl. Math. Lett. 5 (1992), 59–62.

[65] M. E. Gurtin, R. C. MacCamy, E. A. Socolovsky, A coordinate transformation for the porous media equation that renders the free-boundary stationary. Quart. Appl. Math. 42 (1984), 345–357.

[66] M. A. Herrero, A limit case in nonlinear diffusion. Nonlinear Anal. 13 (1989), 611–628.

[67] M. A. Herrero, J. L. Vazquez, The one-dimensional nonlinear heat equation with absorption: Regularity of solutions and interfaces. SIAM J. Math. Anal. 18 (1987), 149–167.

[68] E. Hopf, The partial differential equation $u_t + uu_x = \mu u_{xx}$. Comm. Pure Appl. Math. 3 (1950), 201–230.

[69] K. Höllig, M. Pilant, Regularity of the free boundary for the porous medium equation. Indiana Univ. Math. J. 34 (1985), 723–731.

[70] S. D. Howison, Complex variables methods in Hele–Shaw moving boundary problem. Europ. J. Appl. Math. 3 (1992), 209–224.

[71] H. E. Huppert, The propagation of two-dimensional viscous dravity currents over a rigid horisontal surface. J. Fluid Mech. 121 (1982), 43–58.

[72] N. H. Ibragimov, Transformation groups applied in mathematical physics. Nauka, Moscow 1983. English translation: D. Reidel Publ., Dordrecht 1985.

[73] A. S. Kalashnikov, The occurrence of singularities in solutions of the non-steady see-page equation. U.S.S.R. Comput. Math. Math. Phys. 7 (1967), 269–275; translated from: Zh. Vichisl. Mat. i Mat. Fiz. 7 (1967), no. 2, 440–443.

[74] A. S. Kalashnikov, The nature of propagation of perturbations in processes that can be described by quasilinear degenerate parabolic equations. In: Trud. Sem. Petrovsk., 1975, vol. 1, pp. 135–144 (in Russian).

[75] A. S. Kalashnikov, The effect of absorption on heat propagation in a medium in which the thermal conductivity depends on temperature. U.S.S.R. Comput. Math. Math. Phys. 16 (1976), 141–149; translated from: Zh. Vychisl. Mat. i Mat. Fiz. 16 (1976), 689–696.

[76] A. S. Kalashnikov, Some questions of the qualitative theory of non-linear degenerate parabolic second-order equations. Russ. Math. Surveys 42 (1987), 169–222; translated from: Uspehi Mat. Nauk 42 (1987), 135–176.

[77] S. Kamenomostskaya, The asymptotic behavior of the solution of the filtration equation. Israel J. Math. 14 (1973), 73–87.

[78] S. Kamin (Kamenomostskaya), Similar solutions and the asymptotics of filtration equation. Arch. Rational Mech. Anal. 60 (1976), 171–183.

[79] S. Kamin, Ph. Rosenau, Thermal waves in an absorbing and convecting medium. Phys. D 8 (1983), 717–737.

[80] L. I. Kamynin, On the linear problem of Verigin. Soviet Math. Dokl. 4 (1963), no. 3, 847–851; translated from: Dokl. Akad. Nauk SSSR 150 (1963), no. 6, 1210–1213.

[81] L. I. Kamynin, On the existence of a solution of Verigin's problem. U.S.S.R. Comput. Math. Math. Phys. 2 (1962), no. 5, 954–987; translated from: Zh. Vichisl. Mat. i Mat. Fiz. 2 (1962), no. 5, 838–858.

[82] S. V. Khabirov, Solutions of second-order evolution equations obtained by differential substitutions. In: Dinamika Sploshn. Sredy 85 (1988), 146–161 (in Russian).

[83] S. V. Khabirov, Method of differential substitutions in solution of boundary value problems. In: Differentsial'nye uravneniya s chastnymi proizvodnymi (Partial differential equations), Institut Matematiki SO AN SSSR, Novosibirsk 1984, pp. 113–131 (in Russian).

[84] J. R. King, Exact multidimensional solutions to some nonlinear diffusion equations. Quart. J. Mech. Appl. Math. 46 (1993), 419–436.

[85] J. G. Kingston, C. Rogers, Reciprocal Bäcklund transformations of conservation laws. Phys. Lett. A 92 (1982), 261–264.

[86] E. G. Kirnasov, Wahlquist–Estabook type coverings over the heat equation. Math. Notes 42 (1987), 732–739; translated from: Mat. Zametki 42 (1987), no. 3, 422–434.

[87] B. F. Knerr, The porous medium equation in one dimension. Trans. Amer. Math. Soc. 38 (1977), 375–404.

[88] A. N. Kolmogorov, S. V. Fomin, Elements of Functional Analysis, Nauka, Moscow 1968 (in Russian).

[89] V. A. Kondrat'ev, Boundary-value problems for elliptic equations in domains with conic and angular points. Trans. Moscow Math. Soc.16 (1967), 227–313; translated from: Trudy Moskov. Mat. Obshch. 16 (1967), 209–292.

[90] A. L. Kopa-Ovdienko, On the use of Lagrangian coordinates for modelling light beams with strong spatial deformation. U.S.S.R. Comput. Math. Math. Phys. 26 (1987), no. 1–2, 193–196; translated from: Zh. Vychisl. Mat. i Mat. Fiz. 26 (1986), no. 2, 311–315.

[91] E. R. Kosygina, On anisotropic solutions of the multi-dimensional equation of nonstationary filtration. Comput. Math. Math. Phys. 35 (1995), no. 2, 191–205; translated from: Zh. Vychisl. Mat. i Mat. Fiz. 35, no. 2, (1995), 241–259.

[92] A. S. Kronrod, E. M. Landis, On the smoothness of the level-curve sets of functions of several variables. Dokl. Akad. Nauk SSSR 58 (1947), no. 7, 1269–1272 (in Russian).

[93] S. N. Kruzhkov, Nonlinear parabolic equations in two independent variables. Trans. Moscow Math. Soc. 1967, 355–373; translated from: Trudy Moskov. Mat. Obshch. 16 (1967), 326–346.

[94] R. J. Kunze, D. Kirham, Simplified accounting for membrane impedance in capillary conductivity determinations. Soil Sci. Soc. Am. Proc. 26 (1962), 421–426.

[95] A. A. Lacey, J. R. Ockendon, A. B.Tayler, "Waiting time" solutions of a nonlinear diffusion equation. SIAM J. Appl. Math. 42 (1982), 1252–1264.

[96] O. A. Ladyzenskaya, The mathematical theory of viscous incompressible flow, Gordon and Breach, New York 1969.

[97] O. A. Ladyzhenskaya, V. A. Solonnikov, N. N. Ural'tseva, Linear and quasilinear equations of parabolic type, Transl. Math. Monographs 23, Amer. Math. Soc., Providence, RI, 1968.

[98] O. A. Ladyzenskaya, N. N. Uraltseva, Équations aux dérivées partielles de type elliptique, Dunod, Paris 1968.

[99] E. W. Larsen, G. C. Pomraning, Asymptotic analysis of non-linear Marshak waves. SIAM J. Appl. Math. 39 (1980), 201–212.

[100] E. V. Lensky, On the group properties of equations of motion of nonlinear viscous-plastic medium. Vestnik Moskov. Univ. Ser. I Mat. Mekh., 1966, no. 5, 116–125.

[101] B. A. Magadeev, On the contact symmetries of nonlinear evolution equations. Algebra i Analiz 5, no. 2, (1993), 141–156 (in Russian).

[102] R. E. Marshak, Effect of radiation on shock wave behaviour. Phys. Fluids 1 (1958), 24–29.

[103] V. P. Maslov, V. G. Danilov, K. A. Volosov, Mathematical modelling of heat and mass transport processes., Nauka, Moscow 1987 (in Russian).

[104] A. M. Meirmanov, The Stefan problem. Walter de Gruyter, Berlin-New York 1992.

[105] A. M. Meirmanov, Problem of advance of the surface of contact discontinuity in the filtration of immiscible compressible fluids (the Verigin problem). Siberian Math. J. 23 (1982), no. 1, 65–80; translated from: Sibirsk. Mat. Zh. 23 (1982), no. 1, 85–102.

[106] A. M. Meirmanov, On the solvability of the Verigin problem in an exact formulation. Soviet Phys. Dokl. 25 (1980), no. 7, 535–536; translated from: Dokl. Akad. Nauk SSSR 253 (1980), no. 3, 588–591.

[107] A. M. Meirmanov, Two-dimensional self-similar solutions of the problem on filtration of compressible liquid with free boundaries. Chislennye Metodi Mekhaniki Sploshnoy Sredy 12 (1981), no. 6, 56–64 (in Russian).

[108] A. M. Meirmanov, The unique solvability and asymptotic behavior as $t \to \infty$ of a Stefan type problem arising in hydrodynamics. Dinamika Sploshn. Sredy 30 (1977), 98–111 (in Russian).

[109] A. M. Meirmanov, V. V. Pukhnachov, Lagrangian coordinates in the Stefan problem. Dinamika Sploshn. Sredy 47 (1980), 90–111 (in Russian).

[110] V. Mikhailov, Équations aux dérivées partielles. Éditions Mir, Moscou 1980 (Translated from the Russian).

[111] A. V. Mikhailov, A. B. Shabat, A symmetry approach to classification of integrable equations. In: Integriruemost' y kineticheskie uravneniya dlya solitonov (Integrability and kinetic equations for solitons), Naukova Dumka, Kiev 1990, pp. 213–279 (in Russian).

[112] A. Munier, J. R. Burgan, J. Gutierrez, E. Fialkov, M. R. Feix, Group transformations and the nonlinear heat diffusion equations. SIAM J. Appl. Math. 40 (1981), 191–207.

[113] M. Muskat, The Flow of Homogeneous Fluids through Porous Media. Mc Graw-Hill, New York 1937.

[114] T. Nagai, M. Mimura, Some nonlinear degenerate diffusion equations related to population dynamics. J. Math. Soc. Japan 35 (1983), 539–522.

[115] T. Nagai, M. Mimura, Asymptotic behaviour for a nonlinear degenerate diffusion equation in population dynamics. SIAM J. Appl. Math. 43 (1983), 449–464.

[116] V. I. Nalimov, V. V. Pukhnachov, Nonstationary movement of ideal liquid with free boundary. Novosibirsk State University, 1975, 174 p. (in Russian).

[117] S. A. Nazarov, B. A. Plamenevsky, Elliptic problems in domains with piecewise smooth boundaries, Walter de Gruyter, Berlin-New York 1994.

[118] M. Necati Özişik, Heat Conduction, Wiley-Interscience, New York 1980.

[119] O. A. Oleinik, E. V. Radkevich, Second order equations with nonnegative characteristic form, Amer. Math. Soc., Providence, RI, 1973.

[120] A. Oron, Ph. Rosenau, Some symmetries of the nonlinear heat and wave equations. Phys. Lett. A 118, , no. 4, (1986), 172–176.

[121] L. V. Ovsyannikov, Group analysis of differential equations. Nauka, Moscow 1978. English translation: Academic Press, New York-London 1982.

[122] L. V. Ovsyannikov, Introduction to continuum mechanics. Vol. I, II. Novosibirsk State University, 1974 (in Russian).

[123] L. V. Ovsyannikov. The group properties of nonlinear heat conduction equations. Dokl. Akad. Nauk SSSR 125, no. 3, (1959), 492–495 (in Russian).

[124] A. de Pablo, J. L. Vazquez, Regularity of solutions and interfaces of a generalized porous medium equation in $\mathbb{R}^N$. Ann. Mat. Pura Appl. (4) 158 (1991), 51–74.

[125] R. E. Pattle, Diffusion from an instantaneous point source with a concentration-dependent coefficient. Quart. J. Mech. Appl. Math. 12 (1959), 407–409.

[126] L. A. Peletier, The porous media equation. In: Applications of Nonlinear Analysis in the Physical Sciences, H. Amann, N. Bazley, and K. Kirchgässner, eds., Pitman, Boston 1981, pp. 229–241.

[127] L. A. Peletier, Problems in filtration theory. Appl. Sci. Res. 37 (1981), 145–153.

[128] J. R. Philip, Evaporation and moisture and heat fields in the soil. Journal of Meteorology 14 (1957), 354–366.

[129] J. R. Philip, The theory of infiltration. Adv. Hydrosc. 5 (1969), 215–296.

[130] M. Pierre, Uniqueness of the solutions of $u_t - \Delta\phi(u) = 0$ with initial datum of measure. Nonlinear Anal. 5 (1982), 175–187.

[131] A. V. Pogorelov, Monge-Ampère equations of elliptic type. Khar'kov State University, 1960. English translation: Noordhoff, Groningen 1964.

[132] P. Ya. Polubarinova-Kochina, Theory of groundwater movement. Nauka, Moscow 1977. English translation: Princeton University Press, 1962.

[133] V. V. Pukhnachov, Nonlocal symmetries in nonlinear heat equations. In: Proceedings of the workshop energy methods for free boundary problems in continuum mechanics, S. N. Antontsev, J. I. Diaz, S. I. Shmarev, eds., Kluwer, Dordrecht 1996.

[134] V. V. Pukhnachov, Reciprocal transformations for radial equations of nonlinear heat transfer. Zap. Nauchn. Sem. St.- Petersburg, POMI, 213 (1994), 151–163 (in Russian).

[135] V. V. Pukhnachov, Lagrangian coordinates in free boundary problems for parabolic equations. In: Emerging applications in free boundary problems, Proceedings of the International Colloquium Free Boundary Problems: Theory and Applications, J. M. Chadam, H. Rasmussen, eds., Essex: Longman Sci. Tech., Harlow 1993, pp. 64–77.

[136] V. V. Pukhnachov, Equivalence transformations and hidden symmetry of evolution equations. Soviet Math. Dokl. 35 (1987), no. 3, 555–558; translated from: Dokl. Akad. Nauk SSSR 294 (1987), no. 3, 535–538.

[137] V. V. Pukhnachov, Evolution equations and coordinates of Lagrange. Dinamika Sploshn. Sredy 70 (1985), 127–141 (in Russian).

[138] V. V. Pukhnachov, Exact multidimensional solutions of the nonlinear diffusion equation. J. Appl. Mech. Tech. Phys. 36 (1995), no. 2, 169–176; translated from: Prikl. Mekh. Tekhn. Fiz. 36 (1995), no. 2, 23–31.

[139] V. V. Pukhnachov, S. I. Shmarev, The multidimensional self-similar Stefan problem in linearized statement. VINITI, 1984, no. 4272-84, 25 p. (in Russian).

[140] E. V. Radkevich, On the spectrum of the pencil of the Verigin–Muskat problem. Math. Notes 49 (1991), no. 3–4, 279–288; translated from: Mat. Zametki 49 (1991), no. 3, 77–90.

[141] C. Rogers, On a class of moving boundary problems in nonlinear heat conduction: application of a Bäcklund transformation. Internat. J. Nonlinear Mech. 21 (1986), 249–256.

[142] C. Rogers, R. Broadbridge, On sedimention in a bounded column. Internat. J. Nonlinear Mech. 27 (1992), 661–667.

[143] C. Rogers, W. F. Shadwick, Bäcklund transformations and their applications. Academic Press, New York 1982.

[144] C. Rogers, M. P. Stallybrass, D. L. Clements, On two-phase filtration under gravity and with free boundary infiltration. Nonlinear Anal. 7 (1983), 785–799.

[145] G. Rosen, Method for the exact solution of a nonlinear diffusion. Phys. Rev. Lett. 49 (1982), no. 25, 1844–1846.

[146] G. Rosen, Nonlinear heat conduction in solid H_2. Phys. Rev. B (3) 19 (1979), no. 4, 2398–2399.

[147] L. H. Rowen, Polynomial identities in ring theory. Academic Press, New York 1980.

[148] B. L. Rozhdestvenskii, N. N. Yanenko, Systems of quasilinear equations and their applications to gas dynamics. Nauka, Moscow 1968. English translation: Amer. Math. Soc., Providence, Rhode Island, 1983.

[149] G. A. Rudykh, E. I. Semenov, The construction of exact solutions of the multidimensional quasilinear heat conduction equation. Comput. Math. Math. Phys. 33 (1993), no. 8, 1087–1097; translated from: Zh. Vychisl. Mat. i. Mat. Fiz. 33 (1993), no. 8, 1228–1239.

[150] P. G. Saffman, G. I. Taylor, The penetration of a fluid into a porous medium of Hele–Shaw cell containing a more viscous fluid. Proc. Roy. Soc. London Ser. A 245 (1958), 312–329.

[151] A. A. Samarskii, V. A. Galaktionov, S. P. Kurdyumov, A. P. Mikhailov, Blow-up in problems for quasilinear parabolic equations. Nauka, Moscow 1987. English translation: Walter de Gruyter, Berlin–New York 1995.

[152] A. Sard, The measure of the critical values of differential maps. Bull. Amer. Math. Soc. 48 (1942), 883–897.

[153] O. N. Shablovskii, Transformation of independent variables in solution of Stefan type problems. VINITI, 1986, no. 143-B86, 24 p. (in Russian).

[154] G. Schlichting, Boundary layer theory. 4th edition. McGraw-Hill, New York 1960.

[155] S. I. Shmarev, On a degenerate parabolic equation in filtration theory: monotonicity and C^{∞}-regularity of interface. Adv. Math. Sci. Appl. 5 (1995), 1–29.

[156] S. I. Shmarev, On the free boundary properties for a class of degenerate parabolic equations of filtration theory. Soviet Mathematics. Russian Academy of Sciences. Doklady. Mathematics 46 (1993), no. 2, 296–300; translated from: Dokl. Akad. Nauk 326 (1992), no. 3, 431–437.

[157] S. I. Shmarev, Instantaneous appearance of singularities in solution of a degenerate parabolic equation. Siberian Math. J. 31, no. 4, (1991), 671–682; translated from: Sibirsk. Mat. Zh. 31 (1990), 166–179.

[158] S. I. Shmarev, Self-similar solutions of the multidimensional Stefan problem. Soviet Phys. Dokl. 31 (1986), no. 5, 386–389; translated from: Dokl. Akad. Nauk SSSR 288 (1986), no. 1, 95–99.

[159] S. I. Shmarev, J. L. Vazquez, Lagrangian coordinates and regularity of interfaces in reaction-diffusion equations. C. R. Acad. Sci. Paris Sér. Math. I 321 (1994), 993–998.

[160] P. E. Sobolevskii, Local and nonlocal existence theorems for nonlinear second-order parabolic equations. Soviet Math. Dokl. 2 (1961), 63–66; translated from: Dokl. Akad. Nauk SSSR 136 (1961), no. 2, 292–295.

[161] V. V. Sokolov, On the symmetry of evolution equations. Russian Math. Surveys 43 (1988), no. 5, 165–203; translated from: Uspehi Mat. Nauk 43 (1988), no. 5, 133–163.

[162] E. A. Socolovsky, Lagrangian non-oscillatory and FEM schemes for the porous media equation. Comput. Math. Appl. 15 (1988), 611–617.

[163] E. A. Socolovsky, On the numerical approximation of finite speed diffusion problems. Numer. Math. 53 (1988), 97–105.

[164] M. L. Storm, Heat conduction in simple metals. J. Appl. Phys. 22 (1951), 940–951.

[165] W. Strampp, Bäcklund transformations for diffusion equations. Phys. D 6 (1982), 113–118.

[166] S. I. Svinolupov, Analogs of the Burgers equation of arbitrary order. Theoret. and Math. Phys. 65 (1985), 1177–1180; translated from: Teoret. Mat. Fiz. 62 (1985), 303–307.

[167] D. Swartzendruber, The flow of water in unsaturated soils. In: Flow through porous media , R. J. M. de Wiest, ed., Academic Press, New York 1969, pp. 215–292.

[168] S. R. Svirshchevskii, The group properties of the model of heat transfer with relaxation of the heat flux. Inst. Prikl. Mat. AN SSSR, Preprint no.105 (1988), 16 p. (in Russian).

[169] A. N. Tikhonov, A. A. Samarskii, Equations of mathematical physics. Nauka, Moscow 1976 (in Russian).

[170] S. S. Titov, V. A.Ustinov, Investigation of polynomial solutions of the gas filtration equation with an integer adiabate exponent. In: Approximate methods for the solution of free-boundary problems in continuum mechanics, Sverdlovsk 1985, pp. 64–70, (in Russian).

[171] J. L. Vazquez, Asymptotic behavior and propagation properties of the one-dimensional flow of gas in a porous media. Trans. Amer. Math. Soc. 277 (1983), 507–528.

[172] N. N. Verigin, The forcing of viscous solvents for increasing the strength and water-inpermeability of the bases of hydrotechnical constructions. Izv. Akad. Nauk SSSR Otdel. Tekhn. Nauk 5 (1954), 674–687 (in Russian).

[173] A. M. Vinogradov, I. S. Krasil'shchik, V. V.Lychagin, Geometry of jet spaces and nonlinear partial differential equations. Nauka, Moscow 1986. English translation: Gordon and Breach Science Publ., New York 1986.

[174] A. M. Vinogradov, I. S. Krasil'shchik, A method of computing higher symmetries of nonlinear evolution equations, and nonlocal symmetries. Soviet Math. Dokl. 22 (1980), no. 1, 235–239; translated from: Dokl. Akad. Nauk SSSR 352 (1980), no. 6, 1289–1293.

[175] E. T. Whittaker, G. N. Watson A Course of Modern Analysis. Cambridge University Press, Fourth Ed., 1992.

[176] V. E. Zakharov, A. B. Shabat, Exact theory of two-dimensional self-focusing and one-dimensional self-modulation of wave in non-linear media. Soviet Phys. JETP 61 (1972), 62–69; translated from: Zh. Èksper. Teoret. Fiz. 61 (1971), 118–134.

[177] V. E. Zakharov, S. V. Manakov, S. P. Novikov, L. P. Pitaevskii, Theory of solitons. The inverse skattering method. Nauka, Moscow 1980. English translation: Plenum Press, New York-London 1984.

[178] Ya. B. Zel'dovich, G. I. Barenblatt, The asymptotic properties of self-modelling solutions of the nonstationary gas filtration equations. Soviet Phys. Dokl. 3 (1958), 44–47; translated from: Dokl. Akad. Nauk SSSR 118 (1958), no. 4, 671–674.

[179] Ya. B. Zel'dovich, A. S. Kompaneets, On the theory of propagation of heat with the heat conductivity depending upon the temperature. In: Collection in honour of the seventieth birthday of academician A. F. Ioffe, Izdat. Akad. Nauk SSSR, Moscow 1950, pp. 61–71 (in Russian).

[180] Ya. B. Zel'dovich, Yu. P. Raiser, Physics of shock waves and high temperature hydrodynamics phenomena. Academic Press, New York 1966.

Notation

Assumptions

(A.0)–(A.2) (Ch.3) 109
A.1–A.2 (Ch.4) 273–274
(B.0)–(B.2) 110-111
(B.3) 113
(C.1)–(C.3) 118
(D.0)–(D.2) 123–124
(F.0)–(F.4) 124
(G.1)–(G.3) 131
(H.1)–(H.3) 162
(K.1)–(K.2) 188
(K.3) 194
(N.1)–(N.3) 187
(U.1)–(U.5) 10
(U.6)–(U.9) 17–18

Problems

(E_1)–(E_3) 10
(L_1) (Ch.1) 15
(L_2) 16
(L_3) 33
(E)*, ($E_{m,\lambda}$), (E) 109
(P) 116
(P_1) 117, 185
(L) 123, 186
($L_{m,\lambda}$) 123
(L_1) (Ch.3) 162

Index

Boussinesq's equation in
filtration theory 250
similarity solutions of first and
second type 254
existence and uniqueness theorems
of similarity solutions,
255–256
linearization 266
Burgers's equation 62

Cauchy identity 271
Cauchy–Verigin problem 10, 102
self-similar solution 7
comparison principle 120
contact surface 4–5
continuity equation 2, 47, 85, 213
continuous motions, filtration 3

Darcy's law 2, 104, 271
diffraction problem 16
diffusion-reaction equation 127
Dym's equation 83, 88

Euler variables 12, 85
connection with Lagrange variables
27, 124–125, 272–273
evaporation 105
evolution equations in complex
domain 87

fast diffusion equation 57, 90
self-similar solution 91
filtration equation 3
finite speed of propagation
(FSP property) 106
free boundary 4, 106

gas dynamics equations 85–87
generalized motions, filtration 4
with strong discontinuity 5
gradient-nonlinear equation 69

heat conduction equation in solid
hydrogen 74
hidden symmetry of evolution
equations 43, 53

interface 106, 270
infinite differentiable 177
instantaneous loss
of regularity 144
representation via Lagrange
variable 126, 181, 191

kinematic condition 108, 246
Kingston–Rogers theorem 79
Kirchhoff's transformation 89
Korteweg–de Vries equation 81

L-transformation 43, 48
L-invariant equations
of second order 43, 65
of third order 83
Lagrangian coordinates
mass 13, 47, 86, 122, 180, 186
geometrical 49, 85, 244
canonical 246
in multi-dimensional Stefan's
problem 212
in multi-dimensional PME 270
for equations in divergence form
122

for equations not in divergence form 126
level curves 132
Lie group admitted by
nonlinear heat equation 51
equation of non-Newtonian filtration 54
Boussinesq's equation 251

mass conservation law 104, 271
balance law 107, 121, 185
in the differential form 2, 122
for equations without mass conservation 127

nonlinear filtration equation 106
nonlinear heat conduction equation 51
nonlinear Schrödinger equation 44, 75
nonlocal equation in population dynamics 115, 179
nonlocal-invariant solution 60

one-phase Stefan's problem
one-dimensional 6
self-similar solution 8
explicit solution with a functional arbitrariness 99, 101
degenerate 209
multi-dimensional 210
linearization in Lagrangian coordinates 216
existence and uniqueness theorems for linearized problem 216

parabolic boundary 132
porous medium equation (PME)
one-dimensional 90
generalized 161
multi-dimensional 106, 211, 270
Barenblatt solution 250, 254, 107, 192
exact solutions 248
self-similar solution 91, 107, 115
differentiable asymptotics 119, 194
weak solution 118
weak and generalized solutions of generalized PME 109, 111

quasi-homogeneity 110

reciprocal equation 48

trajectory equation 6, 85, 121, 271

Verigin's problem 1, 5
self-similar solution 7
existence and uniqueness theorems 11
formulation in Lagrange variables 14
von Mises transformation 50
variable 146

waiting time 113

Bei Fragen zur Produktsicherheit wenden Sie sich bitte an:
If you have any questions regarding product safety,
please contact:

Walter de Gruyter GmbH
Genthiner Straße 13
10785 Berlin
productsafety@degruyterbrill.com